THE EARTH'S CLIMATE AND VARIABILITY OF THE SUN OVER RECENT MILLENNIA: GEOPHYSICAL, ASTRONOMICAL AND ARCHAEOLOGICAL ASPECTS

Le Glacier d'Argentière in (*a*) *ca.* **1850–60**, and (*b*) **1966**. The engraving (*a*) and the photograph (*b*) (taken by Madeleine Le Roy Ladurie) are both taken from *Histoire du climat depuis l'an mil* by Emmanuel Le Roy Ladurie, and reproduced courtesy of Libraire Flammarion, Paris.

THE EARTH'S CLIMATE AND VARIABILITY OF THE SUN OVER RECENT MILLENNIA: GEOPHYSICAL, ASTRONOMICAL AND ARCHAEOLOGICAL ASPECTS

PROCEEDINGS OF
A ROYAL SOCIETY AND ACADÉMIE DES SCIENCES
DISCUSSION MEETING
HELD ON 15 AND 16 FEBRUARY 1989

ORGANIZED AND EDITED BY
J.-C. PECKER, MEM., ACADÉMIE DES SCIENCES,
AND S. K. RUNCORN, F.R.S.

LONDON
THE ROYAL SOCIETY
1990

Printed in Great Britain for the Royal Society
by the
University Press, Cambridge

ISBN 0 85403 406 4

First published in *Philosophical Transactions of the Royal Society of London*,
series A, volume 330 (no 1615), pages 395–687

♾ The text paper used in this publication meets the minimum requirements of American National Standard for Information Sciences—Permanence of Paper for Printed Library Materials, ANSI Z39.48–1984.

Copyright

British Library Cataloguing in Publication Data

The Earth's climate and variability of the Sun over recent
  millennia.
  1. Climate, Changes
  I. Pecker, Jean-Claude   II. Runcorn, S. K. (Stanley Keith)
  III. Royal Society   IV. Académie des Sciences
  551.6

ISBN 0-85403-406-4

Published by the Royal Society
6 Carlton House Terrace, London SW1Y 5AG

PREFACE

This volume contains the papers and contributions to discussion of a joint meeting of the Académie des Sciences and the Royal Society held in February 1989. The meeting was an historic one being the first joint Discussion Meeting of the two academies in their over three centuries' existence. It was fitting, therefore, that the President of the Académie des Sciences, Professor J. Aubouin, presided at the opening session and the President of the Royal Society, Sir George Porter, presided at the final session.

It was fitting that this joint Discussion Meeting dealt with a fundamental aspect of a topic, climatic change, that is of increasing importance to humanity. It was also appropriate in this meeting to have strong participation from the Archaeology section of the British Academy.

For decades, a few scientists only have been devoting serious attention to the physics of solar–terrestrial relations; Walter O. Roberts, at the High Altitude Observatory in Boulder, Colorado, is one of these pioneers, and has much stimulated studies in this field.

As organizers privileged to arrange this historic meeting, we would like to express our appreciation to many who gave us scientific advice and particularly to Ms Maureen Hopkinson of the Department of Physics, University of Newcastle upon Tyne, to Miss Christine Johnson and Mr Jonathan Wainwright and other members of the staff of the Royal Society and the Académie de Sciences who gave great help in the organization.

We hope that this important interdisciplinary subject will attract young and enthusiastic workers as a result of the publication of these proceedings.

October 1989

J.-C. PECKER
S. K. RUNCORN

CONTENTS

[Frontispiece, three plates]

CONTENTS

CONTENTS

ix

PAGE

Phil. Trans. R. Soc. Lond. A **330**, 399–402 (1990)
Printed in Great Britain

399

Introductory remarks

BY J.-C. PECKER[1] AND S. K. RUNCORN[2], F.R.S.

[1] *Collège de France, Annexe, 3 Rue d'Ulm, 75231 Paris, Cedex 05, France*
[2] *Department of Physics, The University, Newcastle upon Tyne NE1 7RU, U.K.; University of Alaska Fairbanks, Fairbanks, Alaska 99775, U.S.A.; Imperial College of Science, Technology and Medicine, London SW7 2BZ, U.K.*

The search for periodic or quasi-periodic variations in the solar constant through the analysis of climatic and meterological data has proved elusive. The reason is evident: the atmosphere is a wet gas with much energy stored as latent heat and is in complex interaction dynamically and thermally with the oceans and land areas. This confronts the investigator with a hydrodynamic problem of awesome difficulty and has hitherto frustrated attempts at weather prediction over more than a few days. The instabilities, what we call the weather, cause not only day-to-day but also year-to-year variations so great that many experts have concluded that these would have completely masked possible small changes due to fluctuations of the energy input from the Sun. Yet, as the seasonal changes of solar energy falling on each hemisphere result in such obvious effects, it should not be impossible to detect in the climatic records much smaller changes in the total global input of heat energy into the atmosphere, especially if these are cyclical, by integrating out short-term fluctuations.

That the Sun varies with one or more periods has long been known. Solar physicists noted the time variations of the spotted parts of the Sun as soon as they identified the spots as a solar feature. Scheiner (1630), in the first half of the seventeenth century, used them to determine the solar rotation axis: the spots stayed visible during 14 days after which they disappeared at the west solar limb, eventually reappearing at the east limb. It was also noticed that spots have a short life, a few months, and they differ one from another. It was only much later that the German amateur astronomer, Schwabe (1838), suspected from his observations over the years from 1826 to 1837 and later concluded (Schwabe 1844) that the number of spots on the solar disk was a regular, perhaps periodic, function of time. Since Schwabe's work this clear periodicity of about 11 years has been monitored continuously. It was quickly noted that one 'solar cycle' does not look like another: in some the number of spots rises slowly and the maximum is often of moderate importance, whereas some other cycles, more conspicuously active, have the quickest rise from minimum to maximum.

The magnetic nature of spots was discovered later at Mt Wilson by G. E. Hale (1908), who found the signature of the Zeeman effect in their line spectra. Then he discovered (Hale 1924) a pattern in the magnetic polarities of the spots: that they usually occur in pairs of opposite polarities in the same latitude. In both hemispheres one polarity leads during one 11-year period. In the following period, these magnetic properties are reversed in sign: the leading and following spots of the pairs change from N to S, suggesting that a physical or magnetic cycle of 22 years, rather than one of 11 years, is the basic phenomenon. These researches have been carefully refined over the past 50 years or so: the inequality of the observed cycles suggested strongly that other periods should be looked for. Gleissberg (1958) suggested from the sunspot data a 75–80-year cycle which also appears in auroral frequency data (Gleissberg 1965) and this seems well established. From meteorological data, Link (1968) suggested a 300–400-year cycle and now (see later) we see a 200-year modulation. The obvious difficulty is that one has good magnetic records only since the beginning of this century, complete sunspot records only since the beginning of the nineteenth century, but only uncalibrated and incomplete data in western Europe and China in earlier centuries.

This explains why solar physicists take so much interest in those terrestrial phenomena that might be associated with solar behaviour: if these relations could be demonstrated safely, Earth records (not only of meteorological conditions, but also of human migrations and geological evolution perhaps) could help to understand the processes within the Sun. And this is why the two typical solar signatures – the 11-year periodicity and 27-day rotation – have been carefully searched for in terrestrial data of all kinds. In fact it was precisely with that object that an astronomer, A. E. Douglass, a devoted and curious mind, at the end of the last century and during the first decades of this, noted the regularity in the distribution of a tree-ring thickness. As an astronomer he first searched for a clear 11-year periodicity, each tree ring being the result of the annual growth of the tree, mostly at springtime, with a thickness depending on climatic factors (Douglass 1919). The presence of such a periodicity remained difficult to prove statistically, but the study led Douglass to show how absolute dating could be done by using tree rings.

It also encouraged many scientists to look everywhere on Earth for the 11-year signature. Mitchell (1979), for example, has found correlations between the 22-year solar activity and the recurrence of drought in western U.S.A. Long ago Beveridge in an early application of time series analysis in economics claimed to find the solar cycle in the price of wheat, which, because it is linked with climate, may merit being taken seriously! In such indices as wheat price, or tree-ring thickness, climatic effects, rainfall or sunshine, are averaged, or integrated, over a year. Some meteorological data, and some indices of the higher atmosphere (auroral frequency and geomagnetic indices) have been collected at much more frequent intervals than a year. The wealth of data has thus become much larger, and research of solar–terrestrial relations more securely based: in the tropospheric layers, climatic patterns are now observed for example, as convincingly discussed in this meeting. An early example is R. A. Fisher's statistical studies of rainfall in southern Australia. Another of the same kind is the study of rainfall in polar and moderate latitude stations by Xanthakis (1967). Rain is an element of the daily meteorological patterns: therefore, one might think that meteorologists, and not only climatologists, should be interested in looking for the solar signature in continuous weather records, perhaps for use in their short-term prediction. This has recently been attempted. May & Hitch (1989) of the Meteorological Office conclude from British weather records from 1881 to 1986 that heavy rains occur at the climax of the 11-year cycle of sunspot activity.

One might mention in this context the interesting work by Shapiro (1956), who found a persistence of a meteorological pattern a few days after a large geomagnetic disturbance associated with solar corpuscular emission followed by 10 days of decreasing persistence correlation. Unfortunately, this showed only the difficulty of the forecasting in periods of marked activity, but it has not proved helpful in improving forecasts.

Nevertheless, the search either for a 27-day or the 11-year signature, or for some systematic relation, is one way in which the mechanisms of solar interaction with the climate may be better understood in the future: it is wise to look at these early findings, or hints – be they only that – sceptically, but also with some care, for in spite of the statistical problems some studies may have uncovered real effects and quite recently stronger suggestions of this connection has been published.

Many investigations have been made by suitably averaging climatic data from different parts of the globe to search for periodic variations in the Sun's output: the implicit assumption being made that certain parts of the atmospheric circulation may be especially sensitive to them. To take a recent example: Colacino & Rovelli (1983) have analysed the long series of air temperature measurements made in Rome from 1782 to 1975. The annual mean temperatures, mean maximum and mean minimum temperatures were calculated, the latter increasing due to human activity as the city grew. The natural climatological effects were thus separated and maximum entropy analysis gave a spectrum with a sharp peak at 11 years. Analysis of trends show a decrease in temperature from 1810 to 1880 and an increase from 1880 to 1950 with a decrease since.

The problem with all such investigations is that they are studies of data from small areas of the globe, yet the variations are being attributed to a cause affecting the whole Earth. It is therefore interesting that Russell (1975) has discussed the records of the geomagnetic field, in which certain days feature disturbances different from the quiet day magnetic variation which arises from the tidal motions in the ionosphere. The disturbances arise in the magnetosphere and have long been measured in the celebrated geomagnetic indices. As has long been known, the sunspot cycle is clearly present in these geomagnetic disturbances, but Russell shows that there is a long-term variation of geomagnetic activity: the solar cycle average increases from 1872 to 1950 and decreases since. This correlates with the trend in air temperature in Rome since 1880 described above. Are we therefore in addition to the sunspot cycle seeing a much longer cycle of solar activity with a period of between 150 and 200 years? The absurdity of discussing this question on data available over such a short period is evident. Such a longer period of solar variability had become a matter of debate when the sightings of sunspots before Galileo's observations were collected from historical sources. Maunder (1922) called attention to Spörer's (1889) studies of a marked absence of sunspots in the seventeenth century and earlier such minima have now been established from geophysical studies, suggesting an approximate 200-year modulation of the solar cycle.

We need to study a record spanning recent millenia if progress is to be made. As is so often the case in scientific research, a much more convincing record comes from an unexpected field, tree rings, hitherto entirely a tool for archaeological dating. This relatively new, but so far only partly explored, source of information on solar activity is the spectrum of the ^{14}C variations during the past millenia found from tree rings. As early as in 1956 H. de Vries in Groningen, The Netherlands, found indications of variations in the ^{14}C-content of wood that grew in the seventeenth and fifteenth centuries. These variations could be explained by variations in the cosmic-ray production rate of ^{14}C in the atmosphere. Their existence was soon confirmed by other ^{14}C-laboratories and their significance for the reliability of ^{14}C dating was soon appreciated. However, an explanation of this so called 'de Vries effect' was not possible, except that it seemed probable that it was caused by variations in the modulation of the cosmic-ray flux by magnetic fields from the Sun. This contention was then supported by the observation that the ^{14}C-variations are not random fluctuations, such as produced by 'red' noise, but they are due, as was first shown by Hans E. Suess (1973, 1974), to a consistent line spectrum with a dominant 200-year line that can be recognized to be present with varying intensities for the whole time recorded by tree rings back to about 5000 B.C. It is this ^{14}C record from tree rings of known date which has given a new impetus to the study of variations in the Sun and climate

[3]

over recent millenia. These proceedings therefore begin with an account of the discovery of the Suess wiggles. The early data from which Suess derived the 200-year cycle was noisy and was not at first accepted by other laboratories, although later their reality was generally agreed. Harold C. Urey used to say that true progress in science is only possible if the record of such progress 'is kept straight'. In this case it was only after some years that Suess's discovery was independently verified by another laboratory using German oak tree rings (de Jong *et al.* 1979).

It is significant that since the meeting two papers have been published finding the solar cycle in global temperatures. Newell *et al.* (1989) find from the analysis of global and hemispheric marine temperatures for 1856–1986 a prominent 22-year period which they concluded may be related to the solar magnetic cycle. Barnett (1989) also in studying global sea surface temperatures concludes that the quasi-biennial variation is modulated with the 11-year period. Thus earlier negative conclusions reached by meteorologists – for instance, Pittock (1978), who concluded that 'little convincing evidence has yet been produced for real correlations between sunspot cycles and the weather/climate,' – may need revision.

It would not be right to conclude without reference to the work of D. J. Schove (1983), who over many years, in season and out of season, argued for a connection between sunspot cycles and the climate. His contributions, with many other papers, were brought together by him just before his death. Discussion of some of these questions has recently been reopened by a NATO workshop (Stephenson & Wolfendale 1988).

<h2 style="text-align:center">REFERENCES</h2>

Barnett, T. P. 1989 A solar–ocean relation: fact or fiction? *Geophys. Res. Lett.* **16**, 803–806.

Colacine, M. & Rovelli, A. 1983 The yearly averaged air temperature in Rome from 1782 to 1975. *Tellus* A**35**, 389–397.

de Jong, A. F. M., Mook, W. G. & Becker, B. 1979 Confirmation of the Suess wiggles. *Nature, Lond.* **280**, 48–49.

Douglass, A. E. 1919 Climatic cycles and tree growth. Carnegie Institute of Washington.

Gleissberg, W. 1958 The eighty-year sunspot cycle. *J. Br. astr. Ass.* **68**, 148–152.

Gleissberg, W. 1965 The 80-year solar cycle in auroral frequency number. *J. Br. astr. Ass.* **75**, 227–231.

Hale, G. E. 1908 On the possible existence of a magnetic field in sunspots. *Astrophys. J.* **28**, 315–343.

Hale, G. E. 1924 The law of sunspot polarity. *Proc. natn. Acad. Sci. U.S.A.* **10**, 53–55.

Link, F. 1968 The 400-year cycle. *J. Br. astr. Ass.* **78**, 195–205.

Maunder, E. W. 1922 The prolonged sunspot minimum, 1645–1715. *J. Br. astr. Ass.* **32**, 140–145.

May, B. R. & Hitch, T. S. 1989 Periodic variations in extreme hourly rainfall in the United Kingdom. *Meteorological Mag.* **118**, 45–50.

Mitchell, J. M., Stockton, C. W. & Meko, D. M. 1979 Evidence of a 22-year rhythm of drought in the western United States related to the half solar cycle since the 17th century. In *Solar terrestrial influences on weather and climate* (ed. B. M. McCormac & T. A. Seliga), pp. 125–143. Dordrecht: Reidel.

Newell, N. E., Newell, R. E., Hsiung, J. & Zhongxiang, W. 1989 Global marine temperature variation and the solar magnetic cycle. *Geophys. Res. Lett.* **16**, 311–314.

Pittock, A. B. 1978 A critical look at long-term Sun–weather relationships. *Rev. Geophys. Space Phys.* **16**, 400–420.

Russell, C. T. 1975 On the possibility of deducing interplanetary and solar parameters from geomagnetic records. *Sol. Phys.* **42**, 259–269.

Scheiner, C. 1630 *Rosa Ursina sive Sol.* Bracciani, Italy.

Schove, D. J. 1983 *Sunspot cycles.* Stroudsburg, Pennsylvania: Hutchinson Ross.

Schwabe, H. 1838 Ueber die Flecken der Sonne *Astr. Nachr.* **15**, 243–248.

Schwabe, H. 1844 Solar observations during 1843. *Astr. Nachr.* **21**, 233–236.

Shapiro, R. J. 1956 *Meteorology* **13**, 335.

Spörer, F. W. G. 1889 Ueber die periodicität der Sonnenflecken seit dem Jahre 1618. *R. Leopold–Caroline Acad. Bll. Astron., Halle* **53**, 283–324.

Stephenson, F. R. & Wolfendale, A. W. (eds) 1988 *Secular solar and magnetic variations over the last 10 000 years.* Dordrecht: Kluwer.

Suess, H. E. 1973 Natural radiocarbon. *Endeavour* **32**, 34–38.

Suess, H. E. 1974 Natural radiocarbon: evidence bearing on climatic change. *Colloq. Int. CNRS* **219**, 311.

Xanthakis, J. 1967 Probable values of the time of rise for the forthcoming sunspot cycles. *Nature, Lond.* **215**, 1046–1048.

Phil. Trans. R. Soc. Lond. A **330**, 403–412 (1990)

Printed in Great Britain

The ^{14}C record in bristlecone pine wood of the past 8000 years based on the dendrochronology of the late C. W. Ferguson

By H. E. Suess and T. W. Linick†

Department of Chemistry and Scripps Institution of Oceanography, University of California, San Diego, La Jolla, California 92093, U.S.A.

When, in 1950, Willard Libby and his coworkers obtained their first radiocarbon (^{14}C) dates, C. W. Ferguson at the University of Arizona Tree Ring Laboratory was working on establishing a continuous tree ring series for the newly discovered bristlecone pine *Pinus aristata*. Before his untimely death in 1986, he had extended the series nearly 8000 years into the past. From the Ferguson series I obtained for ^{14}C determinations wood samples grown at various times. Also, two other laboratories obtained such samples. For b.c. times in particular, our measured ^{14}C-values that deviated consistently from those calculated from tree rings, and the deviations increased with age. This general trend was observed by other laboratories, but the presence of deviations from these trends, of the so-called 'wiggles', was questioned by other workers. To me these wiggles indicated the existence of a most interesting geophysical parameter valid for the whole terrestrial atmosphere. Fourier spectra obtained at my request by Kruse in 1972, and by Neftel, demonstrated the consistency of the results, and supported my contention that the secular variations of ^{14}C in atmospheric CO_2 are related to variations of solar activity.

1. Introduction

When I was offered the opportunity in 1952 to set up a routine radiocarbon counting laboratory at the U.S. Geological Survey in Washington, D.C., I contemplated using acetylene ($HC\equiv CH$) as the counting gas. In Germany, more than 10 years before, I had used acetylene to investigate 'hot atom' reactions as they occur in connection with nuclear reactions (Libby 1947; Suess 1939). I therefore expected acetylene to be ideal as a counting gas as it would combine with atoms and radicals that interfere with the counting of radioactive decay. Indeed, I found it to be much less sensitive than CO_2 to the presence of minute quantities of impurities, in particular of parts per million of oxygen in the counting gas. Carbon dioxide is commonly used by physicists as a counting gas. However, it easily can be converted to acetylene by hydrolysing a carbide, such as strontium carbide, obtained from strontium carbonate by reduction with magnesium metal, or lithium carbide obtained by reacting lithium metal with CO_2 gas at sufficiently high temperature. Acetylene has practically no vapour pressure at the temperature of liquid nitrogen. Its purity can be checked easily by measuring its triple point pressure.

From 1952 to 1954, at the U.S. Geological Survey I used acetylene for precision measurements of ^{14}C, dubbed 'radiocarbon' by W. F. Libby in 1952. There I measured the time of the maximum extent of the continental glaciation of North America to be about 20000 radiocarbon years before the present. Also, the dilution of the ^{14}C in the atmospheric CO_2 by the addition of ^{14}C-free anthropogenic CO_2 from the burning of fossil fuels, the so-called Suess effect, was observed.

† Dr Timothy W. Linick died in Tucson, Arizona, on 4 June 1989.

2. RADIOCARBON IN TREE RINGS

When Willard Libby and his coworkers published their first radiocarbon date list (Arnold & Libby 1951) it was considered most desirable to test their method on samples of wood of precisely known ages. The best samples for this purpose were considered wood samples from trees, for which the time of growth could be determined from tree rings. At that time the California *Sequoia sempervirens* offered such wood. Accurately dated samples of wood close to 2000 years old were available from the University of Arizona Tree Ring Laboratory. Another tree, the bristlecone pine, *Pinus aristata*, discovered by Professor Schulman of the same laboratory, made it possible to derive a tree ring series more than 7000 years into the past. This was done by the late C. W. Ferguson (1968) who for the rest of his life concentrated entirely on this work.

Ferguson wisely decided to distribute 10 g samples of his wood, consisting of 10 annual growth rings each, to three ^{14}C-dating laboratories, at the University of Pennsylvania, at the University of Arizona in Tucson, and at the Scripps Institution of Oceanography of the University of California in La Jolla. There, commencing in 1955, I used acetylene as the counting gas. For precision measurements I used two ultra low background counters of the Houtermans–Oeschger type (1955), that Dr Oeschger kindly had ordered made by his institute shop.

It soon became obvious that the measured ^{14}C values indeed deviated consistently from those calculated, using ages derived by Ferguson from tree-ring counting. For the past 2000 years the magnitudes of the deviations were relatively small but increased in B.C. times. ^{14}C values for the oldest dendrochronologically dated wood samples were nearly 10 % too high, i.e. their ^{14}C ages up to 800 years too young. It was therefore necessary to answer the following questions.

1. Are the tree ring ages, the so-called 'Dendro-ages', correct?

2. Do wood samples from different kinds of trees, grown at different geographic locations and having the same tree-ring age, show the same ^{14}C content?

3. Can the ^{14}C content of wood samples change through contamination, irradiation, or in any other way?

Much careful work was done during the following years, and we can trust now that these questions are reliably answered. The ^{14}C in the cellulose present in wood in a given annual ring corresponds remarkably well to that in the CO_2 of the atmosphere at the time of the growth of the ring if one corrects for isotope fractionation during photosynthesis. Other organic compounds, such as the sap and lignin in sapwood, may migrate into older rings. This has been studied by Cain in detail (Cain & Suess 1976), making use of bomb ^{14}C present after 1955 in atmospheric CO_2 as tracer.

If data are normalized as discussed above, one finds that the ^{14}C-content of the cellulose in wood grown at the same time is practically independent of geographic location and altitude. An exception seems to be a slight difference of a few per mille in ^{14}C in air from the two hemispheres. Cellulose from wood from the Northern Hemisphere seems to have a slightly higher ^{14}C content than that from the Southern Hemisphere (see, for example, Baxter & Walton 1971).

In any case, we can assume that the differences between calculated and measured ^{14}C are essentially a unique function of their ages. The general trend of this function, as determined by the La Jolla Laboratory, can be seen in figure 1. This general trend was soon confirmed by other laboratories.

[6]

About half of all the ^{14}C measurements by the La Jolla laboratory during the time of its existence from 1956 to 1981 were carried out for the purpose of determining this function. These measurements showed consistently relatively small deviations from a smooth curve, the so-called 'wiggles' that were superimposed upon the general trends. These wiggles, in general with amplitudes of less than 1%, were not important for radiocarbon dating, but appeared most interesting for ^{14}C-geophysics.

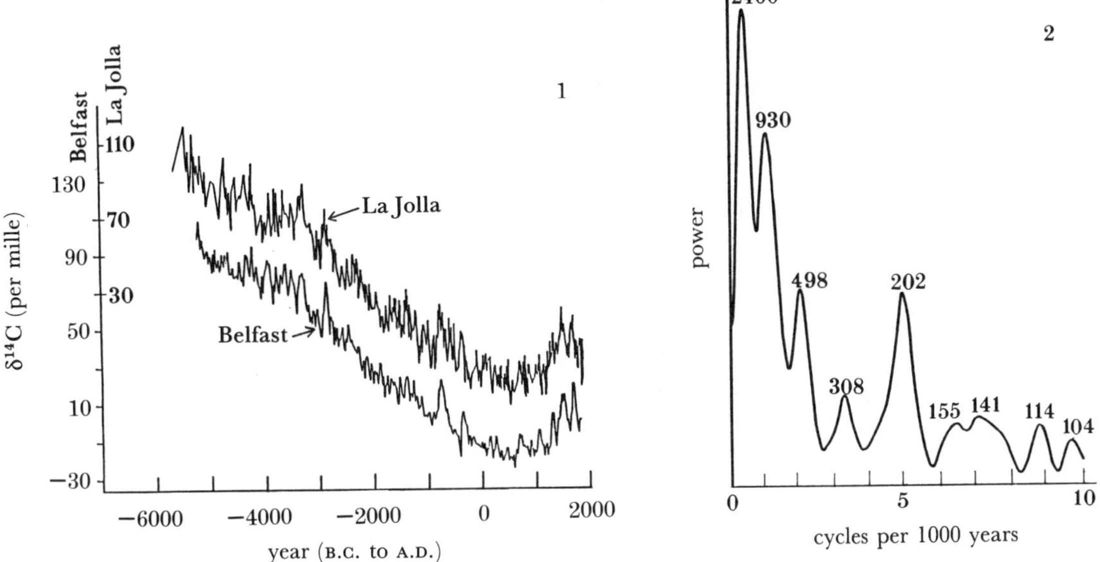

Figure 1. Upper curve: δ^{14}C (normalized) as a function of tree ring date, as measured at La Jolla in Ferguson bristlecone pine wood. Lower curve: same as measured in Belfast in Irish Oak wood.

Figure 2. Fourier spectrum of the δ^{14}C variations during the period from 5300 B.C. to 1500 A.D. as observed by the La Jolla radiocarbon laboratory in samples from bristlecone pine wood dated by Ferguson according to A. Kruse (see Suess 1980).

The first important question was whether the wiggles represented random variations or were, at least in part, periodic deviations from the smooth trend. The Fourier spectrum shown in figure 2 was derived by A. Kruse (see Suess 1980). It confirmed the existence of a conspicuous periodicity of about 200 years.

Unfortunately, the accuracies of the measurements of the other laboratories were insufficient to demonstrate even the existence of 'wiggles'. A statistical analysis by Clark (1975) stated that 'on the basis of all the existing measurements the assumption of wiggles was statistically not justified' (figure 4). Even after a paper by De Jong et al. (1980) had appeared entitled 'Confirmation of the Suess wiggles', other experts did not concern themselves with this interesting phenomenon.

Shortly thereafter the U.S. National Science Foundation, Atmospheric Science Section, terminated the financial support of the La Jolla Laboratory, supposedly, because our measurements were 'not sufficiently accurate for the purpose in question'. With no funds for continuing my own research, I personally contacted a number of experts in time series analyses who might possibly be interested in these observations. I asked for help in data reduction and interpretation of data. One of them, Professor C. P. Sonett of the University of Arizona in Tucson, responded.

3. Normalization of data and comparisons of time series

Most authors publish their laboratory results in 'conventional radiocarbon dates', expressed in radiocarbon years, and give their accuracy as one- (or two-) sigma statistical counting errors. Regarding these counting errors (given in general as plus and minus) widespread misconceptions prevail. Therefore, it should be emphasized that these ± numbers do not indicate the maximum but rather the minimum average error of a number of measurements. This means that the average error of a sufficiently large number of measurements can never be smaller, but may well be much larger than what is indicated by these ± values, because the uncertainty of the result of a measurement not only arises from statistical fluctuations but may result from a multitude of other experimental factors.

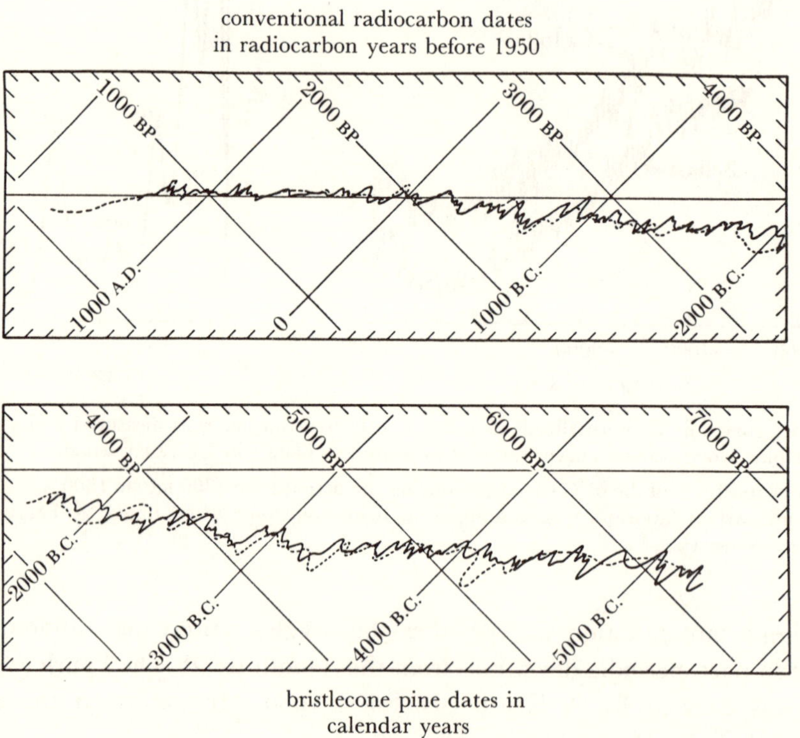

FIGURE 3. Dendrochronological age as a function of radiocarbon age determined in La Jolla, representing a calibration curve. The dotted line was published by Suess in 1970 to indicate the character of the expected curve. The solid line shows the same, but is based on about 700 individual measurements (Suess 1980). The diagram is tilted 45° to conserve space and a spline curve is drawn through the measured points.

In figure 3 we drew a line through the measured points free-hand with what was called cosmic 'Schwung'. This was taken by some of our colleagues as an indication of inaccuracy. This expression, however, was coined by the late eminent Austrian 'cosmic' (sic!) physicist and oceanographer Albert Defant, who, when plotting the amplitudes of ocean tides, used this expression to indicate that the amplitudes resulted from cosmic, extraterrestrial forces, and also, that these forces did not change abruptly. An ancient dictum, *natura non facet saltum* (nature does not make jumps), expresses this appropriately. Today one might say: the second time-derivative of (macroscopic) quantities in nature does not change abruptly. Therefore, unknown parts of a function in nature can best be approximated by spline functions.

For many reasons archaeologists should not expect ^{14}C data (before and after their calibration) to give more than the correct century, at best. Geophysicists, however, require an optimal precision for the interpretation of the observed data. In this case it is not always simple to compare results from different laboratories, because the numerical results depend on a number of normalizations. These are the following.

1. Radiocarbon ages are given by convention as years before the present (BP). Time zero is taken as 1950 A.D.

2. The measured counting rate is compared with that of a ^{14}C sample prepared by using a sample distributed by the U.S. Bureau of Standards.

3. Normalization of δ^{13}C of the sample to δ^{13}C of the standard, assuming mass-dependent isotope fractionation (Libby 1952). (Mass-independent isotope fractionation has not yet been observed for carbon isotopes (Thiemens 1983).)

4. By agreement, to preserve consistency of published ^{14}C-dates, the 'Libby' or 'conventional' half-life of 5568 years (by definition) is used. The more recently determined 'Cambridge' half-life of 5730 ± 40 years should be used in connection with the physics of ^{14}C decay.

Archaeologists usually report conventional Libby dates. Calibrated dates are expressed, of course, in calendar years. Usually, the uncalibrated conventional Libby dates differ only by a few times ten years because of small differences in normalization. This is negligible for archaeological dates but it may interfere with their interpretation for geophysical purposes.

In his book *Radiocarbon dating* Professor R. Ervin Taylor (1987) has presented graphically the results of calibration measurements from several laboratories, including those obtained in La Jolla, and has plotted them as we had done previously (Suess 1970, 1980). As far as we know there exist now two complete series of precision ^{14}C measurements of wood samples with times of growth of the past 7000 years. Figures 3 and 4 show the La Jolla bristlecone pine and the Belfast Irish oak series respectively. The agreement is most remarkable. The two series were obtained completely independently at different times during the past 20 year, using two different methods of ^{14}C determination: (a) acetylene counting of bristlecone pine samples from California and (b) scintillation counting of oak samples from Ireland. What is measured is a steady-state concentration of ^{14}C in wood. This concentration reflects the ^{14}C in the CO_2 of air, which is a function of its cosmic ray production rate and the rate by which it equilibrates with the world's oceans, where most of it decays (Houtermans *et al.* 1973). The similarities of the two time series, as shown in figures 1, 3 and 4, are easily recognized. Errors in the measured data, or in data reduction, can never produce this kind of resemblance.

The curve by Clark (1975) in figure 5 was derived by statistically evaluating the measurements by various authors assuming, however, that all the data have the same accuracy, which is certainly not the case. One sees that with this assumption, no fine structure and no wiggles result and all the irregularities are smoothed out. Also, the smooth curve in figure 6 was drawn in such a way as to eliminate most of the structure.

Perhaps even more convincing are the Fourier spectra of the two time series. They both show the by now well-known prominent spectral line around 205 years, first recognized visually in the La Jolla series as early as 1970, and confirmed by Fourier analyses by Kruse (figure 2) and then by Neftel (Neftel *et al.* 1981). As expected, the 200-year line is present in the Belfast Irish oak spectrum and can also be recognized now in other time series.

According to Sonett, the empirical average standard error sigma of our measurements is 7.8

conventional radiocarbon dates
in radiocarbon years before 1950

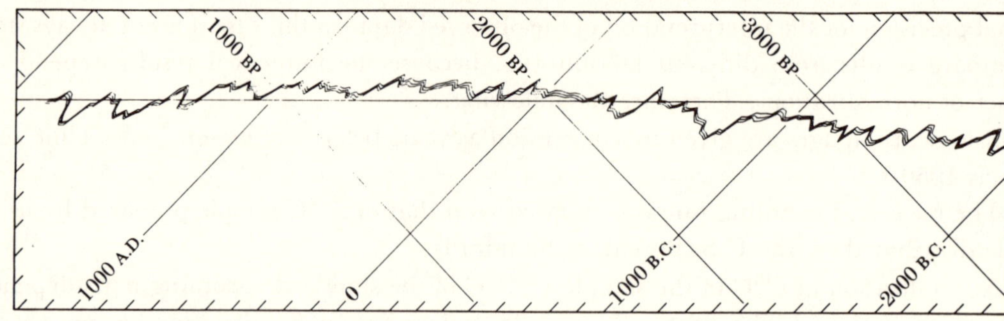

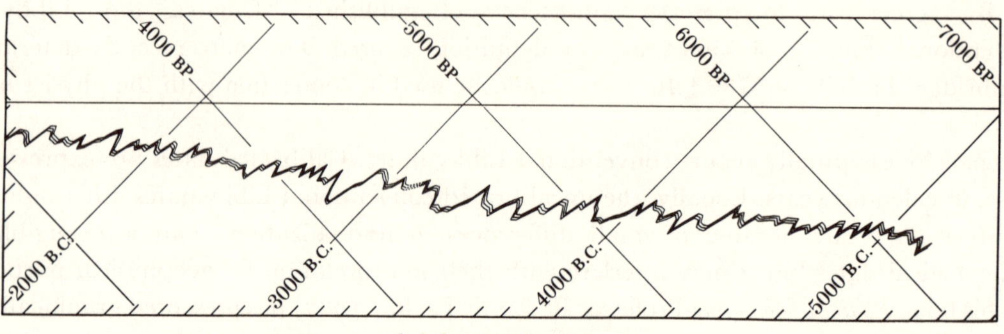

bristlecone pine dates in
calendar years

FIGURE 4. Same as figure 3, but with data from the Belfast laboratory. After Pearson *et al.* (1986), normalized in the same way as figure 3. Empirical values are connected by straight lines. The parallel lines indicate statistical one sigma limits. The graph is adapted from R. E. Taylor (1987) to conform with figure 3 and published here with permission.

radiocarbon dates

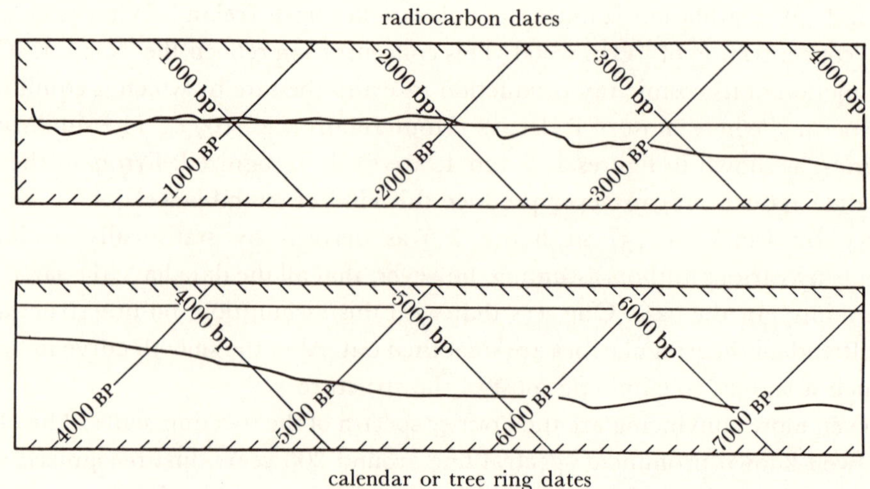

calendar or tree ring dates

FIGURE 5. Calibration lines after Malcolm Clark (1975). The average values of published results of many measurements are calculated assuming they have the same experimental errors. With this assumption the depicted smooth line appears 'statistically justified'. Adapted to conform with the other figures by R. E. Taylor and published here with his permission.

radiocarbon dates

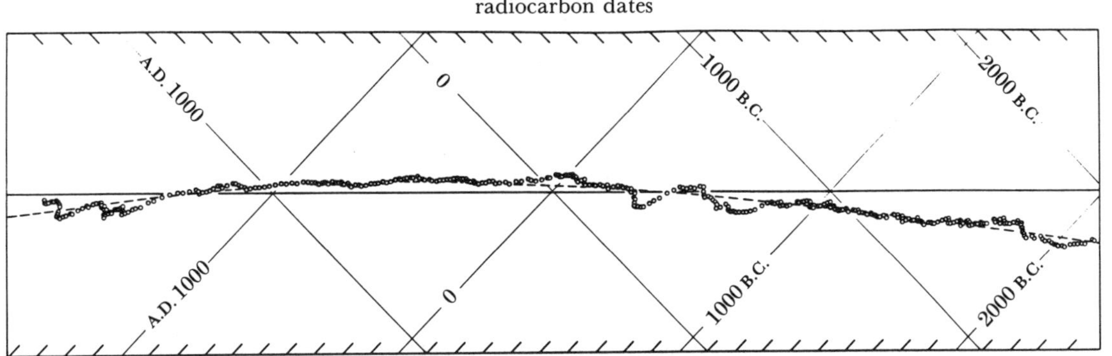

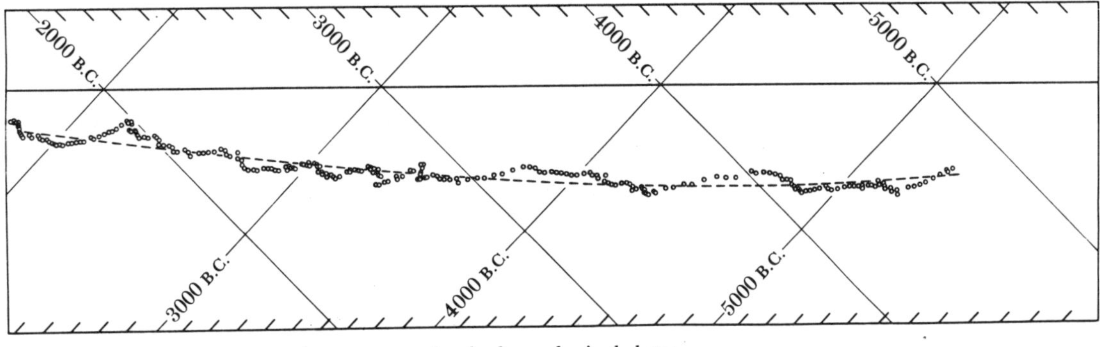

dendochronological dates

FIGURE 6. 'MASCA' calibration curve according to Ralph *et al.* (1973) derived by calculating 'running means' to eliminate the noise and thus the fine structure. Redrawn by R. E. Taylor and published here with his permission.

per mille. This is quite good, considering a statistical counting error for each measurement between 4 and 8 per mille (depending on age) and the fact that the La Jolla measurements were carried out over a period of nearly 20 years at two different locations of the UCSD campus. Professor Sonett's paper (this Symposium) concerns this. We list in table 1 the strongest lines in the spectra obtained by Sonett.

TABLE 1

La Jolla		Belfast	
period/years	power spectrum densities	period/years	power spectrum densities
2266.8	53870	2387.6	40530
504.5	18872	508.6	18464
358.2	9977	352.8	12988
201.5	14953	206.9	12218
158.5	8849	149.5	10143

The measured ¹⁴C values reflect steady-state concentrations, which essentially depend on the cosmic ray production rate and the rate by which ¹⁴C is equilibrated with the world's oceans, where most of it decays. If the production rate varies, then the steady state concentration varies, whereby the ocean ⇌ atmosphere system acts as a low pass filter (see figure 7).

[11]

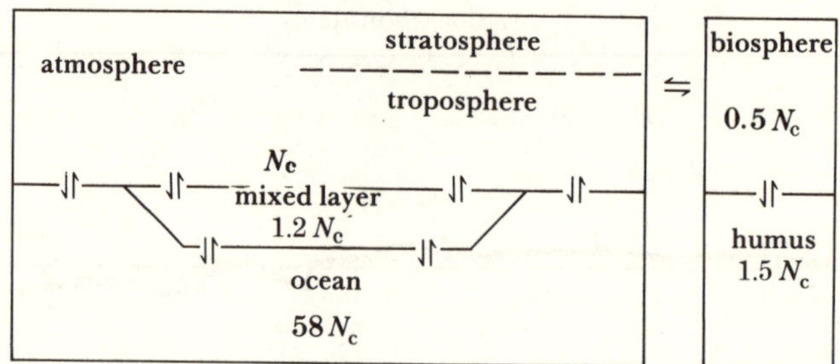

FIGURE 7. Carbon exchanging reservoirs on the surface of the Earth. N_c denotes the relative amounts of exchangeable carbon in each reservoir, with the amount in the atmosphere equal to one. Radiocarbon is produced by cosmic rays in the atmosphere. Most of it decays in the oceans. The concentration of ^{14}C in the total carbon, the $\delta^{14}C$, is measured in wood. It is a function of the $\delta^{14}C$ in the atmospheric CO_2. The ^{14}C in the atmospheric CO_2 roughly depends on (1) the cosmic ray production rate, (2) the relative amounts N_c in the different reservoirs, and (3) the exchange and transfer rate of carbon between the different reservoirs (Libby 1952).

The fact that there exist two time series of radiocarbon measurements obtained by completely different experimental methods, namely, acetylene counting and scintillation counting, of wood ^{14}C from two different kinds of trees that grew at different altitudes and continents makes it possible to recognize reliably worldwide global fluctuations in the ^{14}C-level of atmospheric CO_2.

4. THE THREE CAUSES OF SECULAR ^{14}C FLUCTUATIONS

Professor Sonett and we have looked into the possibilities of explaining these observations. We agree now that most of the spectral lines must be a result of periodic variations of the intensity of the ^{14}C-producing radiation that reaches the Earth. Its important galactic component is attenuated by magnetic fields of the Sun and also by the magnetic dipole moment of the Earth. Changes in the carbon distribution in the ocean $\rightleftharpoons$ atmosphere system on the surface of the Earth are unimportant at the present time but were important during glacial and early postglacial times. To obtain more information on cosmogenic radioisotopes during these periods of time is a most desirable research objective that promises to answer many important questions on solar activity and global climate. The following three causes of ^{14}C variations are being discussed in the literature:

(1) changes in the magnetic field of the Earth;

(2) changes in the size and exchange rates of the carbon reservoirs on the surface of the Earth (figure 7);

(3) changes in the intensity of the components of the galactic (and solar) radiations that give rise to the ^{14}C formation on Earth.

The spectral lines must result from periodic variations of the intensity of the ^{14}C-producing radiation that reaches the Earth.

In addition to the two series of measurements discussed here in detail, many so-called precision, high-precision, or ultra-high-precision measurements have been published during the past few years (see, for example, Kra 1986). However, to our knowledge analyses of the spectra of their time series and comparison with our, or Dr Pearson's, results have not been made yet in most cases. We hope very much that this will be done soon by the respective authors. Many details presented here graphically need explanations.

5. Concluding remarks

When, in 1958, H. de Vries published his paper 'Variation in concentration of radiocarbon with time and location on Earth', it appeared probable, and soon certain, that the overall concentration of this carbon isotope had not remained constant, but had changed both slowly on a timescale of several thousand years and also more rapidly on a timescale of some hundred years. We have considered here the more rapid variations, fluctuations, or 'de Vries wiggles', as they were called by some investigators. Mostly they were considered rare stochastic phenomena, in some way connected with the occurrence of climatic change. However, it was possible to show less than 20 years later that bristlecone pine wood, dated by its tree rings by C. W. Ferguson, showed ^{14}C-variations throughout the investigated time range, albeit with greatly varying amplitudes (Suess 1970). These variations were then shown to exhibit a well-defined time spectrum with a prominent line at slightly more than 200 years.

For more than 10 years now the majority of experimental workers has been most skeptical regarding the existence of periodicities in the natural ^{14}C record and of meaningful spectral lines in the observed secular variations of the cosmic ray produced ^{14}C on the surface of our planet. It was pointed out by Damon *et al.* (1978) that 'no single de Vries type fluctuation (prior to the Medieval Warm Epoch) has been confirmed by two or more laboratories', and also that 'wiggles of the type reported by Suess (1970) have not been confirmed'. This, and probably other factors, then led the U.S. National Science Foundation to deny repeated requests for further financial support. Now, thanks to the outstanding success of the experimental work at Belfast, headed by Dr Gordon Pearson, and the sophisticated theoretical evaluation of the results by Professor Sonett at the University of Arizona, there cannot be any doubt that these wiggles are real and reflect periodic occurrences. They constitute an unexpected, important source for new scientific information. Unfortunately, it will be too late for me to participate in its exploration, but hopefully still in time to enjoy the findings of my colleagues.

References

Arnold, J. R. & Libby, W. F. 1951 *Science, Wash.* **113**, 111–120.
Baillie, M. G. L., Pilcher, J. R. & Pearson, G. W. 1983 *Radiocarbon* **25**, 171–178.
Baxter, M. S. & Walton, A. 1971 *Proc. R. Soc. Lond.* A **321**, 105–127.
Clark, R. M. 1975 *Antiquity* **49**, 251–266.
Damon, P. E., Lerman, J. C. & Long, A. 1978 *A. Rev. Earth planet. Sci.* **6**, 484.
De Jong, A. F. M., Mook, W. G. & Becker, B. 1980 *Nature, Lond.* **280**, 48–49.
de Vries, H. 1958 *Proc. K. ned. Akad. Wet.* B **61**, 1.
Ferguson, C. W. 1968 *Science, Wash.* **159**, 839–846.
Houtermans, F. G. & Oeschger, H. 1955 *Helv. phys. Acta* **28**, 464–466.
Houtermans, J. C. & Oeschger, H. 1958 *Helv. phys. Acta* **31**, 117–126.
Houtermans, J. C., Oeschger, H. & Suess, H. E. 1973 *J. geophys. Res.* **78**, 1898–1908.
Kra, R. S. (ed.) 1986 *Radiocarbon* **28** (Calibration Issue).
Libby, W. F. 1947 *J. Am. chem. Soc.* **69**, 2523–2534.
Libby, W. F. 1952 *Radiocarbon dating.* University of Chicago Press.
Libby, W. F. 1970 *Phil. Trans. R. Soc. Lond.* A **269**, 1–10.
Linick, T. W., Suess, H. E. & Becker, B. 1985 *Radiocarbon* **27**, 1, 20–30.
Neftel, A., Oeschger, H. & Suess, H. E. 1981 *Earth planet. Sci. Lett.* **56**, 127–147.

Pearson, G. W. 1979 *Radiocarbon* **21**, 1–21.
Pearson, G. W., Pilcher, J. R., Baillie, M. G. L., Corbett, D. M. & Qua, F. 1986 *Radiocarbon* **28**, 911–934.
Ralph, E. K., Michael, H. N. & Hahn, M. C. 1973 *MASCA Newslett.* **9**, 1–20.
Sonett, C. P. & Suess, H. E. 1984 *Nature, Lond.* **307**, 141–143.
Suess, H. E. 1939 *Z. Elektrochem. angew. phys. Chem.* **45**, 647–648.
Suess, H. E. 1970 *Proc. 12th Nobel Symp., Upsala 1969* (ed. I. Olsson). Stockholm: Almquist Wiksell-Gebers Forlag.
Suess, H. E. 1980 *Radiocarbon* **22**, 200–209.
Suess, H. E. 1986 *Radiocarbon* **28**, 259–266.
Taylor, R. E. 1987 *Radioactive dating, an archaeological perspective*. New York: Academic Press.
Thiemens, M. H. & Heidenreich, J. E. III 1983 *Science, Wash.* **219**, 1073–1076.

Phil. Trans. R. Soc. Lond. A **330**, 413–426 (1990)

Printed in Great Britain

413

The spectrum of radiocarbon

By C. P. Sonett[1] and S. A. Finney†

*Department of Planetary Sciences ([1]also Lunar and Planetary Laboratory), University of Arizona,
Tucson, Arizona 875721, U.S.A.*

The power spectral density of the specific activity of radiocarbon variations, using an
absolute chronology based on tree ring count, exhibits spectral lines at a number of
periods: 2300, 964, 753, 717, 493, 413, 357, 229 and 208 years as well as recording
a secular variation over the full 9000-year record. These variations appear in both the
La Jolla and Belfast radiocarbon records and some are also detected in the Camp
Century ^{10}Be record, though its secular variation appears to lead that of the
radiocarbon by about 2100 years. Of the total number of nine prominent radiocarbon
features, most are mutually dependent with perhaps only three independent lines.
The 208-year period appears modulated by the long 2300-year period. The evidence
of modulation of the 208- (and possibly 229-) year period(s) by the 2300-year period
suggests a solar source for the latter features, through 200-year spectral features are
also detected in the tree ring spectrum, thus tying Sun and climate together. A solar
source signalled jointly through correlated radiocarbon and atmospheric temperature
variations suggests that solar hydromagnetic and bolometric variations are coupled.
Moreover, as evidence is lacking for variations in ^{10}Be or in the geomagnetic field with
a period of 2300 years, by a process of elimination together with suggestive global
ocean deep water return times of more or less a millennium, a surmise is that chemical
resonances may account for the periodicity in radiocarbon with the variation in
oceanic carbonate concentration recorded in the radiocarbon record as tracer. The
evidence for correlated oscillations in air temperature detected by tree-ring-growth
cyclicity in Campito Mt bristlecone pine trees and the radiocarbon variability is
consistent with a model of atmosphere–ocean resonances underlying the radiocarbon
periodicities.

1. Introduction

That the atmospheric inventory of radiocarbon (^{14}C) is constant was long a basic principle of
radiocarbon dating, though it was shown as early as 1958 by de Vries (1958) that at least for
the short term of a few hundred years this principle is violated (see de Jong & Mook 1980).
The well-known major and possibly secular variation of *ca.* 10 % was also discovered
subsequent to Ferguson's extension of the bristlecone pine chronology (Ferguson 1970; Suess
1965, 1967, 1978; Damon *et al.* 1972). As the principle of reservoir constancy has been slowly
abandoned, the explanation of the time variability of the radiocarbon record has become a
central problem of geophysics and geochemistry and possibly solar physics. Cosmic ray (CR)
flux variations are hydromagnetic and probably involve the Sun though marginally likely to
arise from interstellar variability (Sonett *et al.* 1987). Determination of the various forcing
functions responsible for the variations in the radiocarbon record is a major problem in the
study of carbon on Earth. It is this problem that we explore in this paper.

---

† Present address: FTD/SQDEI, Wright–Patterson Air Base, Ohio 45433-65503, U.S.A.

[15]

Of the eight isotopes of carbon, radioactive (^{14}C) is produced terrestrially primarily by the nuclear reaction

$$^{14}N(n, p) \rightarrow {}^{14}C, \tag{1}$$

where ^{14}N is atmospheric. ^{14}C decays by

$$^{14}C \rightarrow {}^{14}N + v^- + \beta^-, \tag{2}$$

where v^- is the antineutrino and β^- the electron. The half-life of radiocarbon, $\rho_{\frac{1}{2}} = 5730$ years (Lederer *et al.* 1967). The ultimate source of radiocarbon is traceable to the CR flux incident upon the atmosphere from which, by spallations, at atmospheric neutron sea is generated. It is these neutrons that participate in the $N(n, p)$ reaction yielding ^{14}C (Lingenfelter & Ramaty 1970; O'Brien 1979). If the CR flux were constant, the atmospheric ^{14}C inventory would be in secular equilibrium.

Carbon is a geochemically active element, making it difficult to trace through the environment, but at the same time providing an important tracer for the physics and chemistry of the oceans, atmosphere and biosphere. Because of the reservoir capacity of the atmosphere and the relatively long half-life of ^{14}C, the atmosphere act like a low-pass filter with periods *ca.* 100 years attenuated in amplitude by a factor of *ca.* 20 (Houtermans *et al.* 1973; Siegenthaler *et al.* 1980). This tends to increase the difficulty of measurements involving short periods as the signal:noise ratio is small. The specific variable that occupies a central role is the Δ^{14}C or delta radiocarbon defined by $\Delta^{14}C = (^{14}C_{ref} - {}^{14}C_{inv})/{}^{14}C_{inv}$ where $^{14}C_{ref}$ is the chronological reference activity at time t determined from tree rings count and $^{14}C_{inv}$ is the true measured activity. It is customary to multiply this value by 10^3, giving units of 'per mille'. Because of the very complex chemistry of radiocarbon, the certification of spectral features can be materially aided by reference to the Camp Century ^{10}Be record (Beer *et al.* 1983) for which long-period variability is most likely assignable to modulation of the incident CR flux, interplanetary or geomagnetic or both.

The radiocarbon record is obtained almost without exception from measurement of radiocarbon ages in wood from trees for which an absolute chronology exists by virtue of growth ring count. Much of the variability is quasi-periodic, though the long trend (*ca.* 10000–12000 years), discussed next, may be secular. The records often though not exclusively considered by us are those of Suess at the University of California (La Jolla), the Belfast (Pearson) sequence (Pearson *et al.* 1986) and to a slightly lesser extent the sequence from Rhinegraben oak for which the chronology was worked out by Beker and the ^{14}C record by Suess.

Delta radiocarbon records are basically noisy. As putative signal levels are low, it is important to establish the primary statistical properties of these time sequences so that we know just what it is that we are dealing with and what the likelihood is of the appearance of artefacts. Some idea of the moments and of the stationarity is given by dividing the sequences into halves before computation of the moments. Both the La Jolla and Belfast subsequences are normally distributed; Belfast is strongly stationary whereas La Jolla shows non-stationarity at great age as a result of lessened accuracy. It becomes closely stationary when truncated to the length of the Belfast sequence (520 B.C.–1835 A.D.) (stationarity can be strong, weak, or absent depending upon independence of various moments upon time). To what degree the normal distribution is indicative of noise rather than signal has not been established.

2. The long trend

The 'secular' trend is the most obvious feature of the radiocarbon record, a 14% decrease in the atmospheric radiocarbon inventory from *ca.* 6000 B.C. to a broad minimum near 500 A.D. followed by a sharp rebound by about 2% to the modern value. The trend is demonstrated convincingly in figure 1, showing both the La Jolla and Belfast records low-pass filtered to and upper frequency of 0.01 a^{-1} (100 years). Detrending conventionally is by least squares fitting of period, amplitude and phase of a sinusoid (see, for example, Damon *et al.* 1988) and is attributed to modulation of the incident CR flux at the top of the atmosphere by changes in the geomagnetic field (Bucha 1969, 1970; Creer 1989; Barton *et al.* 1979) and by changes in the global carbon reservoir (Lal & Revelle 1984). In this paper we use the alternate of a sine wave and a third-order Legendre polynomial to detrend the record.

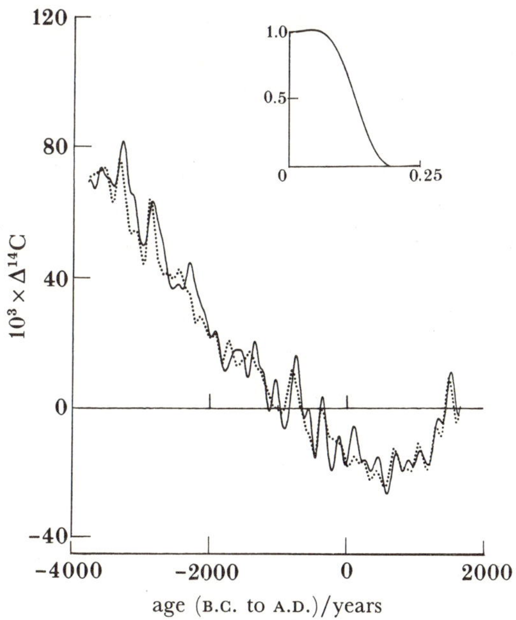

FIGURE 1. La Jolla (solid line) and Belfast (broken line) radiocarbon time sequences filtered to remove periods less than *ca.* 100 years. Inset shows filter amplitude transfer function. Periodicity of *ca.* 200 years appears prominently. (From Sonett (1985).)

By representing the long-term trend as a sinusoid of 8000-year period and using the archaeomagnetic data of Bucha (1970); Suess (1969, 1970) found an attenuation (the box model of Houtermans *et al.* (1973)) of 7 times and a phase shift of 30° between the atmospheric inventory and the radiocarbon production rate. To match the 0.52 power law for geomagnetic modulation of the radiocarbon production (Elsasser *et al.* 1956), the Creer archaeomagnetic data and the Belfast radiocarbon record suggest a phase lag of 500 years and an attenuation factor of 3.2 between production and inventory. The phase shift is consistent with that of Houtermans *et al.* but the attenuation is less. Thus a contribution from a non-magnetic perturbation is suggested. Lal & Revelle (1984) have shown that an increase in atmospheric pCO_2 could result in a decrease of about 7% in $\Delta^{14}C$, making a climatic effect possible.

The trend is also detected in the ^{10}Be record from the Greenland Camp Century ice core

(Beer *et al.* 1988) that is used to serve as a chemistry-free proxy for ^{14}C production. (The atmospheric beryllium inventory follows the production rate closely with little lag (Raisbeck & Yiou 1981).) But the beryllium trend is somewhat difficult to see because of the lack of reservoir-induced high-frequency attenuation and the trend is somewhat obscured by short-wavelength phenomena. A puzzling result from comparing the ^{10}Be and Belfast radiocarbon records is a peak in their correlation for a lag of 2100 years in the ^{14}C record against ^{10}Be. If real, we have no explanation for this time shift, but it may be a computational artefact because of high-frequency oscillations in the ^{10}Be record corrupting the Legendre function fit.

3. Power spectral density

The spectral interval considered here is restricted to periods greater than *ca.* 200 years. Suess (1980) reported previously unpublished calculations of Kruse of the power spectrum of the La Jolla Δ^{14}C record showing spectral lines at 2400, 930, 498, 308, 202, and lesser years. Neftel *et al.* (1981) confirmed the 200-year period in the La Jolla Δ^{14}C spectrum. A more-detailed statistical analysis (Sonett 1984) showed marginal evidence that the *ca.* 200 year period was subject to modulation by a longer period at *ca.* 2000 years. Finney & Sonett (1988) obtained the spectrum of the Belfast radiocarbon record using the new bayesian algorithm (Bretthorst 1988) and Finney (1988) has applied it to both the La Jolla and newer Belfast sequences and shows a high degree of compatibility of the two records and the statistical reliability of the spectral features.

The Δ^{14}C spectrum computed by using the discrete Fourier transform (DFT) (figure 2a) and maximum entropy (MEM) (figure 2b) confirms the complexity indicated earlier by Suess; some 10 spectral lines of differing intensity in the range of 200–2300 years are present. Both the La

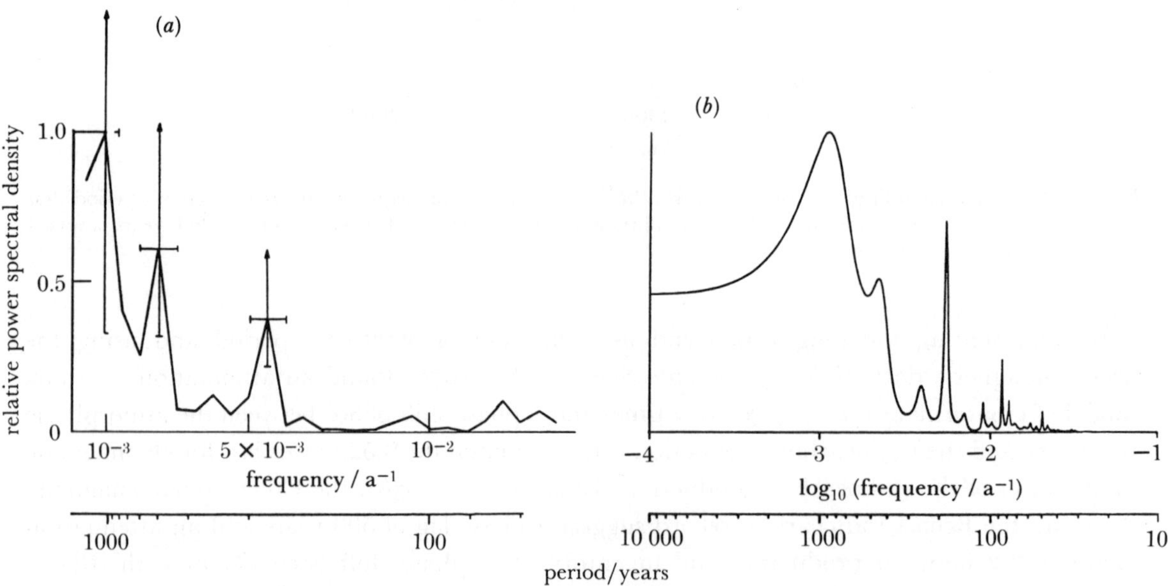

FIGURE 2. (a) Discrete power spectral density (PSD) of La Jolla ^{14}C sequence. Vertical bars are χ^2 amplitude estimates. Horizontal bars are estimated frequency resolution based upon computation steps, i.e. $f_0, f_1, \ldots$, where f_0 is the base frequency. (b) Maximum entropy PSD of La Jolla sequence showing prominent 200-year period. Resolution of both spectra are just sufficient to show major very-long period power, lesser at intermediate periods, and significant power in 200-year neighbourhood.

Jolla and Belfast records show similar lines. With decreasing period, spectral features are increasingly damped because of the atmospheric reservoir capacity. Most power appears at long periods. But there frequency resolution is reduced, and computational points may not be centred on line peaks leading to errors in estimation of amplitude. Moreover, line-frequency estimates are biased by window convolution with the data leading to side lobe and leakage problem. Some relief from the convolutional problem occurs by using the MEM algorithm, but this is counteracted by the lack of an adequate theory for specifying the optimum computational order. Nevertheless, MEM is a useful tool when used judiciously.

In the algorithm for computation of spectra developed by Jaynes and by Bretthorst (1987) one or more model functions are fitted to the data; the choice of an orthogonal coordinate basis set into which the data are projected eliminates side lobes. Additionally, data interpolation is unnecessary. Model probability is computed by projection of the data vector (in a Hilbert space) onto orthogonalized sinusoidal model functions. These provide a model likelihood. Table 1 gives model estimates using the combined La Jolla and Belfast records. Because the Bretthorst algorithm yields frequency probability estimates for a prior model (sines and cosines here) a spectrum is not directly provided. However, a spectrum can be constructed (figure 3) by plotting lines centred on the frequency probability maxima and with widths given by $1/e$ of the maximum probability. Then line amplitudes are made to correspond to model amplitudes.

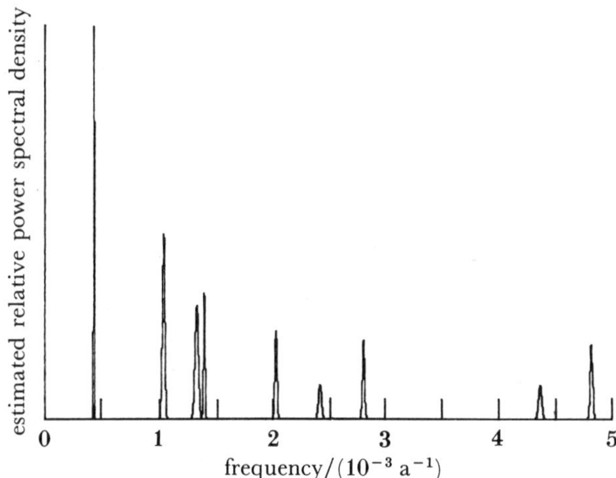

FIGURE 3. Spectral construction of spectrum from sine or cosine model bayesian estimates using both Belfast and La Jolla Δ^{14}C data. Line widths at base correspond to $1/e$ decrease in frequency positional probability from line centre. Amplitudes are estimated from sine and cosine model components.

Many of the lines appear to be linearly dependent upon one another. It is straightforward to show that all nine of the most prominent lines can be explained by linear combinations of a non-unique choice of three of the lines. As noted earlier we make use of the ^{14}Be record to distinguish the direct CR variability. Table 2 shows the frequencies and amplitudes of the beryllium record determined by the Bretthorst algorithm. However, the beryllium sample ages are based upon ice flow modelling and thus are intrinsically of greater uncertainty than indicated merely from the spectral modelling. Therefore, we do not report frequency uncertainties in table 2.

Lines $f_{Be,2}$ and $f_{Be,3}$ are within *ca.* 3 % of $f_{C,5}$ and $f_{C,8}$ in frequency. Because this implies less

TABLE 1. BAYESIAN ESTIMATES OF ^{14}C SPECTRAL LINES

$10^4 \times$ frequency[a]	period	amplitude	phase[b]
a^{-1}	years		
4.3267 ± 0.0891	2311	4.38	-119
1.0372 ± 0.0290	965	2.97	98
1.3276 ± 0.0408	753	2.34	-3
1.3941 ± 0.0186	717	2.46	76
2.0290 ± 0.0256	493	2.07	145
2.4209 ± 0.0357	413	1.27	87
2.8043 ± 0.0267	357	1.94	13
4.3637 ± 0.0400	229	1.28	-146
4.8170 ± 0.0304	208	1.90	129

[a] Errors are 2σ.

[b] Angles are measured $+$ (counterclockwise) and $-$ (clockwise).

TABLE 2. CAMP CENTURY ^{10}BE FREQUENCY AND AMPLITUDE ESTIMATES

(Subscript c means cosine component; subscript s means sine component.)

line	frequency	amplitude
$f_{\mathrm{Be},1,c}$	3.0544×10^{-4}	-0.0923 ± 0.0155
$f_{\mathrm{Be},1,s}$		-0.0484 ± 0.0121
$f_{\mathrm{Be},2,c}$	1.9550×10^{-3}	0.0501 ± 0.0156
$f_{\mathrm{Be},2,s}$		0.0698 ± 0.0207
$f_{\mathrm{Be},3,c}$	4.478×10^{-3}	-0.0631 ± 0.0254
$f_{\mathrm{Be},3,s}$		-0.0352 ± 0.0231

than a full cycle difference between these lines in the two records, a possible common forcing mechanism is suggested. Contrasting this $f_{\mathrm{Be},1}$ and $f_{\mathrm{C},1}$ frequencies differ by 30%. Thus it is unlikely that forcing by the underlying non-dipole component of the geomagnetic field is responsible for these features. $f_{\mathrm{C},1}$ has been suggested as a result of geomagnetic modulation (Damon *et al.* 1990) but the parameter detected in the geomagnetic field at 2300 years is the direction sometimes attributed wholly to variations in the magnitude of the quadrupole moment. But assuming that the whole field participates in modulation, primarily the direction rather than the magnitude of the moment would be involved, and this cannot globally affect the CR flux.

The following assignment rationale assumes that the longest radiocarbon period $f_{\mathrm{c},1}$ is primary. Shuffling linear combinations of frequencies discloses that either $f_{\mathrm{C},2}$ or $f_{\mathrm{C},4}$ and $f_{\mathrm{C},8}$ or $f_{\mathrm{C},9}$ must be primary. Identification of $f_{\mathrm{C},5}$ as a second harmonic of $f_{\mathrm{C},2}$ and its appearance in the ^{10}Be record strengthens the supposition that $f_{\mathrm{C},2}$ is primary. It absence in the ^{10}Be may be because of its low amplitude. The appearance of $f_{\mathrm{C},8}$ in both the radiocarbon and ^{10}Be records suggests it to be primary (Finney 1988). But all these assignments should be regarded as tentative. In making the inferred relation to the ^{10}Be record note should be taken of the very restricted 141-datum ^{10}Be record. Table 3 shows line assignments satisfying these restrictions.

TABLE 3. A PLAUSIBLE ^{14}C LINE SOURCE ASSIGNMENT

symbol	period	source
$f_{c,1}$	2311	primary
$f_{c,2}$	965	primary
$f_{c,3}$	753	$3f_{c,1}$
$f_{c,4}$	717	$f_{c,2}+f_{c,1}$
$f_{c,5}$	493	$2f_{c,2}$
$f_{c,6}$	413	$2f_{c,2}+f_{c,1}$
$f_{c,7}$	357	$2f_{c,2}+2f_{c,1}$
$f_{c,8}$	229	primary
$f_{c,9}$	208	$f_{c,8}+f_{c,1}$

4. MODULATION

Given two functions

$$\alpha(\omega_1, t, \phi_1) = A\sin(\omega_1\tau+\phi_1)$$

and

$$\beta(\omega_2, t, \phi_2 = B\sin(\omega_1 t+\phi_1)$$

a nonlinear operation R jointly upon α and β yields a function that is Taylor expanded into a series of products of α and β to increasing powers. Exponentiation of a periodic signal merely changes the amplitudes of its harmonics. As each term in R is a product it can be reduced to a series of amplitude modulations and variations in harmonic structure, but the symmetry properties of simple amplitude are no longer applicable. However, the resulting spectrum may still contain features that are linear combinations of the original two fundamental frequencies.

In the unified spectrum of table 1 several lines exhibit spacing that is within the statistical uncertainty of other major features. Lines $f_{c,5}$ and $f_{c,6}$ are separated by *ca.* 2300 years suggesting modulation by the 2300-year line of a higher-frequency carrier. (We borrow the term 'carrier' from communication engineering to denote the higher of two interaction periodic functions; which function is the 'carrier' is a matter of choice.) Moreover, the two lines, $f_{c,8}$ and $f_{c,9}$ are also separated within 2σ by 2300 years, indicating modulation between the 2300-year period and two high-frequency periods. However, the side band expected at 3.93×10^{-3} a^{-1} is not detected.

5. THE RADIOCARBON INVENTORY

Carbon is stored in the atmosphere, oceans, biosphere, and in mineral form. Its accessibility varies greatly; the atmospheric time constant is of order 1–2 years, while at the other extreme at least some of the abyssal oceanic component is recycled in time as great as 200 million years (the maximum age of the ocean floor). Major terrestrial reservoirs of CO_2 are the atmosphere and oceans with a lesser addition from the biosphere. Reservoirs are tabulated by Walker (1977). From the standpoint of geochemical carbon, radiocarbon constitutes a tracer, though ^{14}C clearly is a substance of independent importance when considering its variations and their source(s). Variations in the inventory arise from (1) changes in the incident flux of galactic cosmic rays (interplanetary magnetic field modulation), (2) changes in solar flare associated cosmic rays, (3) modulation of the incoming CR by changes in the Earth's magnetic field, and finally (4) possible variations in the take-up and release of CO_2 by the oceans. The

[21]

latter is important; later we discuss qualitatively how the long-period spectrum might be forced by oceanic currents.

Of the 10 features identified separately in the La Jolla and Belfast spectra, eight are common to the two (within 2σ). The resulting spectrum of the combined data confirms the earlier statistically tentative surmise that the lines in the 200-year neighbourhood are primary features modulated by the 2300-year line. Because $f_{C,8}$ and $f_{Be,8}$ are common features, it is unlikely that they arise from terrestrial forcing as no evidence exists for geomagnetic variations of the required magnitude. The magnetic record is of insufficient length to clearly rule out magnetic forcing, but neither paleomagnetic or present epoch geomagnetic data provide any evidence (Cox 1969). Chemical forcing is ruled out for Be. The conclusion that *ca.* 200-year periods are extraterrestrial in origin together with the evidence for modulation leads to the likelihood that the 2300-year period is due to terrestrial forcing of the radiocarbon inventory. If radiocarbon is regarded as a tracer for carbon generally, then the global inventory of carbon itself is being forced at 2300 years.

Though $f_{C,2}$ (960 years) does not appear in the beryllium spectrum, the second harmonic ($f_{C,5}$) does. As the former is the weaker of the two it suggests that $f_{Be,2}$ may exist but obscured by noise in the beryllium spectrum. A geomagnetic origin cannot be tested for as the period is in excess of the geomagnetic record. However, there are no linear side band combinations between $f_{C,2}$ and $f_{C,8}$, implying that the two lines do not interact. This is contrary to what would be expected if $f_{C,2}$ were the result of geomagnetic forcing.

6. The joint air temperature Δ^{14}C correlation

Evidence for a correlation exists between the radiocarbon spectrum and the spectrum of variations in the growth of bristlecone pine trees from Campito Mt in the White Mts of eastern California. Campito Mt bristlecone (from the same general region as the trees which supplied the wood for the La Jolla radiocarbon dating) supplies ring growth information that in the arid environment near timberline, is sensitive to air temperature (La Marche 1973). Sonett & Suess (1984) have shown a close similarity between the autospectra of the La Jolla radiocarbon and growth of these trees. This relation appears also by the cross-MEM spectral, coherence, and phase shown in figure 4. The radiocarbon and tree ring data correlation peaks with a relative shift in lag favouring the tree ring data, i.e. the tree response leads by *ca.* 80 years. Furthermore, the two records are inversely correlated in the sense that thicker tree rings (more rapid growth), if associated with higher temperature and greater solar activity also infers greater interplanetary modulation and thus lessened production of radiocarbon. These are surmises and depend upon the idea that solar activity is positively correlated with atmospheric temperature. The lag is consistent with an expected fast tree response to temperature against a longer atmospheric response to the radiocarbon source.

7. Feedback and oscillators

Line $f_{C,1}$ has been surmised to be associated with variations in the higher-order (non-dipolar) component of the Earth's magnetic field because an apparent cyclicity exists in paleomagnetic directions (Lund 1983; Creer 1983). However, globally averaged radiocarbon and ^{10}Be production rates are unlikely to be sensitive to changes in field direction alone. Moreover, the

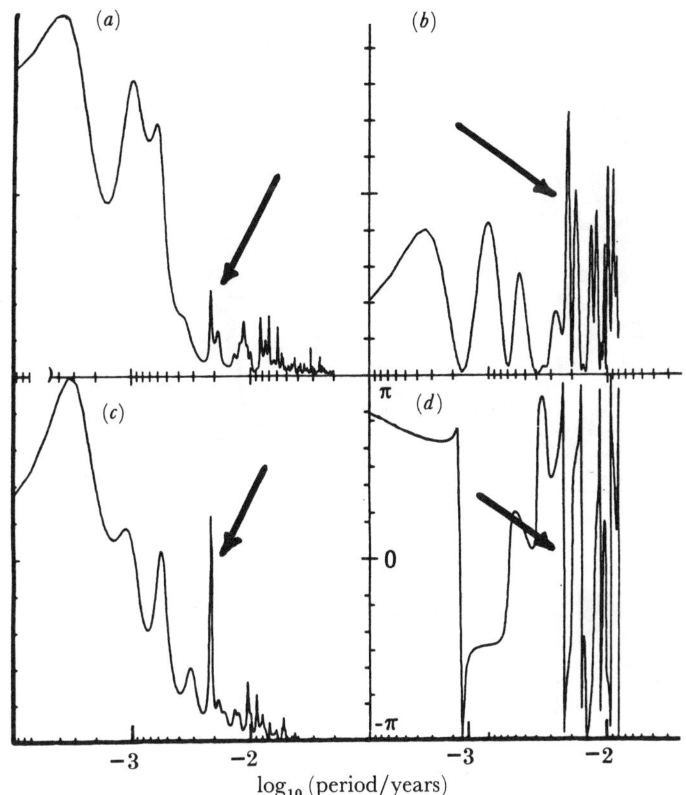

Figure 4. mem cross spectrum of Campito Mt bristlecone pine growth ring sequence and La Jolla Δ^{14}C sequence. (a) Campito autospectrum, (b) coherence, (c) La Jolla autospectrum, (d) relative phase. (Modulo 2π; mem order = 75, arrows point to a ca. 200-year line.)

^{10}Be record lacks a ca. 2000-year spectral feature. Because a vague suggestion exists that the Maunder-like minima are amplitude modulated with a period of ca. 2000 years a solar origin cannot be ruled out with confidence. Nevertheless, as the ^{10}Be record shows a continuous ca. 200-year period and the longer ca. 2000-year) period is absent in this record, the argument for a terrestrial source for the ca. 2300-year period is strengthened. By a process of elimination, the major candidate appears to be a 'chemical' oscillation.

The tree ring record (aside from possible arcane photosynthesis effects) involves an air-temperature–radiocarbon-production relation at several periods within the radiocarbon spectral range. Can oscillatory oceanic forcing exist and if so can periods be inferred within the radiocarbon range but independent of the ^{10}Be record? It is known that the global deep water cycle times vary from fractional millennia to 1000–2000 years. Such a return path in effect stores carbon except for vertical diffusion and sedimentation (Broecker & Li 1970). If an increase of atmospheric carbon is destablizing with respect to respiration of CO_2 from surface waters, then the deep water return path could function as a kind of delay system. The electrical analogy is a dispersive delay line that for certain periods depending upon circuit constants will shift phase sufficiently to sustain oscillations. Under such conditions the deep-water radiocarbon record becomes a tracer for a global oscillation of that part of the carbon not locked into the geological reservoir. A fanciful example of the return paths is given by Broecker & Peng (1982) from which figure 5 is adapted. The importance of this to climate would rest

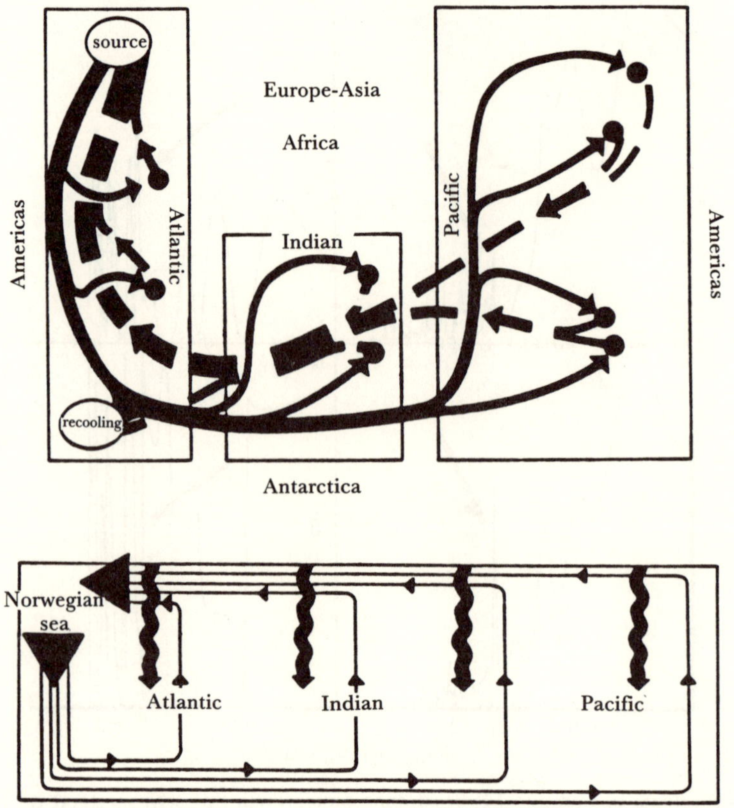

FIGURE 5. A fanciful schematic of the deep water return paths for the world ocean. Return circuit times have been variously reported from 700 to 2000 years based upon deep water radiocarbon ages. Return path corresponds to a frequency sensitive (dispersive) delay line. Combined with positive gain and sufficient phase shift, self-excitation of the system is conjectured. The heavy wave lines represent particle sedimentation, which would have the effect of adding younger radiocarbon, thus decreasing inferred deep water return times. (From Broecker & Peng (1982).)

upon the amplitude of the atmospheric reservoir oscillations at 2300 years that were linked to the oceanic system.

That the Sun can support oscillations reflected in the terrestrial record must rest basically upon solar hydromagnetics for the transit time for waves across the Sun is not more at most of order 1 h. The long terrestrial record is from radiocarbon, but shorter records, still long relative to the transit time come from auroras (Schröder 1988) and from study of sunspots. The Gleissberg period (*ca.* 80–90 years) is obvious in the sunspot index autocorrelation (figure 6). However, reservoir attenuation of the atmosphere spectral line amplitudes of radiocarbon decrease by large factors (*ca.* 200 for 11-year period) (Houtermans, *et al.* 1973). (See also Castagnoli *et al.* 1984.)

Other periods, e.g. 100–150 years, are also present in the radiocarbon record from time to time (de Jong & Mook 1980), but may be non-stationary (Damon *et al.* 1990). Stuiver & Quay (1980) have shown that variability in the radiocarbon record extends downward in period through the Maunder, Spörer, etc., minima (figure 7), whereas Kocharev (1987) has shown that the variability seems to exist even in the 11-year cycle (see also Beer *et al.* 1983). Kocharev's Δ^{14}C record furthermore implies the continuation of solar activity through the

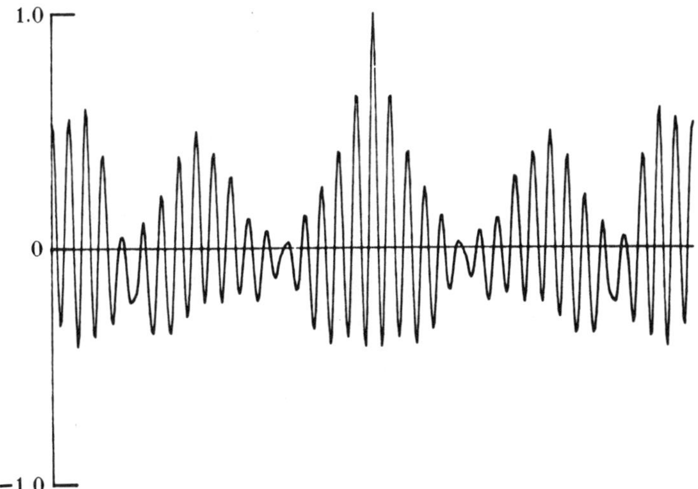

FIGURE 6. Autocorrelation of Wolf sunspot index numbers, showing prominent 80–90-year amplitude modulation (Gleissberg period) of the basic 11-year solar activity cycle.

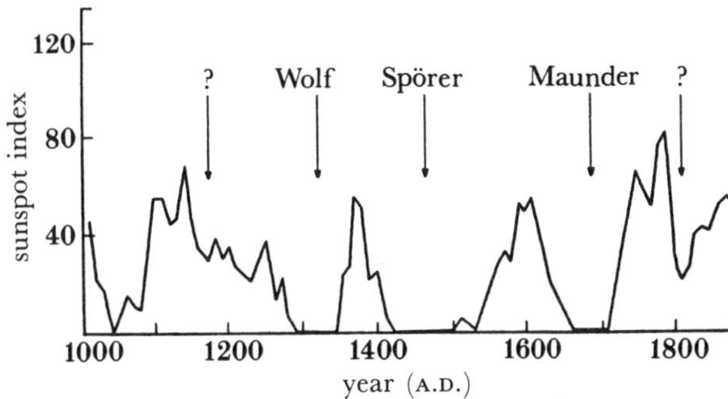

FIGURE 7. Reconstruction of the sunspot index for the past millennium from Pacific northwest radiocarbon (From Stuiver & Quay (1980).)

Maunder Minimum with a change in the primary period from 11 to 22 years during the interval!

The combination of a small fossil magnetic field in the solar core added to the dynamo field, all modulated by a field having the Gleissberg period, yields a spectrum similar to the observed sunspot index spectrum (Sonett 1984). Substitution in this model of a very-long-period field for the fossil (or in addition to it) may provide power in the frequency range of interest for the radiocarbon spectrum. In the Sonett model the substitution is made for the fossil field so that now the sunspot index has the general form (restricted to amplitude modulation) of

$$S = [(1 + \alpha \cos \omega_{m1} t) (\cos \omega_c t + \cos \omega_{m2} t)]^2 + n(t)^2, \qquad (3)$$

where α is the amplitude modulation factor, ω_{m1} the long-period modulation frequency, ω_c the Hale period, ω_{m2} an additional longer period, and $n(t)$ noise. Such a model is *ad hoc* in the sense of the addition of ω_{m2} and a physical model must be developed to support this conjecture.

The major evidence that the 200-year period is solar comes from the suspected modulation by the longer *ca.* 2300-year period, inferring a terrestrial modulation of a periodic extraterrestrial source, but even this is not unique. The evidence that line features are non-stationary in amplitude and perhaps even in frequency increases the difficulty in making assignments of sources. It should be noted that the general shape of the radiocarbon spectrum is red, i.e. increasing power with period. Together with the transient stability of the spectra (Sonett 1984) the suggestion is arguable that the underlying dynamical system responsible for the variations is chaotic or at least quasi-period (as seen in the time period discussed in this paper) (see, for example, Thompson & Steward 1982). However, as noted earlier the low-pass filter properties of the atmosphere may explain the 'redness'. Moreover, tests for fractional dimensionality require, at the least, evenly spaced data (Grassberger & Procaccia 1983), which is not satisfied by the La Jolla sequence. Although the Belfast sequence is nearly so, its length is only half that of the La Jolla sequence that already, in view of its low signal:noise ratio, is marginal for fractional dimensionality tests. Lastly, Feynman & Gabriel (1990) have studied evidence that the Gleissberg cycle and Maunder minima are tied together through quasi-periodicity or chaotic behaviour. Thus this matter remains unsettled.

We thank H. E. Suess for a continuing dialogue on radiocarbon problems. This research was supported by the solar–terrestrial programme of the National Science Foundation.

8. References

Barton, C. E., Merrill, R. T. & Barbetti, M. 1979 Intensity of the earth's magnetic field over the last 10000 years. *Phys. Earth planet. Interiors* **20**, 96.

Beer, J., Andree, M., Oeschger, H., Stauffer, B., Balzer, R., Bonani, G., Stoller, M., Suter, M., Wölfli, W. & Finkel, R. C. 1983 Temporal ^{10}Be variations in ice. *Radiocarbon* **25**, 269.

Beer, J., Siegenthaler, U., Bonani, G., Finkel, R. C., Oeschger, H., Suter, M. & Wölfi, 1988 Information on past solar activity and geomagnetism from ^{10}Be in the Camp century ice core. *Nature, Lond.* **331**, 675–679.

Bretthorst, G. L. 1988 *Bayesian spectral analysis and parameter estimation. Lecture notes in statistics, vol. 48* (ed. J. Berger, S. Fienberg, J. Gani, K. Krickeberg & B. Singer). Berlin: Springer-Verlag.

Broecker, W. S. & Li, Y.-H. 1970 Interchange of water between the major oceans. *J. geophys. Res.* **75**, 3545–3552.

Broecker, W. S. & Peng, T.-H. 1970 *Tracers in the sea.* New York: Lamont-Doherty Geological Observatory.

Bucha, V. 1969 Changes of the Earth's magnetic moment and radiocarbon dating. *Nature, Lond.* **224**, 681.

Bucha, V. 1970 Influence of the earth's magnetic field on radiocarbon dating. In *Radio-carbon variations and absolute chronology. Nobel Symp.* **12**, 595.

Castagnoli, G. C., Bonino, G., Attolini, M. R., Galli, M. & Beer, J. 1984 Solar cycles in the last centuries in ^{10}Be and δ^{18}O in polar ice and in thermoluminescence signals of a sea sediment. *Neuvo Cim.* C **7**, 235–244.

Cox, A. 1969 Geomagnetic reversals. *Science, Wash.* **163**, 237.

Creer, K. M. 1989 Geomagnetic field and radiocarbon activity through Holocene time. In *NATO advanced research workshop on secular, solar, and geomagnetic variations through the last 10000 years* (ed. F. R. Stephenson & A. W. Wolfendale). Reidel: NATO. (In the press.)

Creer, K. M., Tucholka, P. & Barton, C. E. 1983 *Geomagnetism of baked clays and recent sediments.* Amsterdam: Elsevier.

Damon, P. E. 1970 Solar induced variations of energetic particles at one AU. In *The solar output and its variations* (ed. O. R. White). Colorado Associated University Press.

Damon, P. E. 1988 Production and decay of radiocarbon and its modulation by geomagnetic field-solar activity changes with possible implications for global environment. In *Secular solar and geomagnetic variations in the last 10000 years* (ed. F. R. Stephenson & A. W. Wolfendale). Kluwer.

Damon, P. E., Cheng, S. & Linick, T. W. 1990 Fine and hyperfine structure in the spectrum of secular variations of atmospheric ^{14}C. *Radiocarbon* (In the press.)

Damon, P. E., Lerman, J. C. & Long, A. 1978 Temporal fluctuations of atmospheric C: causal effects and implications. *A. Rev. Earth planet. Sci.* **6**, 457.

Damon, P. E., Long, A. & Wallick, E. I. 1972 Dendrochronologic calibration of the carbon-14 time scale. In *Int. Conf. carbon dating, 8th Proc.* (ed. T. A. & G. Taylor), pp. A28–71. Royal Society of New Zealand.

de Jong, A. M. F. & Mook, W. G. 1980 Medium-term atmospheric variations. *Radiocarbon* **22**, 267.

de Vries, H. 1958 Variation in the concentration of radiocarbon with time and location on Earth. *Proc. K. ned. Akad. Wet.* B **61**, 94.

Elsasser, W., Ney, E. P. & Winckler, J. R. 1956 Cosmic ray intensity and geomagnetism. *Radiocarbon* **28**, 266.

Ferguson, C. W. 1971 Dendrochronology of Bristlecone pine, Pinus Aristata: establishment of a 7484 year chronology in the White Mts. of eastern-central California. In *Radio-carbon variations and absolute chronology, Proc. 12th Nobel Symp.* (ed. I. U. Olsson), pp. 237–259. New York: Wiley.

Feynman, J. & Gabriel, S. B. 1990 Period and phase of the 88-year solar cycle and the Maunder Minimum: evidence for a chaotic Sun. *Sol. Phys.* (In the press.)

Finney, S. A. 1988 The spectrum of radiocarbon. M.S. thesis, University of Arizona, U.S.A.

Finney, S. A. & Sonett, C. P. 1988 High resolution spectral analysis of the Irish oak radiocarbon record. *Proc. 19th Lunar and Planetary Science Conf.*

Grassberger, P. & Procaccia, I. 1983 Measuring the strangeness of strange attractors. *Physica* D 9, 189.

Houtermans, J. C., Suess, H. E. & Oeschger, H. 1973 Reservoir models and production rate variations of natural radiocarbon. *J. geophys. Res.* 78, 1897.

Kocharev, G. E. 1987 Nuclear processes in the solar atmosphere and the particle acceleration problem. *Astrophys. Space Phys.* 6, 155.

Lederer, C. M., Hollander, J. M. & Perlman, I. 1967 *Table of isotopes*, 6th edn. New York: Wiley.

Lingenfelter, R. E. & Ramaty, R. 1970 Astrophysical and geophysical variations in ^{14}C production. *Radiocarbon variations and absolute chronology, 10th Nobel Symp.* (ed. I. U. Olsson), pp. 513–537. New York: Wiley.

Lal, D. & Revelle, R. 1984 Atmospheric pCO_2 changes recorded in late sediments. *Nature, Lond.* 308, 344.

La Marche, V. C. Jr. 1973 Holocene climatic variations inferred from treeline fluctuations in the White Mts., California. *Quaternary Res.* 3, 632–660.

Lund, S. P. 1983 Quaternary secular variations recorded in central North American wet lake sediments. In *Geomagnetism of baked clays and recent sediments* (ed. K. M. Creer, M. P. Tucholka & C. E. Barton). Amsterdam: Elsevier.

Neftel, A., Oeschger, H., Staffelbach, T. & Stauffer, B. 1988 CO_2 record in the Byrd ice core 50 000–5000 years BP. *Nature, Lond.* 331, 609–611.

O'Brien, K. O. 1979 Secular variation in the production of cosmogenic isotopes in the Earth's atmosphere. *J. geophys. Res.* 84, 423–431.

Pearson, G. W., Pilcher, J. R., Baillie, M. G. L., Corbett, D. M. & Qua, F. 1986 High precision ^{14}C measurement of Irish oaks to show the natural ^{14}C variations from AD 1840 to 5210 BC. *Radiocarbon* 28, 911.

Raisbeck, G. M. & Yiou, F. 1981 Cosmogenic $^{10}Be/^{7}Be$ as a probe of atmospheric transport processes. *Geophys. Res. Lett.* 8, 1015–1018.

Schröder, W. 1988 Aurorae during the Maunder minimum. *Meteorology atmos. Phys.* 38, 246–251.

Siegenthaler, U., Heimann, M. & Oeschger, H. 1980 ^{14}C variations caused by changes in the global carbon cycle. *Radiocarbon* 22, 177–191.

Sonett, C. P. 1982 Sunspot index spectrum from square law modulation of the Hale cycle. *Geophys. Res. Lett.* 9, 1313.

Stuiver, M. & Quay, P. D. 1980 Changes in atmospheric carbon-14 attributed to a variable Sun. *Science, Wash.* 207, 11–19.

Sonett, C. P. 1982 Sunspot index form square law modulation of the Hale cycle. *Geophys. Res. Lett.* 9, 1313.

Sonett, C. P. 1984 Very long periods and the radiocarbon record. *Rev. Geophys. Space Phys.* 22, 239.

Sonett, C. P. 1985 Suess 'wiggles' – a comparison between radiocarbon records. *Meteoritics* 20, 383–394.

Sonett, C. P. & Suess, H. E. 1984 Correlation of bristlecone pine ring widths with atmospheric ^{14}C variations: a climate-Sun relation. *Nature, Lond.* 307, 141.

Sonett, C. P., Morfill, G. E. & Jokipii, J. R. 1987 Interstellar shock waves and ^{10}Be from ice cores. *Nature, Lond.* 330, 458–460.

Stuiver, M. & Quay, P. D. 1980 Changes in atmospheric ^{14}C attributed to a variable Sun. *Science, Wash.* 9, 1–20.

Suess, H. E. 1978 La Jolla measurements of radiocarbon in tree ring dated wood. *Radio-carbon,* 20, 1.

Suess, H. E. 1970 Bristlecone pine calibration of the radiocarbon time scale 5300 BC to the present. In *Radiocarbon variations and absolute chronology, 12th Nobel Symp.* (ed. I. U. Olsson), pp. 595–612. New York: Wiley.

Suess, H. E. 1980 The radiocarbon record in tree rings of the last 8000 years. *Radiocarbon* 22, 200.

Thompson, J. M. T. & Stewart, H. B. 1986 *Nonlinear dynamics and chaos.* New York: Wiley.

Walker, J. C. G. 1977 *Evolution of the atmosphere.* London: Macmillan.

Discussion

A. Berger (*Département de Physique, Université Catholoque de Louvain, Belgium*). Professor Sonett's discovery of a 2300-year cycle is very interesting. Such a quasi-period has also recently been found in ^{18}O deep sea cores with high sedimentation rates (Pestiaux *et al.* 1987). This was tentatively related to the ocean circulation and the astronomical theory. Can Professor Sonett comment on a possible relation between what causes this 2300-year period in both his data and in mine.

C. P. Sonett. Evidence for a spectral feature in the neighbourhood of 2300–2500 years in various natural time series ranges from the ^{18}O record through radiocarbon, and the fascinating sea core data reported by Pestiaux *et al*. I think it is far too early to attempt to assign an overall mechanism, but as is well known the oxygen record is related to rainfall. Some evidence for this period in the bristlecone pine record also appears. If real it indicts temperature as these trees are at timberline and only thermally stressed (C. W. Stockton, personal communication). That the period also appears in the radiocarbon record is especially difficult to understand because it could be due to solar wind modulation, perhaps deep ocean circulation, or even the Earth's magnetic field which displays a directional 'coning' (precession) of 2400-year period (though I doubt whether this could affect the radiocarbon inventory.

Additional reference

Pestiaux, A., Berger, A. & Duplessy, J. C. 1987 Paleoclimatic variability at frequencies ranging from 1 cycle per 10000 years to 1 cycle per 1000 years: evidence for nonlinear behavior of the climate systems. *Climate Change* **12** (1), 9–37.

Phil. Trans. R. Soc. Lond. A **330**, 427–439 (1990)

Printed in Great Britain

427

Environmental information in the isotopic record in trees

By S. Epstein and R. V. Krishnamurthy

*Division of Geological and Planetary Sciences 170-25, California Institute of Technology,
Pasadena, California 91125, U.S.A.*

Twenty-three trees from widely different geographic locations and different environments were analysed for the δD and $\delta^{13}C$ records. The δD values suggested that the temperature of the Earth's surface rose over the past 100 years and probably for the past 1000 years. The rate of warming appears to be latitude dependent, greatest in the cooler areas. The $\delta^{13}C$ record, obtained for seven of the 23 trees, contain the $\delta^{13}C$ decrease due to the anthropogenic effect, the addition of CO_2 from coal and petroleum burning. This effect appears to be twice as high in the Northern Hemisphere as in the Southern Hemisphere.

Introduction

The direct relation between climatic temperatures and the $D:H$ and $^{18}O:^{16}O$ ratios of precipitation is well documented. Generally, these ratios correspond quite faithfully to the worldwide climatic distribution (Daansgaard 1964; Siegenthaler & Oeschager 1980), and as shown in numerous examples of isotopic records in snow and rain, also respond to seasonal variations (Epstein *et al.* 1959). There is a linear relation between the $D:H$ and $^{18}O:^{16}O$ ratios of precipitation and the mean annual temperature of the locations of the samples. The vigorous activity on the isotopic analysis of ice cores from the polar ice caps is a good example of using isotopic analyses of precipitation to obtain climatic information (Arnason 1981).

More recently, efforts have been made to extract climatic information from the isotopic composition of hydrogen and oxygen in plants and more specifically in tree rings. It has been shown that the hydrogen and oxygen isotopic composition in plants is determined by the isotopic composition of the water used by the plants (Epstein *et al.* 1976; DeNiro & Epstein 1979; Ramesh *et al.* 1986; Gray & Song 1984). Thus in principle the isotopic compositions of plants should permit the measurement of the climatic temperatures of their locations.

There are some complications associated with measuring the isotopic composition of hydrogen, oxygen and carbon in plants, due to the isotopic heterogeneity of the various chemical components. For example it has been shown that the δD^* and $\delta^{13}C^*$ of the lipids in the plants can be $100‰$ and $10‰$ respectively lower than the δ values of the total plants (Smith & Epstein 1970). Thus the variation of the lipid fraction in a plant may introduce large isotopic composition changes independent of those resulting from the water it used. Consequently, it is necessary to use a single chemical component in plants to have the best chance of determining uniquely the isotopic composition of the water it used during its growth.

It has now been established that whereas cellulose should be used for the $\delta^{13}C$ and $\delta^{18}O$ analyses, it cannot be used for the δD analyses. The cellulose monomer contains 10 hydrogens, three of which are in the OH group and hence exchangeable. The exchangeable hydrogen will acquire the δD value of its immediate environmental water and will modify the original δD of the cellulose. This problem can be overcome by nitration to replace the OH hydrogen with NO_2 to form nitrocellulose, which then can be used for the δD analysis.

$$\delta D^* = (R_{sample}/R_{standard} - 1) \times 1000, \quad R = {}^{13}C/{}^{12}C, \quad D/H, \quad {}^{18}O/{}^{16}O.$$

Standard is mean ocean water for O_2 and H_2 and Pee Dee belemnite for C.

A relation between the δD of the nitrocellulose from plants and the environmental water was determined (Epstein *et al.* 1976) using 25 different species of plants including marine turtle grass, fresh water lakes or pond plants, and terrestrial plants. For the non-aquatic plants the δD of the environmental waters were estimated from the analysis of lakes or rivers in close proximity to the plants. This relation, with a standard error of $\pm 10\%_0$, was found to be of the form $\delta D(\text{plants}) = \delta D(\text{water}) - 20$, the water media concentrating deuterium. Considering the large variety of plants used and the possible biochemical and biophysical differences that may exist among them, this relation is good. It is reasonable to expect that isotopic data from a series of tree rings from a single tree will reflect more faithfully the variation of the isotopic composition of the water it uses. Probably the most serious modifying effect would be the evaporative transpiration of the leaf water that would enrich the deuterium in the tree relative to the environmental water source. Thus it is possible that warm, dry climate would show a higher δD value and a higher temperature than actually exists. However, we really do not understand this effect and its magnitude. In many cases, the fixation of CO_2 is inhibited when the tree is subjected to unusually warm dry weather. In addition, Sternberg & DeNiro (1983) have shown that at least some of the oxygen in the wood may be in part determined by the water in the tree trunk rather than the leaf water.

It was necessary to demonstrate that the δD of the nitrocellulose from the trees can be correlated with the environmental climatic temperatures. Hydrogen isotope analyses were made on wood from trees from 25 different localities in the North American continent (Yapp 1980; Yapp & Epstein 1982*a*) that covered a wide climatic range. These locations included Alaska, Texas, southwestern and eastern United States. In addition the δD values of eight time-equivalent five-year periods of seven trees in North America that grew in areas for which the climatic temperatures were well documented were compared with the mean annual temperature of their locations. The spatial temperature distribution versus the δD values gave a linear relation expressed by the equation (T is in degrees Centigrade)

$$\delta D(\text{cellulose nitrate}) = 7.7 \times T - 150.$$

This large coefficient of $8\%_0/°C$ can be useful because δD can be measured to within $1\%_0$ and very precise temperature data should be attainable from the isotopic data.

EXPERIMENTAL PROCEDURES

The experimental techniques involved in the extraction of H_2 and CO_2 for isotope analyses is described in detail elsewhere (Epstein *et al.* 1976; Yapp & Epstein 1982*b*; DeNiro 1981). Basically a cross-sectional strip of the tree is subdivided into five- or three-year intervals. The wood is ground into sawdust, oxidized with sodium chlorite to produce clean cellulose and the cellulose is nitrated with fuming nitric acid to produce nitrocellulose. The nitrocellulose is isolated by taking advantage of its solubility in acetone. The nitrocellulose is combusted and the water produced is passed over uranium to produce H_2. The $\delta^{13}C$ is measured in the CO_2, combustion product of cellulose.

With the knowledge that the trees can record δD values that can be interpreted as environmental temperatures our object was to analyse a series of temporal δD records in trees from many different environments to determine if we can get a consistent record of the Earth's past climatic temperature.

Sections from 23 trees, spanning at least 30 years, randomly selected, including the 12 dealt with by Yapp (1980), were analysed for their δD records. In several cases, $\delta^{13}C$ analyses were also made on the same tree rings. There were no special criteria set for selecting the tree sites except that we preferred that they should have grown on relatively flat land and as much as one can determine to have had a normal growth and represent a wide range of climatic conditions. The δD of 12 of the 23 trees are reported on by Yapp (1980) and Yapp & Epstein (1982 a). Yapp (1980) has provided the most thorough description of the location of the 12 trees he dealt with as well as the information to compare the δD and the local climatic data. The 23 trees allowed us to determine if a random sampling of locations would still permit the extraction of meaningful environmental information from the isotopic record in the trees *per se*, because it is not always possible to obtain wood samples whose conditions of growth are available.

RESULTS AND DISCUSSION

The location of the trees and the isotopic data are shown in table 1 and figure 2, respectively. Figure 1 shows 25-year running averages of the δD data of the bristlecone pines located in Shulman Grove in White Mountain, California. The running averaging of the data provides a better opportunity to determine the trends in the data by eliminating the short-time

TABLE 1. THE IDENTIFICATION AND APPROXIMATE LOCATION OF THE TREE SAMPLES
USED IN THIS STUDY

sample no.	sample label	sample identity and location
1	Red-2	redwood (sequoia sempervirens); Miller Creek, California
2	AW-BO-1	bur oak (Quercus macrocarpa); Albion, Wisconsin
3	MO-O-2	oak from Owensville, Missouri
4	IE-10	juniperous phoenicea; Gebel Halal, N. Sinai
5	MNY-GA-2	green ash (Fraxinus pennsylvania); Montezuma National Wild Life refuge, New York
6	OPW-DF-1	sitka spruce (picea sitchensis); Olympic peninsula, Washington
7	UCI-1	cedar (Lubocedrus decarrens); Great Grove Sequoia National Forest
8	TAS	huon pine (Lagarostrobos franklini); Maydena S.W. Tasmania
9	ORE-1	douglas fir; Illinois valley, Oregon
10	SNO	red gum (Eucalyptus camaldulensis); Snowy Mountains, New South Wales, Australia
11	NEZ	rimu (Dracrydium cupressinum); Westland National Park, New Zealand
12	RE-CO-2	chestnut oak (quercus prinus); Reston, Virginia
13	MNY-BO-7	bur oak (quercus macrocarpa); Montezuma National Wildlife refuge, New York
14	COL-DF-1	douglas fir (Pseudotsuga menziesii) Mt. Vernon canyon; Colorado
15	MNY-RM-5	red maple (acer rubrum); Montezuma National Wildlife refuge, New York
16	SNF-LP-1	lodgepole pine (pinus contorta); Sierra National forest, California
17	Brpn	bristlecone pine; White Mts; California
18	BCT-12	pine near Seeley Lake; British Columbia
19	NMK-1	juniperous procera; Mau Narok, Kenya
20	FCA-WS-1	white spruce (picea glauca); Alberta, Canada
21	KSA-AS-1	aspen; Anchorage, Alaska
22	FAAS1	aspen; Fairbanks, Alaska
23	SAS-JP-1	jackpine (pinus banksiana); Near Porter Lake, North West Territories, Canada

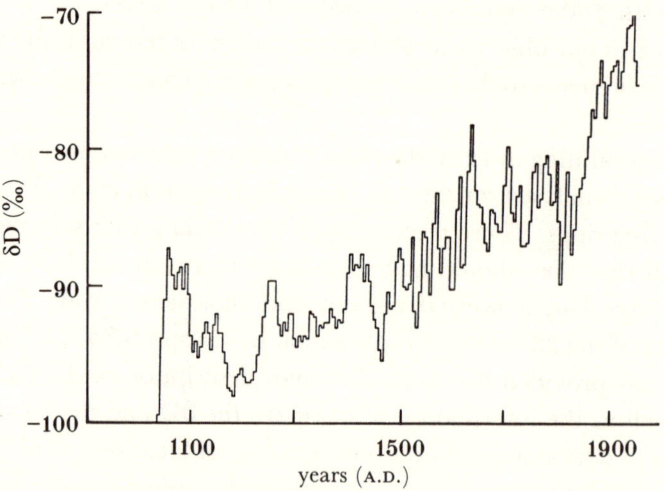

FIGURE 1. δD (25-year running average) against time of the bristlecone pine from White Mountain California.

oscillations. There are several interesting features regarding this isotopic record, the most obvious of which is the overall increase in the δD record with time indicating that the temperature of the White Mountain area and possibly of the Earth as a whole has been rising for the last 1000 years. We cannot discount a possible decrease in relative humidity associated with this temperature rise. Both of these factors can cause increase in the δD values (Epstein *et al.* 1977; Yapp & Epstein 1982 *b*). A temperature rise independent of the anticipated rise due to the greenhouse effect would have some serious consequences as far as our civilization is concerned. This interesting result focused our attention on the necessity to examine whether similar increasing trends in the δD records are present in the other trees that came from many parts of the world.

It should be possible to test the temporal temperature record in tree rings by comparing the δD values in many trees with the temperature records of the area. This was attempted by Yapp (1980) on 12 trees. He showed that wherever such a comparison is possible the δD records sometimes relate to the mean annual temperature, sometimes to the mean summer temperatures and sometimes to the amount of summer precipitation. In two cases (samples 18 and 20), the sources of moisture for the trees were so variable that it was not possible to get a good temporal relation between the recorded temperatures and the δD values in the tree rings.

The time relations between the δD and the recorded temperatures are sensitive to the variations in the δD of the soil water and to any uncertainty in assigning a recorded temperature of an area. For example in some high-latitude areas the relation between the δD of the soil water and the temperature can become complicated if during the growing season two reservoirs of water are available for the soil. One source could be water from melting snow having a low δD value and the other source could be rain water with a higher δD value. A warm dry spring season could introduce low δD melt-water into the soil whereas a cool rainy spring can introduce more high δD rain into the soil. The δD of the tree rings that grew under the warmer conditions would be lower than that recorded in cooler conditions. The relation between the recorded temperature and the δD of the tree rings would be contrary to the expected trend. An exaggerated increase of δD in tree rings with temperature can also be accomplished if mild weather before the growing season removes the snow cover and rain water

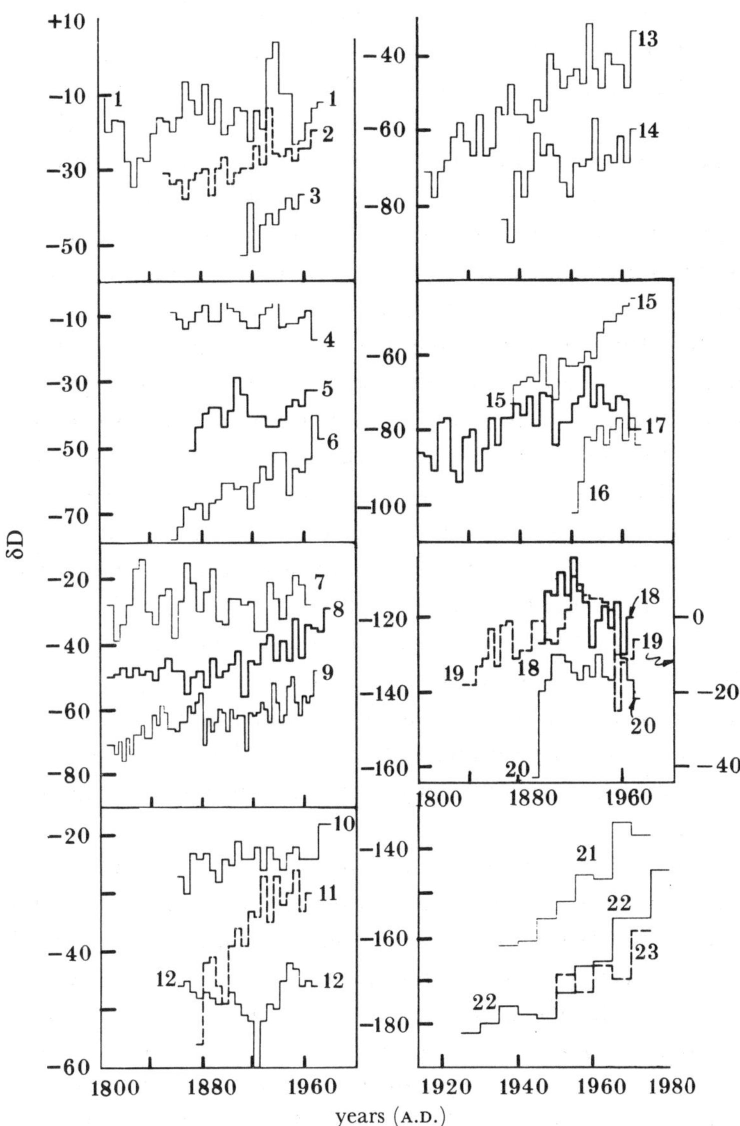

FIGURE 2. δD against time plots of trees from 23 different locations.

with high δD values then dominates the source of water in the soils when the tree grows. These erratic trends appear in short-time records in the tree samples from the high latitude. Over a long range of time such fluctuations in δD tree records would probably average out and result in an overall trend that reflects the average δD of the water and a correct temporal climatic trend of the area.

In spite of these possible difficulties the verification of the presence or absence of a good climate record in the δD of tree rings could be obtained by re-examining the isotopic record of the 12 trees from Yapp & Epstein (1982 a) and the isotopic record of an additional 11 trees from other areas, a total of 23 trees. All totalled, the 23 trees come from three different continents and cover a climatic range from the equatorial and arid areas of Africa to the cold environments, such as found in Tasmania and the Canadian Yukon Territories. All but one of the trees has a δD record greater than 50 years. In the majority of cases the records cover more than 100 years, and two (bristlecone pine and juniper from the Sinai) cover 1000 and 400 years, respectively. Figure 2 shows these records.

It is obvious that 21 of the 23 trees show an increase in δD with time indicating a warming trend of various magnitudes in their respective locations. The two exceptions are in the high-latitude areas we discussed previously. The warming trends are not always smooth and are disrupted by introduction of anomalous surges in the δD records. A good example of an anomaly that modified the trend in the δD record without markedly changing its overall δD characteristic values is illustrated in the 150-year δD record of the juniper tree in Maunarok in Kenya, East Africa (Krishnamurthy & Epstein 1985). The δD in this tree rises from -20 to $+20‰$ from 1830 to 1960, at which time there was a strong decrease of δD of about $40‰$ that lasted for about five years, after which the rise in δD resumed to the present time. The data suggested that the temperature in Kenya rose until 1960 and dropped quickly. It is obvious that such a drop in δD could not be due to world-wide cooling, but to a local perturbation caused by unusual rainfall for several seasons. The compelling evidence that this was the case lies in the measurements made of the level of Lake Victoria, which is in the catchment area of the forest, and which rose by 2 m in 1960, a circumstance that has not been observed in the 50-year record of the measurements of the level of Lake Victoria. This local effect will modify the slope of the line particularly if only the last 50 years of the record are used. Because local effects can modify the δD records in trees, the fact that such a high percentage of trees randomly sampled over a wide geographic area have such a consistent δD record must add credence to the global climatic significance of this record. This must be especially true for the bristlecone pine δD record, because of its length and its continuous trend over and above any possible local perturbations.

Although the majority of the 23 δD records shown in figure 2 show a warming over the period of growth, it is obvious that the degree of warming is not the same for the different trees. The unbiased value for the degree of warming could be estimated by carrying out a linear regression of the δD records in the trees with time. The slope expresses the specific warming as the change of δD per year. The δD value of the youngest tree ring gives the most recent climatic temperature of the area. A plot of the slope against this δD value reveals that there is an approximate linear relationship between the two quantities (figure 3). If all the data points are included, a linear regression gives a correlation coefficient of 0.60 (significant at 0.01 level). If two of the least fitting points (samples 18 and 20) are removed this coefficient rises to 0.83 (significant at 0.01 level).

Yapp (1980) noted that the δD temporal record in samples 18 and 20 was not correlated with the climate record of the area because of the interplay of the different air masses in their areas that produced rain patterns that was not characteristic of the mean annual or average summer temperature but rather of very restricted temperature ranges. The location of the data point for the bristlecone pine (no. 17) in figure 3 appears to be somewhat removed from the trend observed for the 20 other trees. Actually, the trend becomes similar to those of the other trees if we discount the last 30 years of its growth and use the period 1800–1940 in the linear regression. After 1940 the data shows a cooling trend.

The plot in figure 3 alerts us to several complications when we discuss climatic records. It is obvious that in low-latitude areas, where trees have δD values that are on the average of 0 to $-30‰$, a global warming will not be as pronounced as those occurring in the higher-latitude areas. If the appropriate trees of great age can be found in these areas, the effect of

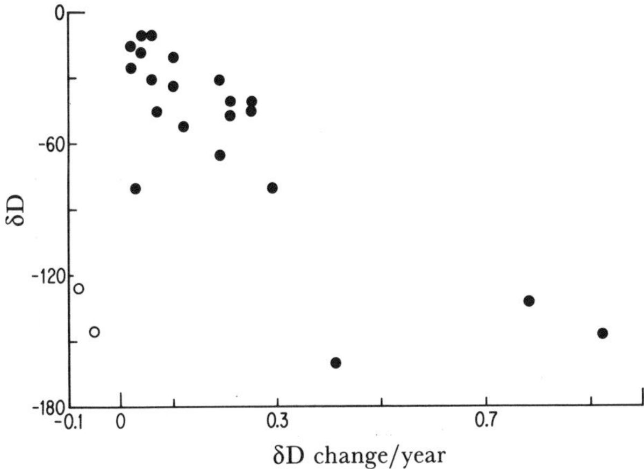

FIGURE 3. Plot showing δD values that are the actual temperatures of the past 10–20 years of growth against the change in δD per year (i.e. rate of warming) of the 23 samples. The correlation coefficient is 0.6. This coefficient improves to 0.83 if the open circled data (samples 18 and 20) are omitted.

global warming on the δD rise should be enhanced. It is reasonable to suppose that the warming trends of the Earth would effect the climatic temperature of the higher latitude to a greater degree than the equatorial latitude, creating smaller latitudinal gradients of temperature. For example, during the Cretaceous period the temperatures of the Earth were much warmer than they are now, and a more uniform temperature existed (Douglas & Woodruff 1981). The historical data that is presented to indicate the presence or absence of global warming or cooling should take this into account (Jones *et al.* 1986).

Our results suggest that isotope analyses of the bristlecone pine or of the trees growing in even cooler climates in Alaska or in the northern latitudes of Europe and Asia would represent a reasonable approach to measure the prehistoric climatic record. In the meantime the δD record in the bristlecone pine may be a good indicator of the relative magnitude of the change in global climate in the past 1000 years.

There are climatic changes reported in the literature (Lamb 1982) that can also be found in the bristlecone pine. There was a cold Arctic period associated with a loss of colonies in southern Greenland about 1100 A.D. There was a parallel drop in the δ value of about 10‰ between 1090 and 1100 A.D. There was a cooling period around 1780 in Europe, and the δD drops around that time.

The relation between the δD in trees and temperature has been estimated to be about 7–8‰ $°C^{-1}$ (Yapp & Epstein 1982a). This would suggest that a temperature rise between 1800 and 1940 is about 2 °C. This value may be too high for a global average but, as pointed out above, the response to world warming or cooling would be magnified in the higher latitudes and altitudes.

The rising trend in global temperature over the past 150 years is present in the 21 samples we analysed. In addition, the bristlecone pine record shows that this trend may have been in effect for the past 1000 years, and thus even if we find a way to stop the greenhouse warming due to anthropogenic activity, there is still an additional warming we might have to deal with if the Earth warming represents a danger to mankind. Considerable thought must be given to that problem as well.

[35]

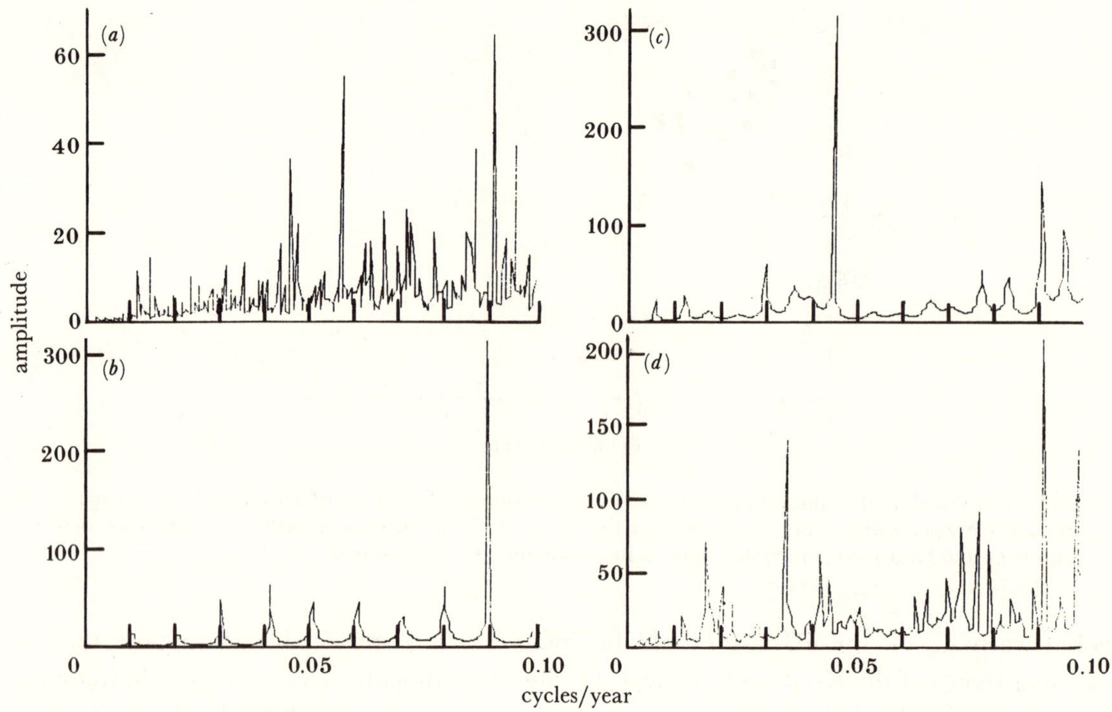

FIGURE 4. Fast Fourier spectra of the tree δD values that show either a 11- or 22-year cycle. The trees that show such cycles are those with a relatively smaller climatic trend; (a) and (b) sample 17, (c) sample 7, (d) sample 4. The time-periods covered in the Fourier analysis are (a) 970–1975 A.D.; (b) 1650–1750 A.D.; (c) 1805–1970 A.D.; (d) 1540–1970 A.D.

It has been previously shown that the δD and $\delta^{18}O$ in trees are affected by the relative humidity of the environment (Epstein *et al.* 1977). The $\delta^{18}O$ disproportionately increases with lowering of humidity, thus simultaneous $\delta^{18}O$ analyses of the cellulose should allow the estimation of the relative humidity on the δD values, and thus provide an appropriate correction on the δD to eliminate the effect of the humidity.

Periodicities in the δD *record in the trees*

It was of interest to subject the hydrogen isotopic record of the trees to a Fourier analysis to determine if significant cycles such as the solar 11-year or 22-year cycles were present in the data. Presumably, such cycles would indicate the influence of the sunspot activity on the Earth's climate. Because it has been shown that a number of factors, such as humidity, anomalous rain patterns, soil drainage (Yapp 1980), and other unknown factors affect the detailed δD record in trees, it is unlikely that Fourier analysis of the data will show striking cycles. It is possible that there are localities where the temperature variations are affected by solar radiation. An effort to find these special trees will be needed. Of the 23 trees that we have analysed, there are three trees that contain in them either 22- or 11-year cycles (figure 4). It is perhaps noteworthy that the cycles are present in trees that suggest a modest warming trend. The δD values of many of the other trees show such a strong trend with time that it probably interferes with the extraction of cycles from their isotopic data. The establishment of the connection of solar cycles with climatic cycles is an important problem. The use of isotopic records in trees to search for these cycles appears to be a viable possibility.

The $\delta^{13}C$ record in trees

One of the important questions that has attracted our attention in recent years is the role that CO_2 concentration in the atmosphere plays in affecting our environment. Here we briefly discuss the $\delta^{13}C$ record of the seven trees analysed. The isotopic compositions of carbon in the cellulose of different trees vary. The $\delta^{13}C$ values that we have observed range from about -17.5 to $-24\%_0$. This large variation reflects the environmental condition of growth rather than the differences in the isotopic composition of carbon in the global atmospheric CO_2. However, for a single tree, the variations are much smaller and vary from ring to ring by significant amounts and reflect the isotopic composition of its immediate atmospheric CO_2 or in some cases the $\delta^{13}C$ of the global CO_2.

The 1000-year comprison of the $\delta^{13}C$ and the δD records of the bristlecone pine is shown in figure 5. There are obvious similarities in their variations with time up to 1850. At this time

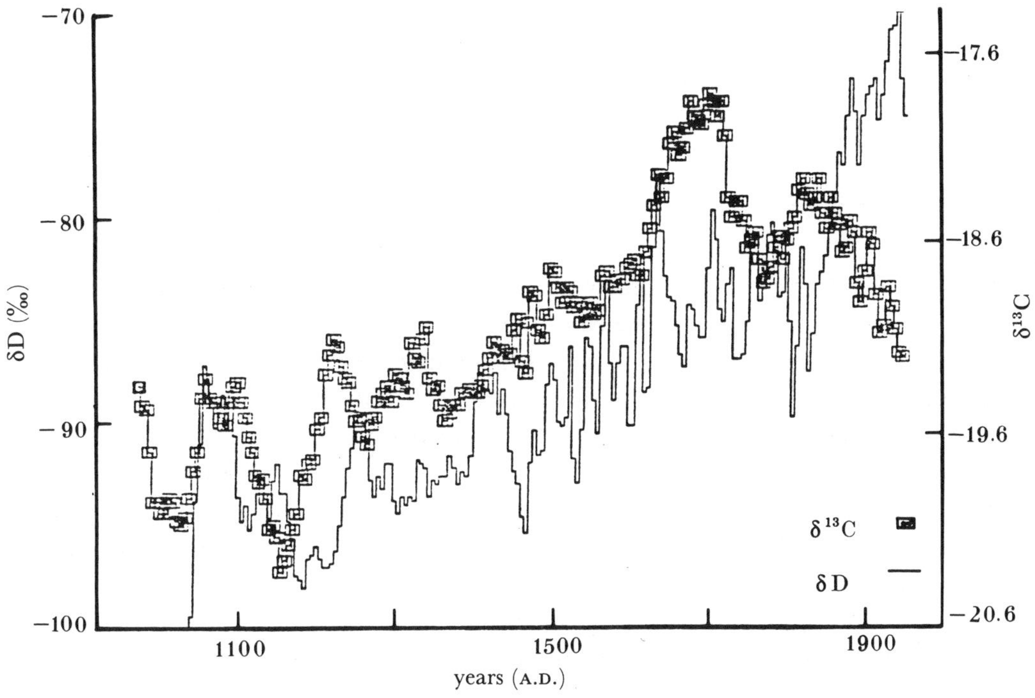

FIGURE 5. A comparison of the δD and $\delta^{13}C$ records of the bristlecone pine covering the past 1000 years.

the $\delta^{13}C$ values drop continuously whereas the δD record continues to climb to its maximum 1940 value. The drop in the $\delta^{13}C$ value is obviously as a result of the introduction of anthropogenic CO_2 of much lower $\delta^{13}C$ into the atmosphere. This signal is present in all of the seven trees we have analysed (figure 6), although in different degrees (Francey 1982; Freyer 1986). These trees are from the Australian mainland, New Zealand, Kenya, Tasmania, the Sinai, and California. Perhaps the most interesting result that we have obtained is shown in figure 7 where the $\delta^{13}C$ value of the bristlecone pine is superimposed on the $\delta^{13}C$ record of the juniper located in the Sinai. These *ca.* 400-year records of $\delta^{13}C$ variations are identical within experimental error. It is of interest to note that the δD records of the two trees do not match as well as the $\delta^{13}C$ record. Therefore, the synchronous nature of the δD–$\delta^{13}C$ record in the

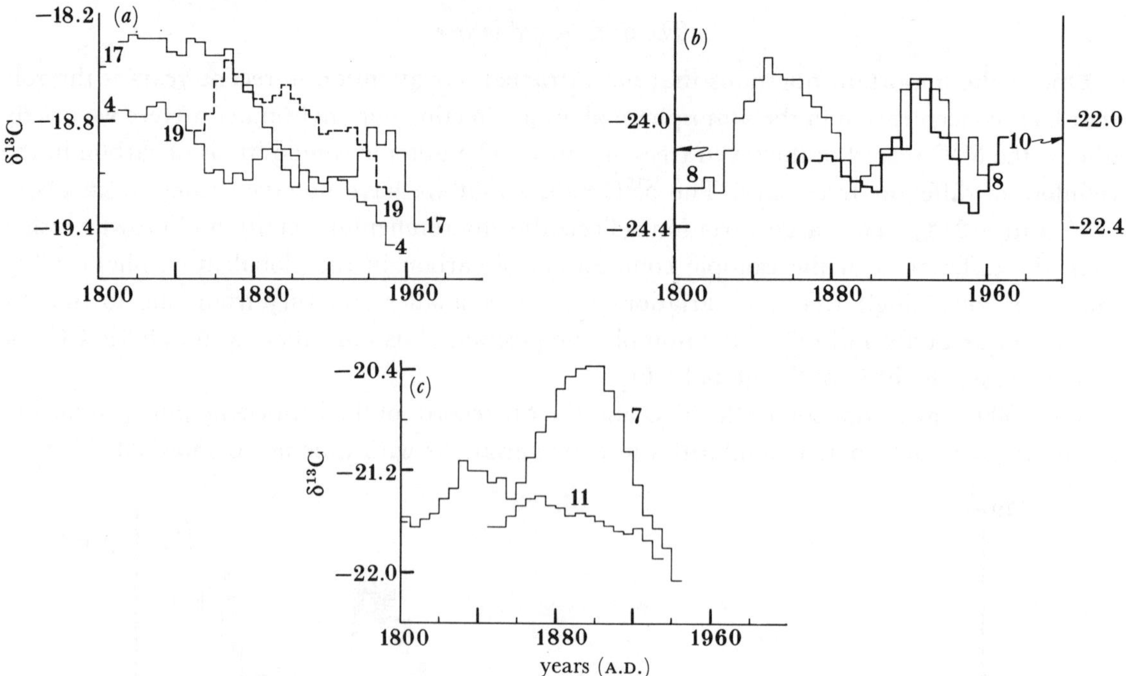

FIGURE 6. $\delta^{13}C$ (25-year running average) against time, of the seven samples analysed.

bristlecone and its absence in the juniper enhance the likelihood that those two trees have in their carbon isotope composition a record of the $\delta^{13}C$ variation in the global CO_2. Otherwise the difference in temperature might very well have affected the $\delta^{13}C$ of the juniper.

We can compare the degree of the anthropogenic $\delta^{13}C$ effect in the different trees we have analysed. Such a comparison based on a simple linear regression model is shown in table 2. It is interesting to note that the $\delta^{13}C$ anthropogenic effects in the Northern Hemisphere trees is about twice that obtained for the trees from the Southern Hemisphere (Tasmania, New Zealand and Australia). Such a difference is reasonable because the major source of industrial CO_2 is the Northern Hemisphere and this CO_2 probably loses much of its $\delta^{13}C$ signature by partial equilibration with the ocean surface before being fixed by plants in the Southern Hemisphere. The possible difference between the anthropogenic ^{13}C effect between the Southern and Northern Hemispheres suggests that there might be an interhemispheric difference in the $\delta^{13}C$ values of the two atmospheres.

In as much as the $\delta^{13}C$ record in the bristlecone pine and the juniper is taken to represent a Northern Hemisphere 'global' trend, it is tempting to ascertain if the $\delta^{13}C$ effects due to deforestation, which would lower the $\delta^{13}C$ value, unusual volcanic activity that would increase the $\delta^{13}C$ value, as well as effects arising out of human activities in the past 1000 years, are present in the bristlecone pine record. The overall increasing trend in the $\delta^{13}C$ in the bristlecone pine must be because of some major world-wide effect, probably the continuous warming of the world's oceans. This would release ^{13}C-enriched CO_2 into the atmosphere and could contribute to the gradual rise in its $\delta^{13}C$. It would be very useful to compare long records from additional trees whose environments are similar to the Sinai and White Mountain area. Sparsely wooded areas at high elevations as well as those away from industrial CO_2 contributions or perhaps even high-wind-prone areas that help in the rapid dissipation of industrial CO_2 might provide good sampling locations.

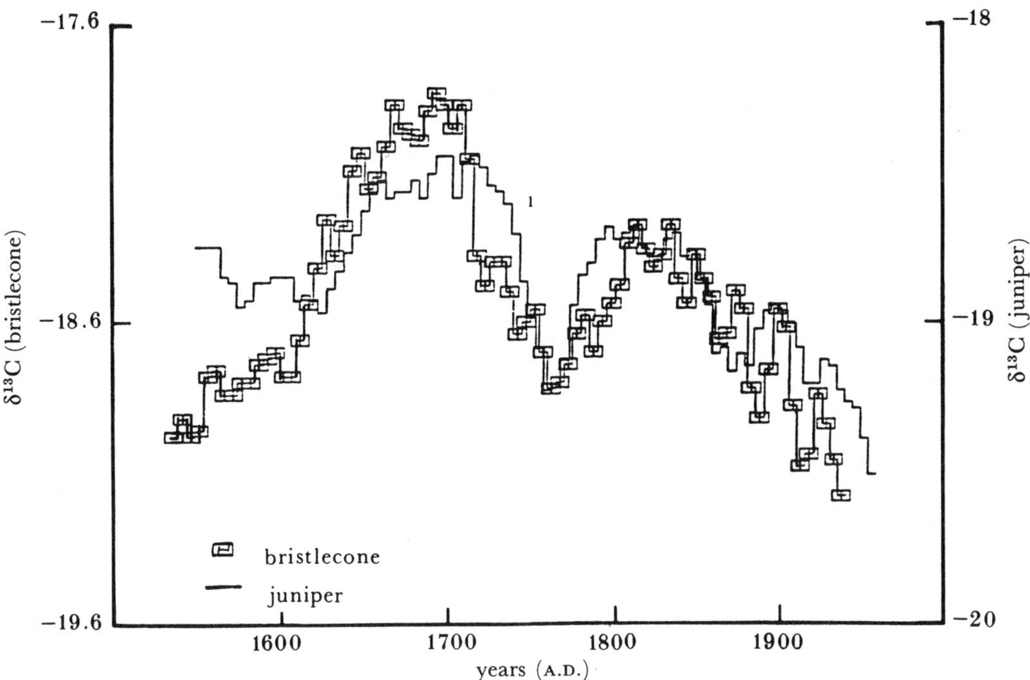

FIGURE 7. Comparison of the $\delta^{13}C$ records of the bristlecone pine and the juniper from the Sinai peninsula.

TABLE 2. THE $\delta^{13}C$ CHANGE ($\Delta^{13}C$) DUE TO ANTHROPOGENIC EFFECT BETWEEN
1850 AND THE PRESENT

(The $\Delta^{13}C$ values were estimated by using a linear regression of the $\delta^{13}C$ record of the individual seven trees.)

sample no.	sample label	$\Delta^{13}C$ (‰)
8	TAS	−0.26
10	SNO	−0.08
11	NEZ	−0.5
19	NMK-1	−0.78
17	Brpn	−0.84
4	IE-10	−0.75
7	UCI-1	−1.08

A well-confirmed 'global' record of the $\delta^{13}C$ in trees would provide the opportunity to determine the effect of local conditions on the $\delta^{13}C$ record of individual trees. For example, a forest fire contributes a great deal of CO_2 of low $\delta^{13}C$ in the atmosphere. The carbon isotope composition of the tree growth for several years subsequent to the fire might provide a pattern characteristic of the event and its magnitude so that such prehistoric events could be identified and its magnitude estimated. Similar environmental ^{13}C perturbations due to volcanic activity and deforestations could also be located and their magnitude estimated.

SUMMARY

Tree rings from 23 trees from various locations have been analysed for the δD and $\delta^{13}C$ values. Twenty-one of these trees give a consistent spatial and temporal pattern suggesting a world-wide warming trend for the past 100 and possibly for the past 1000 years. The degree of warming is latitude dependent.

The $\delta^{13}C$ record in the trees shows the $\delta^{13}C$ decrease in the past 100 years resulting from the anthropogenic contribution of low-^{13}C CO_2 to the atmosphere. The identical $\delta^{13}C$ temporal record of the juniper tree from the Sinai Peninsula and the bristlecone pine in the White Mountains of California suggests that both of these trees probably record the $\delta^{13}C$ of Northern Hemisphere CO_2. The Northern Hemisphere $\delta^{13}C$ effect is twice that for the Southern Hemisphere, and consequently there may be a small steady-state difference of about 0.3‰ in their atmospheric CO_2.

A strong argument has been made for the use of δD values of non-exchangeable hydrogen in cellulose extracted from trees to determine the past history of the Earth's climate and how the temperature was distributed on the surface of the Earth. We perceive this argument as a basis for expanding this effort to search for trees that are especially suited for this purpose. It is obvious that the bristlecone pine trees whose dendrochronology span thousands of years with the possibility of going back to the transition between the glacial–interglacial period are particularly attractive for study. A similar investigation of the $\delta^{13}C$ in trees can go back thousands and millions of years. A sample of glacial–interglacial transition in the $\delta^{13}C$ for ^{14}C-dated tree samples has already been attempted (Krishnamurthy & Epstein 1990).

We thank the late C. J. Ferguson for the dated bristlecone pine sample. We are grateful to the many colleagues (too numerous to list) who have contributed samples from different parts of the world. We acknowledge the help of C. Kendall, L. J. Randolph, E. Dent, S. Newman and J. Ruth for their technical assistance. We have benefited greatly from discussions with Professor M. J. DeNiro and Professor C. J. Yapp. Professor Yapp's contribution to this paper was particularly important in our discussions of his data. We are grateful for financial support from the NSF Grant No. EAR-8504096 and NSF Grant ATM80-8830. This is Contribution No. 4742 from the California Institute of Technology, Division of Geological and Planetary Sciences.

References

Arnason, B. 1981 *Ice and snow hydrology (stable isotope hydrology)* (ed. J. R. Gat & R. Gonfiantini). Vienna: I.A.E.A.
Daansgaard, W. 1964 *Tellus* **16**, 436–468.
DeNiro, M. J. 1981 *Earth planet. Sci. Lett.* **54**, 177–185.
DeNiro, M. J. & Epstein, S. 1979 *Science, Wash.* **204**, 51–53.
Douglas, R. G. & Woodruff, F. 1981 In *The sea* (ed. C. Emiliani), vol. 7. New York: Wiley Interscience.
Epstein, S., Sharp, R. P. & Gow, A. J. 1959 *Trans. Am. geophys. Un.* **40**, 81–84.
Epstein, S., Thompson, P. & Yapp, C. J. 1977 *Science, Wash.* **198**, 1209–1215.
Epstein, S., Yapp, C. J. & Hall, J. H. 1976 *Earth planet. Sci. Lett.* **30**, 241–251.
Francey, R. J. 1981 *Nature, Lond.* **290**, 232–235.
Freyer, H. D. 1986 In *The changing carbon cycle: a global analysis* (ed. J. R. Travalka & D. E. Reichle). London: Springer-Verlag.
Gray, J. & Song, J. S. 1984 *Earth planet. Sci. Lett.* **70**, 129–138.
Jones, P. D., Wigley, T. M. L. & Wright, P. B. 1986 *Nature, Lond.* **322**, 430–434.
Krishnamurthy, R. V. & Epstein, S. 1985 *Nature, Lond.* **317**, 160–162.
Krishnamurthy, R. V. & Epstein, S. 1990 *Tellus* (In the press.)
Lamb, H. H. 1982 *Climate, history and the modern world.* London: Methuen.
Ramesh, R., Bhattacharya, S. K. & Gopalan, K. 1986 *Earth planet. Sci. Lett.* **79**, 66–74.
Siegenthaler, U. & Oeschager, H. 1980 *Nature, Lond.* **285**, 314–317.
Smith, B. N. & Epstein, S. 1970 *Pl. Physiol.* **46**, 738–742.
Sternberg, L. S. L. & DeNiro, M. J. 1983 *Geochim. cosmochim. Acta* **47**, 738–742.
Yapp, C. J. 1980 The variations and climatic significance of D/H ratios in the carbon-bound hydrogen of cellulose in trees. Ph.D. thesis, California Institute of Technology, U.S.A.
Yapp, C. J. & Epstein, S. 1982*a* *Nature, Lond.* **290**, 232–235.
Yapp, C. J. & Epstein, S. 1982*b* *Geochim. cosmochim. Acta* **46**, 955–965.

Discussion

H. Oeschger (*Physics Institute, Bern, Switzerland*). How much do changes in plant physiology affect the isotopic signature in tree rings? For example, the increase in CO_2 (25% in the past 200 years) might have lead to a decrease of evaporation (less stomata opening) that could simulate a δD shift independent of the δD in the environmental water.

S. Epstein. If the evaporation of water from the leaves would decrease as a result of the constriction of the stomata opening, then the δD values in the tree rings would decrease. Because the δD has risen over the past 1000 years in the bristlecone pine, a rise substantiated in the other trees for the past several hundred years, it would appear that the stomata effect would not be important. I would guess that the effect of temperature and possibly the relative humidity would be more profound on the δD hydrogen in the bristlecone pine.

J. A. Eddy (*UCAR, Boulder, Colorado, U.S.A.*). How does Professor Epstein's δD record of the past 900 years obtained from measurements of δD in bristlecone pine from the White Mountains of California compare with the temperature reconstruction derived for the same area and the same period by Valmore C. LaMarche? LaMarche derived a temperature history for simple ring *widths*, having found that trees at this altitude are temperature sensitive in their pattern of growth.

S. Epstein. The tree ring width record in the bristlecone pine determined by LaMarche is reasonably compatible with the δD of the bristlecone pine over the past 500 years. Before that time (between 1000 and 1500 years), there is a considerable difference in the records of the δD and ring width. I see no reason why the bristlecone pine would not record the δD of precipitation before 1500 years ago. The best answer for your question would be the analysis of additional tree records in other locations for both the δD and ring width for this time period. It might prove interesting to check ring width records in younger trees in other locations and see if they match the δD records.

J.-C. Pecker (*Collège de France, Paris*). Is it conceivable to introduce a 'global' (planetary) 'tree-ring index', in the way people have introduced 'planetary' geomagnetic indices, which give many useful ways to study the solar–terrestrial relations? Is it conceivable to use the 'petrified forests' trees (such as in Arizona) to derive tree-ring indices?

S. Epstein. It is conceivable that a global isotope tree-ring index may eventually be developed. This will involve a world-wide cooperative effort. As we have already indicated (figure 4) solar–terrestrial relations may be present in the isotopic record in trees. I cannot comment on the potential of silicified wood. We have not tried to extract cellulose from such materials.

Phil. Trans. R. Soc. Lond. A **330**, 441–443 (1990)

Printed in Great Britain

441

On the need for further isotopic measurements from tree rings

By M. G. L. Baillie

Palaeoecology Centre, Queen's University, Belfast BT7 1NN, U.K.

Archaeological pressure for better chronology has provided the scientific community with long tree-ring chronologies and high-precision radiocarbon calibration curves. Physicists are now using the calibration curves as the only available proxy measure of past solar variation. The underlying tree-ring chronologies can, in theory, offer three lines of research potential: (1) the analysis of other isotopes on a scale of years, (2) the possibility of climatic data on a time resolution compatible with the calibration and (3) possible refinement of the ice-core timescales, by linking related (volcanic) events in both records.

The bringing together of solar physicists and archaeologists to discuss possible relations between past solar activity and the Earth's climate produced some interesting undertones. Not least among these was the implied request from the physicists 'that the archaeologists should provide dated information on environmental change' for comparison with the precisely dated solar variations implicit in the radiocarbon calibration curves.

Now here indeed was a somewhat ironic set of circumstances. Physicists, with 'their' record of past variations in atmospheric radiocarbon, asking archaeologists 'What happened at this date and that date where we see rapid depletions or enrichments in the radiocarbon record?'. It seems not to have occurred to the physicists that 'their' proxy solar record had in fact been provided for them by the archaeological community. In the U.K. archaeological money had funded the construction of the Irish oak tree-ring chronology (Pilcher *et al.* 1984) against which Pearson, also funded from archaeological sources, had performed the first, long, continuous high-precision calibration of the radiocarbon timescale (Pearson *et al.* 1986). Pearson's calibration, combined with that of Stuiver, provides the internationally accepted standard now being used as a proxy solar record.

In a nutshell, archaeologists, in providing the stimulus for the radiocarbon measurements, have provided some of the only chronologically sound data available from the past. As soon as one strays into archaeology proper, time control decreases dramatically in quality. Few archaeological sites are dateable by dendrochronology, which is essentially the only method that would allow exact comparison with the isotopic variations seen in the tree rings. Apart from the problems of archaeological dating, which the calibration was originally intended to address, it is extremely difficult to infer climatic alterations from most archaeological evidence. Even those palaeoenvironmental methods associated with archaeological research, such as pollen or insect studies, suffer badly from the inferior dating control offered by radiocarbon.

In theory it may prove impossible to ever provide environmental information that is well-enough dated to compare with the variations in past radiocarbon concentration! In this vein, we have to remember that the wiggles in the calibration curve, of interest to solar physicists, are precisely dated against the tree-ring standard. All archaeological and palaeoenvironmental dating uses the calibration curve in an attempt to relate raw radiocarbon dates to the dendrochronological timescale! By definition the use of the calibration curve tends to widen the age band. For example, if we take one of the best-dated events in the Neolithic of Britain and Ireland – the elm decline – we can assign to it, on the basis of some 20 relevant radiocarbon determinations, an average radiocarbon date of almost exactly 5000 before present (BP) (Edwards 1985). Let us imagine that there was an error of only ± 20 years associated with this date. When we come to calibrate 5000 ± 20 BP at 95 % confidence we find that the elm decline – if it was a unique event – took place somewhere between broadly 3940 B.C. and 3700 B.C. This is to all intents and purposes the best-dated 'archaeological' event in British and Irish prehistory. It becomes essentially impossible to relate this event to any particular 'solar variation' within the calibration curve. Such considerations make it extremely unlikely that the original physicists' question – 'What happened environmentally at the times of this or that radiocarbon depletion or enrichment?' – can be answered sufficiently accurately to be meaningful.

Again, by a stroke of irony, some of the only environmental events in prehistory that can be precisely dated by tree rings; the possible volcanic-dust-veil events specified by frost rings in the bristlecone pine (LaMarche & Hirschboeck 1984) and by 'narrowest-ring events' in the Irish bog oaks (Baillie & Munro 1988) are events unlikely in themselves to be related to solar activity variations. In the case of the Irish bog oaks, these postulated volcanic events at 4375 B.C., 3195 B.C., 1628 B.C., 1159 B.C., 207 B.C. and A.D. 540 represent the worst general conditions experienced in the tree-ring record. Examination of the calibration curve at these points in time shows no particular pattern between volcanic-induced climatic downturns and radiocarbon variations.

Although in the future further precisely dated environmental events may be forthcoming from the tree-ring record itself, one immediate use of the dates of the apparent volcanic-dust-veil events may lie in applying corrections to the ice-core record. If the dates listed above genuinely relate to Hammer's Greenland acidity peaks at 4400 ± 100 B.C, 3250 ± 80 B.C., 1645 ± 20 B.C., 1100 ± 50 B.C., 210 ± 30 B.C. and A.D. 540 ± 10 (Hammer *et al.* 1980, 1987), then the errors in the ice core dating can be reassessed.

The difficulties and limitations in attempting to relate radiocarbon-inferred solar activity to climate are clearly shown in a recent paper by Wigley (1988). Holocene mean-temperature fluctuations deduced from chronologies of glacial advance or retreat are based on a radiocarbon timescale. To compare the temperature fluctuations with the tree-ring derived solar variations Wigley had to calibrate the radiocarbon-dated temperature curve. Wigley admits that in spite of the general consistency of the temperature record, 'it is doubtful that the dates of the major temperature minima are better than ± 100 years (2σ limits)'. So this generally good temperature record is seen to have severe chronological deficiencies when attempts are made to find out exactly when environmental changes took place.

We can add to this the fact that the temperature record, as deduced from glacial advances and retreats, is 'not a perfect palaeoclimatic record' and the additional fact that there may be 'lags effected by the damping effect of the ocean's chemical or thermal inertia'. With all these provisos it is plain that any ability to infer causal relations is going to be severely impaired.

The net result of these chronological problems is that some other precisely dated variables need to be measured. Most obvious targets are the concentrations of isotopes of hydrogen, carbon and oxygen, directly measurable in tree-ring cellulose. At least some of these isotopic concentrations should be directly related to environmental factors. A concerted effort is required by the scientific community to exploit the long tree-ring records available in Europe and America.

In particular there are replicated data-sets represented by several continuous chronologies, each of the order of 7000 years minimum length, from Ireland and Germany (Pilcher *et al.* 1984; Leuschner & Delorme 1984), England (Baillie & Brown 1988) and California (Ferguson 1969). In the case of these projected isotopic studies the stimulus will no longer come directly from archaeologists but must involve the wider scientific community.

References

Baillie, M. G. L. & Brown, D. M. 1988 An overview of oak chronologies. *Br. archaeol. Rep.* (Br. Ser.) **196**, 543–548.

Baillie, M. G. L. & Munro, M. A. R. 1988 Irish tree-rings, Santorini and volcanic dust veils. *Nature, Lond.* **332**, 344–346.

Edwards, K. J. 1985 The elm decline. In *The quaternary history of Ireland* (ed. K. J. Edwards & W. P. Warren), pp. 288–289. Academic Press.

Ferguson, C. W. 1969 A 7104-year annual tree-ring chronology for bristlecone pine, *Pinus aristata*, from the White Mountains, California. *Tree-Ring Bull.* **29**, 2–29.

Hammer, C. U., Clausen, H. B. & Dansgaard, W. 1980 Greenland ice sheet evidence of post-glacial volcanism and its climatic impact. *Nature, Lond.* **288**, 230–235.

Hammer, C. U., Clausen, H. B., Friedrich, W. L. & Tauber, H. 1987 The Minoan eruption of Santorini in Greece dated to 1645 B.C.? *Nature, Lond.* **328**, 517–519.

LaMarche, V. C. Jr & Hirschboeck, K. K. 1984 Frost rings in trees as records of major volcanic eruptions. *Nature, Lond.* **307**, 121–126.

Leuschner, von H. H. & Delorme, A. 1984 Verlängerung der Göttingen Eichenjahrringchronologien für Nord- und Süddeutschland bis zum Jahr 4008 v. Chr. *Forstarchiv* **55**, 1–4.

Pearson, G. W., Pilcher, J. R., Baillie, M. G. L., Corbett, D. M. & Qua, F. 1986 High-precision ^{14}C measurement of Irish oaks to show the natural ^{14}C variations from A.D. 1840 to 5210 B.C. *Radiocarbon* **28**, 911–934.

Pilcher, J. R., Baillie, M. G. L., Schmidt, B. & Becker, B. 1984 A 7272-year tree-ring chronology for western Europe. *Nature, Lond.* **312**, 150–152.

Wigley, T. M. L. 1988 The climate of the past 10000 years and the role of the Sun. In *Secular solar and geomagnetic variations in the last 10000 years* (ed. F. R. Stephenson & A. W. Wolfendale), pp. 209–224. Kluwer Academic Press.

Discussion

G. H. Schleser (*Biophysical Chemistry Group, Nuclear Research Centre, Jülich, F.R.G.*). My comment concerns the 206-year cycle that has been mentioned frequently in this Symposium. I have measured stable isotopes with depth in a north German peat bog. From the variations observed, two cycles can be calculated by spectral analysis, leading to 187 and 1014 years, the first value being interestingly close to the 206-year cycle.

M. G. L. Baillie. The similarity may have some basis if one considers that the growth of peat bogs may be led by variations in solar output.

Phil. Trans. R. Soc. Lond. A **330**, 445–458 (1990)

Printed in Great Britain

Precambrian cyclic rhythmites: solar-climatic or tidal signatures?

By G. E. Williams

Department of Geology and Geophysics, University of Adelaide, G.P.O. Box 498, Adelaide, South Australia 5001, Australia

[Plates 1 and 2]

For more than 60 years geologists have sought evidence of solar-climatic cyclicity in rhythmically laminated sedimentary rocks, but claims in general have not been persuasive. Three Precambrian rhythmite sequences in Australia that comprise varve-like laminae recently have received attention, however, as their conspicuous cycles of *ca.* 10–14 and/or 20–25 laminae have been ascribed a sunspot-cycle origin. They are the 2500 Ma old Weeli Wolli Formation, the 1750 Ma old Wollogorang Formation, and the 650 Ma old Elatina Formation. New observations for the Weeli Wolli Formation, a siliceous banded iron-formation, suggest a cycle period exceeding the 23 microband couplets proposed by Trendall, casting doubt on the solar interpretation. The Weeli Wolli cyclicity may record Earth-tidal rhythms that modulated the discharge and composition of silica- and iron-bearing fumarolic waters. The structure of the cycles of silty dolomite and mudstone in the Wollogorang Formation does not support a sunspot-cycle origin, and a tidal control on sedimentation should be considered.

The Elatina sequence of cyclic sandstone and siltstone laminae displays several empirical similarities to the sunspot series. The discovery of thicker, more complex lamina-cycles in the correlative Reynella Siltstone has, however, caused reappraisal of the solar interpretation of the Elatina rhythmites. The Reynella cycles consistently contain 14 or 15 laminae, with many laminae comprising pairs of 'semi-laminae'; the overall structure of the cycles is similar to the mixed (semidiurnal and diurnal) fortnightly tidal growth patterns of modern bivalves. Also, recent observations that fine material can be transported in suspension to deeper waters offshore by ebb-tidal currents provides a tidal mechanism that can explain the deposition of thin, laterally extensive, graded laminae whose thickness would be a measure of tidal range and current speed. By this tidal model, the Elatina laminae are regarded as diurnal increments, and the lamina-cycles as commonly abbreviated fortnightly groupings. The Elatina series so interpreted may encode unique information on lunar orbital periods and the Earth's palaeorotation: the data indicate *ca.* 30.5 days per lunar month, 13.1 lunar months and *ca.* 400 days per year, and lunar apsides and lunar nodal cycles of 9.7 and *ca.* 19.5 years respectively some 650 Ma ago.

Evidence for significant modulation of terrestrial climate by the solar activity cycle in the geological past may prove as elusive as have convincing indications of solar signals in modern patterns of weather and climate.

Introduction

A review of evidence for solar variability in the geological past is relevant to the theme of this Discussion Meeting. Confirmation that solar cyclicity similar to that of today has influenced sedimentary processes millions of years ago would (i) demonstrate the durability of the solar activity cycle, (ii) imply that such cyclicity is capable of influencing the Earth's climate, and

(iii) encourage modelling of future solar activity and its possible influence on future terrestrial climate.

Since the early works of Brooks (1926), Udden (1928), De Geer (1929) and Bradley (1930), numerous geologists have sought solar signals, particularly the *ca.* 11-year sunspot period, in regularly laminated sedimentary rocks ('rhythmites') whose laminae were presumed to be annual ('varves') (see Anderson 1961; Richter-Bernburg 1964; Duff *et al.* 1967; Schove 1983). Such studies have used the principle that for a particular deposit varve thickness is a function of climate and that any influence of solar variability on climate should therefore be recorded by changes in varve thickness. Rhythmites of Phanerozoic age (less than *ca.* 570 Ma) have been most studied, and although solar signals commonly have been claimed the patterns in the rocks are not conspicuous and rigorous statistical analysis (see, for example, Anderson & Koopmans 1963) has failed to reveal significant periodicites.

In recent years, however, three cyclic rhythmites of Precambrian age (greater than *ca.* 570 Ma) have received attention in solar physics (see, for example, Bracewell 1986; Sonett & Trebisky 1986) because their cyclicity is conspicuous and in some ways shows empirical similarity to sunspot cyclicity. These rhythmites occur in the following.

1. The Weeli Wolli Formation, a 2500 Ma old banded iron-formation in the Hamersley Basin, Western Australia (Trendall 1973).

2. The Wollogorang Formation, a 1750 Ma old dolomite–mudstone sequence in the McArthur Basin, northern Australia (Jackson 1985).

3. The Elatina Formation, a 650 Ma old sandstone–siltstone sequence deposited in the Adelaide Geosyncline, South Australia, during the Marinoan Glaciation (Williams 1981, 1985; Williams & Sonett 1985).

The origin of these Precambrian cyclic rhythmites is critically reviewed here in the light of new observations. It is concluded that the rhythmites are all better interpreted as recording tidal rather than solar rhythms. Consequently, strong doubts remain as to whether solar cyclicity has significantly affected terrestrial climate in the geological past.

Weeli Wolli Formation

The early Proterozoic banded iron-formations (BIFs) typically display a thin microbanding of chert and iron oxide couplets that usually are regarded as varves (see, for example, Trendall & Blockley 1970; Garrels 1987), although Walter (1972) suggested that some microbanding may be diurnal and possibly related to the modulation of hot-spring activity by earth tides. The microbanding of the Weeli Wolli Formation differs from that of other BIFs in the Hamersley Basin (Trendall & Blockley 1970) in commonly being much thinner and typically displaying a conspicuous cyclicity (figure 1, plate 1). The cyclicity arises from regular variations in both the absolute and relative thickness of chert and iron oxide portions of the microband couplets, giving the Weeli Wolli BIF a characteristic 'striped' appearance (Trendall 1973). The microbands usually are very thin (0.05 mm thick, or less) and only the cyclic stripes are readily discernible (figure 1*b*, *c*). Local pods of chert have better resisted diagenetic compaction, however, allowing the microbands to be more easily seen and counted (figure 1*a*). Trendall & Blockley (1970) noted that the Weeli Wolli cycles contain about 30–38 microband couplets, an estimate later changed to 25 microband couplets (Trendall 1972). Trendall (1973) subsequently determined an average of 23.3 (range 21–28) microband couplets for 23 such

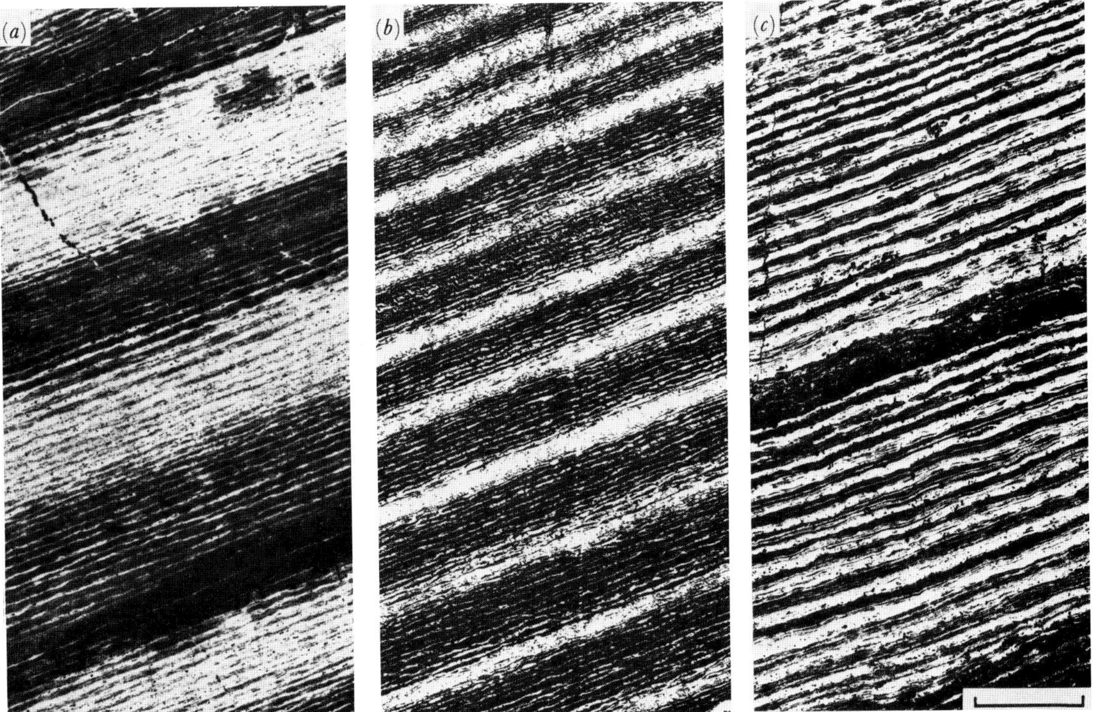

FIGURE 1. Early Proterozoic cyclic banded iron-formation from the Weeli Wolli Formation, Western Australia. Scale bar 5 mm. (*a*) Chert pod containing discernible microband couplets of chert (white) and iron oxide (black); up to about 30 microband couplets occur between the centres of the cyclic 'stripes'. (*b*) Moderately compacted iron-formation with thinner microbands and cycles. (*c*) Strongly compacted iron-formation in which only the cyclic stripes are readily discernible.

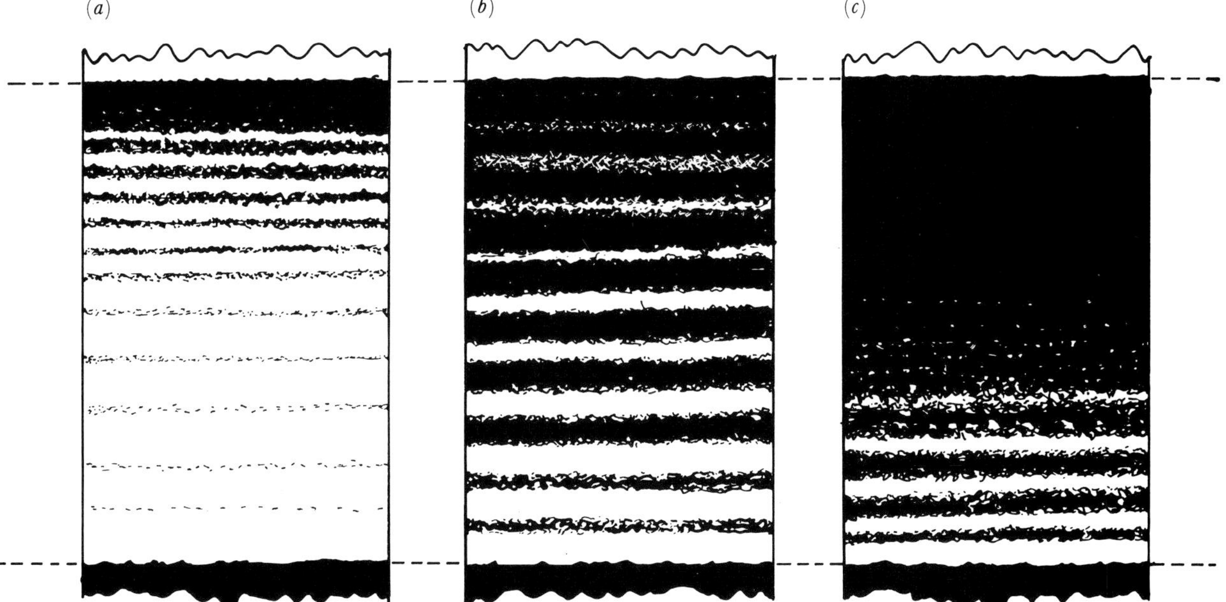

FIGURE 2. Range of cycle types in the mid Proterozoic Wollogorang Formation, northern Australia. The cycles commonly are about 1 cm thick. (*a*) Dolomite (white) dominant over mudstone (black). (*b*) Dolomite and mudstone in about equal proportion. (*c*) Mudstone dominant. (After Jackson 1985.)

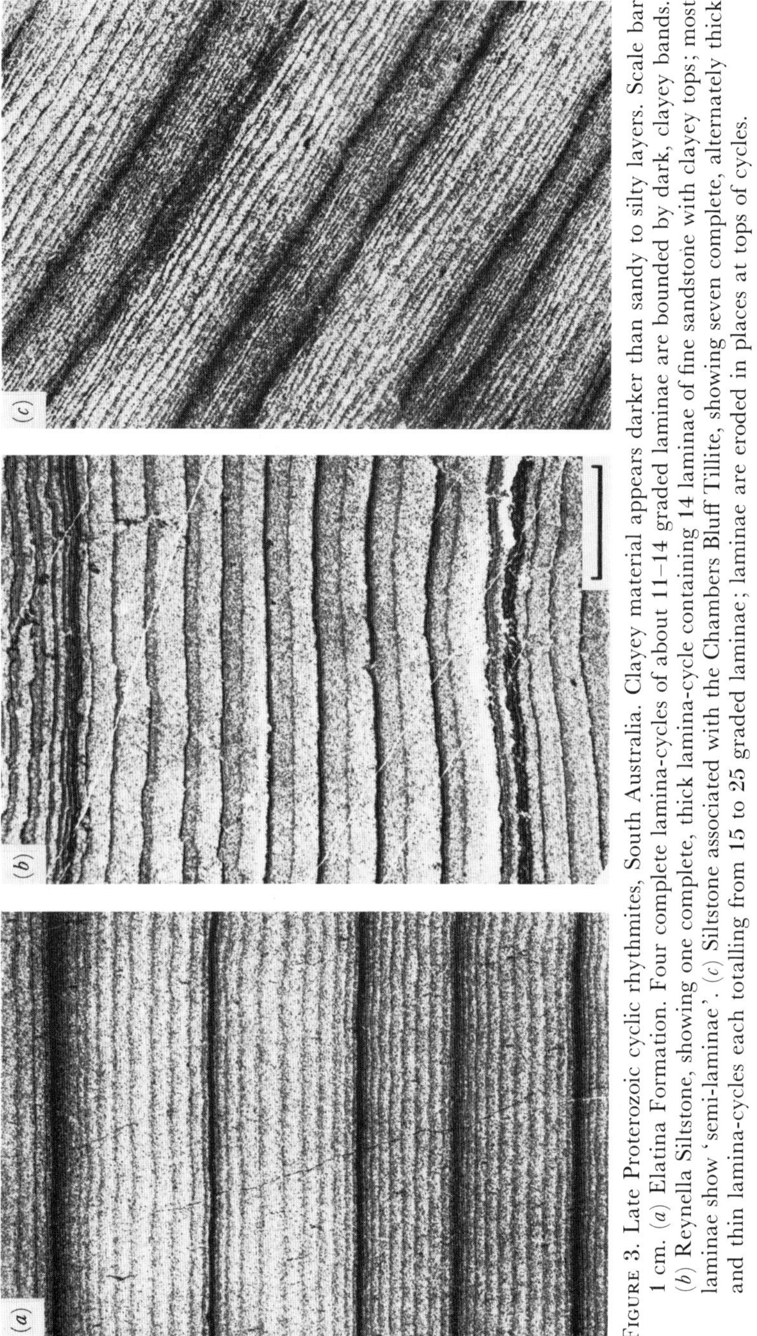

Figure 3. Late Proterozoic cyclic rhythmites, South Australia. Clayey material appears darker than sandy to silty layers. Scale bar 1 cm. (*a*) Elatina Formation. Four complete lamina-cycles of about 11–14 graded laminae are bounded by dark, clayey bands. (*b*) Reynella Siltstone, showing one complete, thick lamina-cycle containing 14 laminae of fine sandstone with clayey tops; most laminae show 'semi-laminae'. (*c*) Siltstone associated with the Chambers Bluff Tillite, showing seven complete, alternately thick and thin lamina-cycles each totalling from 15 to 25 graded laminae; laminae are eroded in places at tops of cycles.

cycles; he interpreted the couplets as varves and suggested the cyclicity may reflect the influence on the basinal environment of the double sunspot cycle, which today has a mean period near 22 years.

Counts I have carried out on thin-sections of newly collected chert pods indicate as many as 28–30 microband couplets per cycle. Cycles containing fewer microband couplets usually show evidence that some adjacent microbands have amalgamated; counts for such cycles probably underestimate the true cycle period. These observations might suggest a cycle period near 30 microband couplets, casting doubt on the solar interpretation.

A submarine fumarolic or volcanic-exhalative origin has been proposed for numerous BIFs (see, for example, Simonson 1985; Fralick 1987); such an origin for the Weeli Wolli BIF is supported by the local presence of beds of volcanic ash (Trendall & Blockley 1970) and aspects of the BIF geochemistry. Today, some geyser activity (Rinehart 1972 a, b) and the composition of gas bubbles in mineral-spring water (Sugisaki 1981) are modulated by earth tides. The question is raised, therefore, whether the Weeli Wolli cyclicity records earth-tidal rhythms that modulated the discharge and composition of silica- and iron-bearing fumarolic waters. As the primary components of the solid earth tide are semidiurnal and fortnightly, the microband couplets could perhaps be semidiurnal increments arranged in fortnightly groups or fortnightly tidal increments grouped in annual cycles through seasonal influences on sedimentation, among other possibilities. The second suggestion may be preferred because geothermal areas usually are so sluggish mechanically that the semidiurnal and diurnal components are filtered out, whereas the activity of geysers may be influenced by the fortnightly tidal component (Rinehart 1974). Furthermore, a yearly time value for the Weeli Wolli cyclic stripes gives sedimentation rates for the compacted facies (figure 1c) that are comparable to presumed rates for other BIFs in the Hamersley Basin whose microbanding is regarded as annual (see Trendall & Blockley 1970).

WOLLOGORANG FORMATION

The middle Proterozoic Wollogorang Formation (Jackson 1985) contains a dolomitic black shale facies that displays laminae of carbonaceous mudstone and silty dolomite arranged in conspicuous cycles. The cycles (figure 2, plate 1) are 1–20 mm thick, and typically comprise a lower portion of mainly dolomite and subordinate mudstone laminae and an upper portion of mainly mudstone with thin dolomite laminae. Jackson (1985) interpreted the dolomite–mudstone lamina couplets as distal lacustrine varves deposited below effective wave base. He suggested that the cycles may reflect an influence of sunspot cyclicity on climate.

The solar interpretation of the Wollogorang cycles must be viewed with reservation, however, for the following reasons.

1. The markedly asymmetric structure of the cycles (figure 2) bears little resemblance to that of the typical sunspot cycle which, although positively skewed, shows a gradual rise to a maximum over some 4–5 years.

2. Jackson (1985) determined a mean period of 7 (± 2 standard deviation) lamina couplets per cycle for a sequence of 100 cycles, and a mean of 11 (± 2 s.d.) couplets for 12 thicker cycles within that sequence. The greater number of laminae in the thicker cycles suggests that many of the cycles are abbreviated and that the counts underestimate the true cycle period.

Asymmetrical sedimentary cycles might form in a tidal setting by algae and mud coating coarser sediment during neaps and raising the threshold current speed for movement of sediment in the neap to spring part of the tidal cycle (see de Boer 1981; Allen 1982). Thus the lamina couplets of the Wollogorang Formation could be diurnal increments arranged in skewed fortnightly groups whose lower, neap-to-spring (mudstone to silty dolomite) portion commonly is missing or abbreviated. The carbonaceous character of the mudstone is consistent with this interpretation. The periods of 14.8 and 28.8 couplets determined by spectral analysis of a short sequence (Sonett & Trebisky 1986) also are consistent with the fortnightly and monthly grouping of diurnal increments.

Further study of the Wollogorang cycles is required, although available evidence favours a tidal origin and by implication a marine setting. Confirmation of periodicities near 15 and 30 lamina couplets would lend support to the tidal interpretation.

Elatina Formation

Several late Proterozoic formations in the Adelaide Geosyncline in South Australia locally display conspicuous cyclic rhythmites of siltstone and fine sandstone. Such rhythmites are best exposed in a 10 m thick member of the Elatina Formation (*ca.* 650 Ma old) at Pichi Richi Pass in the Flinders Ranges. Similar rhythmites have been recognized more recently in the correlative Reynella Siltstone at Hallett Cove near Adelaide, and in siltstones overlying the Chambers Bluff Tillite (between 800 and 650 Ma old) in the Officer Basin in northern South Australia.

Periodicities in the Elatina series

The cyclicity and method of study of the Elatina series are described in detail elsewhere (see Williams 1985, 1988, 1989a, b). Essentially, graded laminae 0.2–3.0 mm thick are grouped in conspicuous 'lamina-cycles' (figure 3a, plate 2) containing on average about 12 laminae (range 8–16). Lamina-cycles are bounded by darker, clayey bands where thinner laminae crowd together. The plot of lamina thickness (figure 4a) shows a characteristic alternation of high- and low-amplitude lamina-cycles. On average, high-amplitude cycles tend to have sharper crests, and contain fewer laminae, than low-amplitude cycles. Also, cycles have a tendency to positive skewness.

Longer periods are evident as variations in lamina-cycle thickness. The most conspicuous long-term rhythm, the 'Elatina Cycle' (figures 5a, b and 6a, b) occurs 59 times in a measured sequence of 1580 lamina-cycle thicknesses obtained from drill core. The Elatina Cycle is marked by first-order peaks in thickness that occur on average every 26.2 ($\pm$0.2 s.d. derived from fast Fourier transform (FFT) spectrum; see Williams 1989a, b) lamina-cycles; most Elatina Cycles also display a second-order peak.

A sawtooth pattern reflecting an alternation of thick and thin lamina-cycles is superimposed on the Elatina Cycle (figures 5a and 6a). This pattern displays a succession of envelopes with 180° phase reversals (a reversal of the thick–thin alternation) at or near the necks between the envelopes (figures 5c and 6c). The interval between phase reversals averages 14.6 ($\pm$0.1 s.d. from FFT spectrum; see Williams 1989b) lamina-cycles, and between 360° changes of phase averages 29.2 ($\pm$0.2 s.d.) lamina-cycles. These and other long-term periods, including several harmonics of the Elatina Cycle, are confirmed by spectral analysis (Williams 1985, 1989a, b). As well, the amplitude of the second-order peak of the Elatina Cycle varies with a period of 19.5 ($\pm$0.5) Elatina Cycles (Williams 1989a, b).

The Elatina rhythmites have no known modern counterpart. Their interpretation therefore

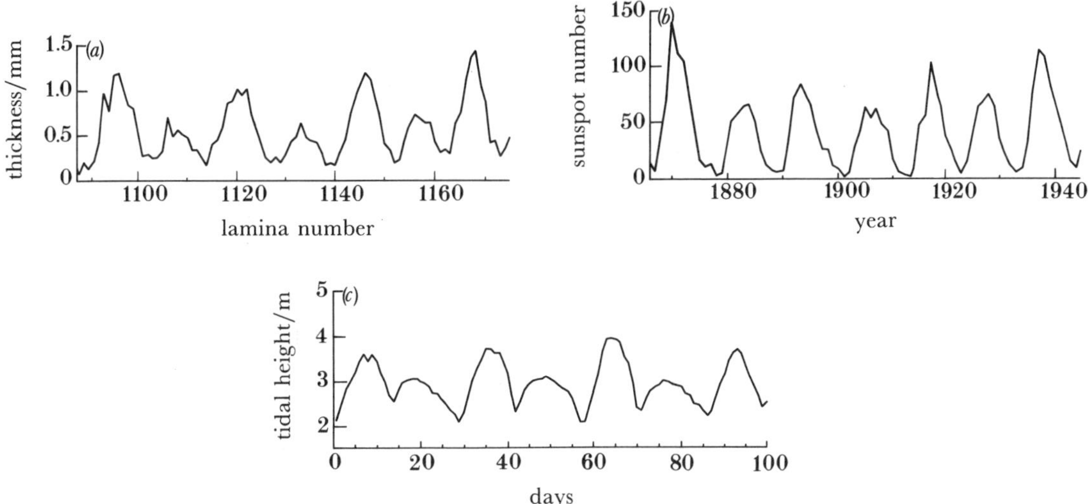

FIGURE 4. (a) Thickness of laminae from the Elatina series, showing lamina-cycles of alternate high and low amplitude. Lamina number increases up-sequence. (b) Mean annual sunspot numbers for sunspot cycles 11–17. (c) Maximum daily tidal heights from 1 January to 10 April 1966, for Townsville, Queensland, showing fortnightly tidal cycles of alternate high and low amplitude.

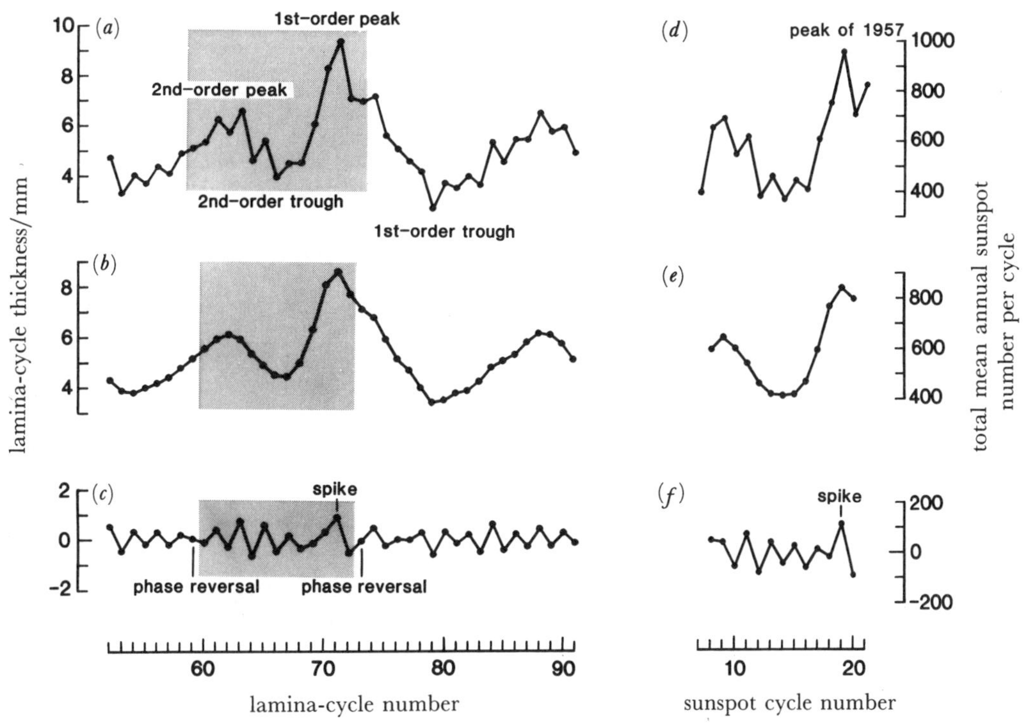

FIGURE 5. Curves comparing the thickness of lamina-cycles in the Elatina series with the magnitude of sunspot cycles for the continuous sunspot record since 1830. (a) Unsmoothed curve extracted from the Elatina series, showing the two peaks and troughs of an 'Elatina Cycle'. (b) The curve shown in (a), smoothed by a 3-point filter (weighted 1, 2, 1). (c) The residual sawtooth curve (curve (a) minus curve (b)). (d) Unsmoothed curve showing the total mean annual sunspot number per cycle for sunspot cycles 7–21. (e) The curve shown in (d), smoothed by a 3-point weighted filter. (f) The residual sawtooth curve (curve (d) minus curve (e)). Correlation coefficients between Elatina (shaded) and sunspot curves are 0.93 (unsmoothed), 0.98 (smoothed) and 0.94 (residual).

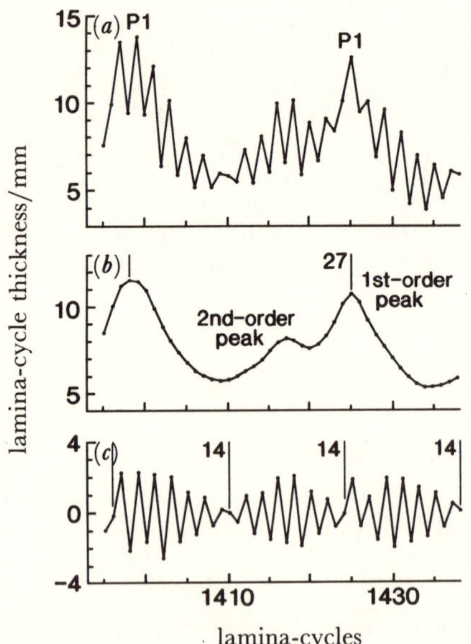
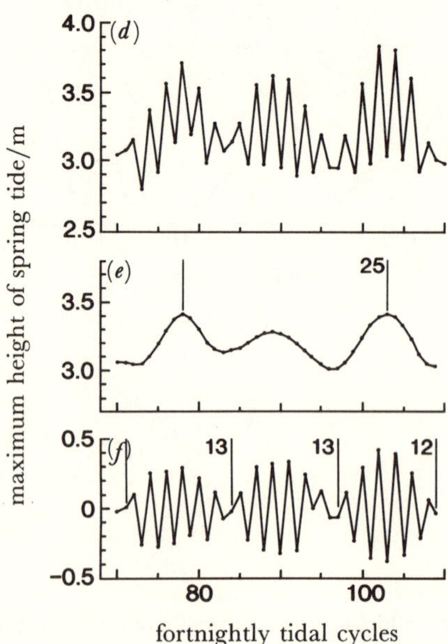

FIGURE 6. (a)–(c) Thickness of lamina-cycles from the Elatina series; lamina-cycle number increases up-sequence. (a) Unsmoothed curve showing first-order peaks (P1) that define the 'Elatina Cycle'. (b) Smoothed curve (5-point filter weighted 1, 4, 6, 4, 1); the number gives lamina-cycles between maxima. (c) Residual sawtooth curve (curve (a) minus curve (b)); the vertical lines mark positions of 180° phase-reversals in the sawtooth pattern, and the numbers give lamina-cycles between phase-reversals. (d)–(f) Tidal patterns for Townsville, Queensland, that are comparable with curves of lamina-cycle thickness. (d) Maximum height of the fortnightly tidal cycle from 19 October 1968 to 3 June 1970. (e) Smoothed curve (5-point filter weighted 1, 4, 6, 4, 1); the number gives fortnightly tidal cycles between yearly maxima that in part reflect seasonal effects (see Pariwono *et al.* 1986). (f) Residual sawtooth curve (curve (d) minus curve (e)); the vertical lines mark positions of 180° phase-reversals in the sawtooth pattern, and the numbers give fortnightly tidal cycles between phase reversals. (Tidal data for figures 4c and 6d, e, f supplied by the Tidal Laboratory of the Flinders Institute for Atmospheric and Marine Sciences, Flinders University of South Australia, copyright reserved.)

depends critically on the time-values ascribed the various sedimentary structures and cycles. A strongly seasonal, arid climate during deposition of the Elatina Formation is indicated by periglacial structures in a penecontemporaneous fossil permafrost horizon bordering the Adelaide Geosyncline (Williams 1986; Williams & Tonkin 1985); a yearly signature therefore should be sought in the rhythmites as a temporal milestone.

The solar-climatic hypothesis

An early interpretation, quite plausible given the periglacial setting, was that the laminae are varves that record spring or summer discharge of turbid meltwater into a deep, periglacial lake (Williams 1981, 1985). The laminae indeed bear resemblance to the 'intermediate distal facies' in the varve classification of Smith (1978). Lamina thickness in the Elatina Formation thus might provide a relative measure of annual meltwater discharge and hence of summer or mean annual temperature, and the Elatina record might be read as an interweaving of climatic cycles.

The varve interpretation underlies the hypothesis that the Elatina periodicities ultimately reflect solar variability (Williams 1981, 1985; Williams & Sonett 1985). The Elatina periods of 12 ± 4, 22–25 and *ca.* 103 'varve-years' may be equated with cycles in the sunspot record, which comprise a basic cycle of *ca.* 11 ± 3 years, the Hale magnetic cycle of *ca.* 22 years, and a longer period between 90 and 110 years. In both the Elatina and sunspot series the basic cycles (figure 4a, b) are of variable length, alternate cycles of relatively high and low amplitude are common, high-amplitude cycles tend to have sharper crests and to be of relatively short duration, and the cycles on average show positive skewness.

Furthermore, the plot of lamina-cycle thickness may be compared with that of the sum of yearly sunspot number per sunspot cycle (figure 5). Running cross-correlations show that the all-too-brief continuous sunspot record since about 1830 is comparable with that portion of the Elatina Cycle that includes the first- and second-order peaks and the intervening second-order trough; the first-order peak in the Elatina Cycle may be equated with the sunspot peak of 1957. Cross-correlations between respective Elatina (shaded) and sunspot curves shown in figure 5 yield correlation coefficients of 0.93–0.98. The strong correlations between elements of the Elatina and sunspot series permit an excellent simulation of the sunspot cycle (Bracewell 1986).

The empirical similarities between elements of the Elatina and sunspot records might arise through a direct connection, by way of climatic temperature variability, between late Proterozoic 'varve' thickness and solar activity. A solar signal, albeit a very weak one, has possibly been recorded in this way by recent varves deposited in Skilak Lake, Alaska (Sonett & Williams 1985). None the less, a mechanism whereby solar cyclicity might strongly influence the late Proterozoic climate has yet to be demonstrated (see Gérard & François 1987).

New data from the late Proterozoic of South Australia

New observations from other late Proterozoic rhythmites in South Australia throw new light on the origin of the Elatina series (Williams 1988, 1989a, b).

The Reynella Siltstone at Hallett Cove displays lamina-cycles like those of the Elatina Formation, but also contains some lamina-cycles up to six times thicker (figure 3b) that allow the detailed structure of individual cycles to be determined accurately. The thick cycles consistently give counts of 14 or 15 laminae per cycle, which exceed the mean count of 12 laminae per cycle in the Elatina series. A further distinctive feature of the Reynella Siltstone is the presence of *two* graded layers or coarse–fine couplets in many laminae (figure 3b). Where such 'semi-laminae' are of about equal thickness, the cycles appear to contain up to 26 or more 'laminae'. Semi-laminae are common even in the thin lamina-cycles in the Reynella Siltstone, but are rare in the Elatina Formation.

The facies at Hallett Cove (Dyson & von der Borch 1986) suggest fluctuations in water depth and distance from the sediment source in a marginal marine or estuarine setting, causing the interbedding of continental and shallow-water marine facies with proximal and distal lamina-cycles of offshore, deeper water origin. An overall greater and more stable depth of water appears to have been maintained at Pichi Richi Pass, where only distal lamina-cycles were deposited.

The lamina-cycles associated with the Chambers Bluff Tillite (figure 3c) also show an overall regular change in thickness of laminae and a pattern of alternate thick and thin cycles with phase reversals. Synsedimentary scouring, however, occurs at the tops of cycles and the thinner laminae in such places commonly are truncated. In a measured sequence of nine cycles the number of laminae per cycle ranges from 15 to 25, averaging 20.0. Because of the scouring, this mean must be viewed as a minimum estimate of cycle period.

The presence of up to 25 laminae per cycle suggests affinity with those cycles in the Reynella Siltstone that contain 26 or more semi-laminae; the fewer laminae in some cycles of the

Chambers Bluff Tillite is explained by contemporaneous scouring. A lamina of the Elatina Formation thus equates with a pair of semi-laminae in the Reynella Siltstone and with two laminae in the Chambers Bluff Tillite. The question is raised, therefore, whether the lamina-cycles at Pichi Richi Pass commonly are abbreviated through non-deposition of clastic laminae at the clayey ('dark band') boundaries between cycles. A reappraisal of the Elatina data and the solar-climatic hypothesis, viewed in the light of these new findings, is warranted.

A tidal model for late Proterozoic rhythmite deposition

The three late Proterozoic rhythmites described here comprise a new sedimentary facies with, as noted above, no matching cyclicity yet recognized in recent laminated sediments. Any new 'facies model' proposed for deposition of the rhythmites must take account of the following observations.

1. Laminae in the Elatina Formation including the very thinnest (*ca.* 0.2 mm) can be correlated in drill cores from holes 200 m apart. Laminae usually are graded within the range of fine sand to clayey fine silt. Deposition likely occurred mainly from suspension in quiet waters below wave base.

2. Palaeoslope directions for the Elatina rhythmites indicated by slumps in drill core are to the east, i.e. directly away from the nearby western margin of the Adelaide Geosyncline. Isolated ripples indicate eastward movement of sediment down this palaeoslope by unidirectional currents. Regional palaeocurrent directions for the Elatina Formation also are to the east (Preiss 1987).

3. A tidal influence is suggested by the clayey drapes within isolated climbing ripples that pass laterally into the Elatina laminar rhythmites, and by bimodal palaeocurrent directions in the Reynella Siltstone (Alexander 1984) and locally in the Elatina Formation. The isolated ripples in the Elatina rhythmites have similarity to the much thicker sand-wave deposits with mud drapes of subtidal origin (see Allen 1982).

These observations and the complex cyclicity displayed by all the rhythmites are explicable by deposition from turbid ebb-tidal currents in an offshore marine setting (Williams 1987, 1988, 1989a, b). In this tidal facies-model, fine sediment is entrained by ebb-tidal currents in tidal inlets and is transported mainly in suspension by turbulent ebb-tidal jets to deeper water offshore (see Özsoy 1986). There, settling suspended sediment would form graded laminae and local small-scale ripples (see Rees 1966) in a distal ebb-tidal delta, estuarine or marine shelf setting below wave base. Sediment deposited offshore in this way is relatively undisturbed by flood-tidal currents, which usually are weak and converge radially toward the tidal inlet (Özsoy 1986).

Effectiveness of the tide as an agent of such entrainment and deposition would be determined by the speed of the ebb-tide and the tidal range (see, for example, FitzGerald & Nummedal 1983; Boothroyd 1985); the potential sediment load of ebb-tidal currents may increase linearly with tidal range and current speed. Hence, relatively thick laminae on the distal delta or shelf would normally be associated with fast ebb-tidal currents and large tidal ranges, such as occur during the spring phase of the tidal cycle. Thin clayey caps could form on sandy laminae during slack water between tides. As movement of sediment by tides may cease below a threshold tidal range and current speed (see Allen 1982; FitzGerald & Nummedal 1983), pauses in deposition of sandy laminae in the distal setting might occur for small, neap-tidal ranges; the quieter waters at such times would allow the settling of further fine material and the deposition of distinct clayey bands between fortnightly groups of laminae. Similarly, deposition of cross-bedded, cyclic 'bundle' sequences of shallow-water, subtidal deposits is explained by unidirectional or strongly asymmetrical tidal currents effecting deposition, a direct relation between clastic cross-bed thickness and tidal range and current speed, and the association of mud layers and drapes with slack water and neaps (see, for example, Visser 1980; Allen 1981, 1982).

The tidal rhythms so recorded by the Elatina Formation might be superimposed on an annual signature of sea-level variation or summer influx of turbid periglacial meltwaters. Annual high sea levels (see, for example, Pariwono *et al.* 1986) might combine with seasonal abundances of sandy and silty material in suspension to produce annual intervals of thicker laminae in the ebb-tidal deposits.

Applying this tidal model to the Elatina series, again the question arises regarding the time values to be ascribed the various sedimentary structures and cycles. In view of the prevailing strongly seasonal climate, a regular pulse should be sought in the sediments, other than the individual lamina, that might record annual (summer) peaks of rapid deposition. By these criteria, the first-order peak of the Elatina Cycle (figures 5*b* and 6*b*) stands clear; it is the most regular of all the oscillations, Fourier transform yielding a strong, narrow spectral peak (Williams 1985, 1989*a*, *b*), and invariably is associated with the thickest laminae and therefore the most rapid deposition. The smoothed curve of the Elatina Cycle (figures 5*b* and 6*b*) is indeed similar to the annual curves of sea level for numerous localities (see figure 6*e* and Pariwono *et al.* (1986)).

Late Proterozoic palaeotidal and palaeorotational periods

If the first-order peak of the Elatina Cycle is regarded as a yearly signal, periodicities displayed by the Elatina series can be equated readily with tidal parameters (see Williams 1987, 1988, 1989*a*, *b*).

1. The late Proterozoic year would be represented by 26.2 ($\pm$0.2) lamina-cycles. The lamina-cycle would thus represent the late Proterozoic fortnightly tidal cycle.

2. The graded laminae of the Elatina Formation and Reynella Siltstone would represent diurnal increments recording the lunar day; the semi-laminae of the Reynella Siltstone and the laminae of the Chambers Bluff Tillite would be semidiurnal increments of the lunar day. The characteristic tidal patterns for these formations would thus be diurnal, mixed and semidiurnal respectively (figure 7); the mainly diurnal signature for the Elatina series may reflect a mixed pattern with the semidiurnal signal filtered out by sedimentary processes. Features of lamina-cycles for the Elatina series (figure 4*a*) – alternate cycles of high and low amplitude, the shorter duration and sharper crests of high-amplitude cycles, and the tendency for some cycles to be positively skewed – are all displayed in the daily pattern of fortnightly tidal cycles for Townsville, Queensland (figure 4*c*) (although, as discussed above, a positive skewness of lamina-cycles may also reflect a higher threshold current speed for sediment movement in the neap to spring part of the tidal cycle). The alternation of high and low spring tides and the shorter duration of high springs result from the eccentric lunar orbit and the Moon's more rapid orbital motion at perigee. (Maximum tidal heights are plotted in figures 4*c* and 6*d*, *e*, *f* as they are a relative measure of tidal range.) Furthermore, the lamina-cycles of the Reynella Siltstone that display semi-laminae (figure 3*b*) have a structure similar to the mixed fortnightly tidal growth patterns of modern bivalves (see Evans 1972).

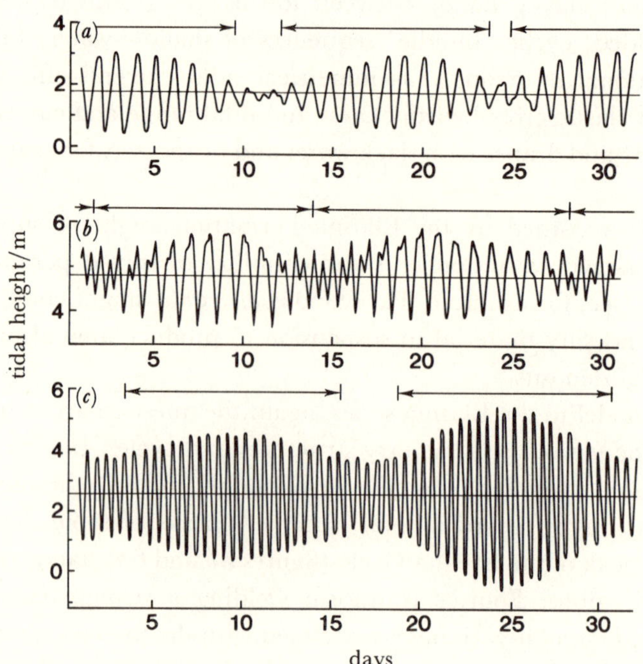

Figure 7. Modern tidal patterns (from Godin 1972; Lisitzin 1974). (*a*) Diurnal (as displayed by the Elatina Formation, see figure 3*a*). (*b*) Mixed (as displayed by the Reynella Siltstone, see figure 3*b*). (*c*) Semidiurnal (as displayed by the Chambers Bluff Tillite, see figure 3*c*). The arrows show schematically the tidal ranges for which semidiurnal and/or diurnal tidal cycles effected the deposition of clastic laminae for respective late Proterozoic rhythmites.

A maximum value of about 16 diurnal tides per late Proterozoic fortnightly tidal cycle implies that up to half the laminae are missing from certain lamina-cycles through their non-deposition at neaps when tidal ranges and maximum current speeds fell below threshold values for sand–silt transport to the area studied. The 'dark bands' of clayey material bounding many lamina-cycles would mark horizons of arrested clastic deposition. Such abbrevation explains the general uniformity of clastic lamina thickness at minima between lamina-cycles (see figure 4*a*). Strongly abbreviated cycles are not necessarily bounded by thicker clayey bands, however, suggesting that the deposition of clayey material also ceased temporarily in the distal setting during protracted neaps of small tidal ranges. Figure 7 shows in a general way the tidal ranges that effected deposition of clastic laminae in the rhythmites.

3. Pairs of lamina-cycles of high and low amplitude (figure 4*a*) would represent the lunar monthly tidal cycle. There would be 13.1 ($\pm$0.1 s.d. from FFT spectrum; see Williams (1989*a*, *b*)) lunar months per late Proterozoic year, compared with 12.37 today. The number of lunar days per late Proterozoic lunar month cannot as yet be directly measured accurately because of the common abbreviation of lamina-cycles. The greatest number of laminae for two successive lamina-cycles in the Elatina series is 29, and the thick lamina-cycles of the Reynella Siltstone contain up to 15 diurnal laminae; these figures suggest 29–30 lunar days per lunar month and in turn imply around 30–31 solar days per lunar month.

4. The second harmonic of the Elatina Cycle, with a period of 13.1 lamina-cycles, would represent the half-yearly tidal signal. The half-yearly tide, by its modulation of tidal range, influenced the number of laminae deposited per lamina-cycle in the Elatina Formation (Williams 1989*a*).

5. The alternation of high and low spring tides at Townsville shows a pattern of 'sawtooth envelopes' and phase reversals (figure 6f) like that in the Elatina series (figures 5c and 6c). This pattern also results from the eccentricity of the lunar orbit. Modern tides, as exemplified by 20 years of data for Townsville, average 27.9 spring tides for a 360° change of phase; the mean value some 650 Ma ago was 29.2 spring tides. These periods, which may be termed the 'tidal year', are slightly longer than respective solar years because of prograde rotation of the Moon's perigee. The late Proterozoic data indicate that the period of the lunar apsides cycle (rotation of the Moon's perigee) was then $29.2 \pm 0.2/(29.2 \pm 0.2 - 26.2 \pm 0.2) = 9.7 \pm 0.1$ years (see also Williams 1989a, b).

6. The amplitude modulation of the second-order peak of the Elatina Cycle indicates a long-term period of 19.5 (± 0.5) Elatina Cycles or years, which is interpreted as that of the palaeolunar nodal cycle (Williams 1989a, b).

Palaeotidal and palaeorotational data for late Proterozoic time *ca.* 650 Ma ago as determined from the rhythmites of the Elatina Formation and Reynella Siltstone, and those calculated for that time by Lambeck (1978, 1988) based on palaeontological data, are compared with their modern equivalents in table 1. The data imply an average equivalent phase lag (the angle between the Earth–Moon axis and the Earth's tidal bulge, derived from the response of the solid earth and ocean tides; see Lambeck 1980) near 3° since late Proterozoic time rather than the present value of 6°.

TABLE 1. LATE PROTEROZOIC (*CA.* 650 Ma) AND MODERN TIDAL AND ROTATIONAL PERIODS

	late Proterozoic			
	Lambeck (1978, 1988)[a]			
parameter	(*a*)	(*b*)	this study[b]	modern
solar days in lunar month	*ca.* 30.7	*ca.* 30.2	30.5 (± 0.5)	29.53
lunar months in year	*ca.* 14.4	*ca.* 13.2	13.1 (± 0.1)	12.37
lunar apsides cycle/years	—	—	9.7 (± 0.1)	8.85
lunar nodal cycle/years	—	—	19.5 (± 0.5)	18.61
days in year	*ca.* 440	*ca.* 400	400 (± 7)	365
length of day/h	*ca.* 20.1	*ca.* 21.9	21.9 (± 0.4)	24.0

[a] The values from Lambeck (1978, 1988) are derived from Phanerozoic palaeontological data. They are based on an average equivalent phase lag angle of (*a*) 6° (the present value), and (*b*) 3°, and assume that tidal friction is the only phenomenon responsible for secular changes in the Earth's rotation and the Moon's revolution.

[b] Periods indicated by the rhythmites of the Elatina Formation and Reynella Siltstone (see Williams 1987, 1988, 1989a, b), with revised error estimates based on FFT spectra where applicable, as discussed in the text.

Overall, the similarities between the structure and relative periods of cycles in the Elatina series and modern tidal cycles are evident in figures 4 and 6, figures 3 and 7, and table 1. Such agreement argues strongly that the late Proterozoic rhythmites in South Australia indeed record a full spectrum of palaeotidal cycles that provide a unique set of palaeorotational and palaeotidal periods for *ca.* 650 Ma ago.

CONCLUSIONS

The Precambrian cyclic rhythmites whose conspicuous cyclicity has been compared with sunspot cyclicity – the Weeli Wolli Formation, the Wollogorang Formation, and the Elatina Formation – are, in the light of new geological data and critical re-examination, better interpreted as encoding tidal rhythms. Despite the varve-like features of the Elatina Formation and the strong correlations between elements of the Elatina and sunspot series, an ebb-tidal model of deposition is much preferred because it can explain the complex patterns of the Elatina series over the *full* range of frequencies as well as new observations from other late Proterozoic rhythmites in South Australia. Furthermore, the ebb-tidal model alone appeals to rhythmic processes operating strongly at the Earth's surface today, and requires stability of the depositional system for only about 60–70 years rather than the 20 000 years necessitated by the varve-solar interpretation. The tidal interpretation of the Elatina series implies that graded strata carrying tidal signals should be preserved in modern, laminated subtidal deposits built by ebb-tidal currents, although bioturbation by benthic fauna may hinder preservation.

In view of these reinterpretations, it would seem that the pre-recent geological record has yet to provide unequivocal evidence for significant modulation of terrestrial climate by the solar activity cycle. Such evidence may prove as elusive as have convincing indications of solar signals in modern patterns of weather and climate. Hence, we cannot conclude, on present geological evidence, that the Sun displayed an activity cycle like that of today during the early eras of Earth history.

On the credit side, the Elatina series evidently provides the first firm benchmark for Precambrian palaeotidal periods and palaeorotation. The study of such rhythmites promises to greatly illuminate the Precambrian history of the Earth's rotation and the Moon's revolution.

I thank Professor K. Lambeck, Professor G. W. Lennon, Dr C. J. Durrant and Dr W. V. Preiss for assistance and helpful discussions.

References

Alexander, E. M. 1984 *Sedimentology of the Marinoan type section, Marino rocks to Hallett Cove area, South Australia.* B.Sc. (Hons) thesis, Department of Geology and Geophysics, The University of Adelaide, Australia.

Allen, J. R. L. 1981 Palaeotidal speeds and ranges estimated from cross-bedding sets with mud drapes. *Nature, Lond.* **293**, 394–396.

Allen, J. R. L. 1982 Mud drapes in sand-wave deposits: a physical model with application to the Folkestone Beds (Early Cretaceous, southeast England). *Phil. Trans. R. Soc. Lond.* A **306**, 291–345.

Anderson, R. Y. 1961 Solar–terrestrial climatic patterns in varved sediments. *Ann. N.Y. Acad. Sci.* **95**, 424–439.

Anderson, R. Y. & Koopmans, L. H. 1963 Harmonic analysis of varve time series. *J. geophys. Res.* **68**, 877–893.

Boothroyd, J. C. 1985 Tidal inlets and tidal deltas. In *Coastal sedimentary environments* (ed. R. A. Davis), pp. 445–532. New York: Springer.

Bracewell, R. N. 1986 Simulating the sunspot cycle. *Nature, Lond.* **323**, 516–519.

Bradley, W. H. 1930 The varves and climate of the Green River epoch. *U.S. geol. Surv. Prof. Paper* no. 158, pp. 87–110.

Brooks, C. E. P. 1926 *Climate through the ages.* London: Ernest Benn. (439 pages.)

de Boer, P. L. 1981 Mechanical effects of micro-organisms on intertidal bedform migration. *Sedimentology* **28**, 129–132.

De Geer, G. 1929 Solar registration by pre-Quaternary varve-shales. *Geogr. Annlr* **11**, 242–246.

Duff, P. M. D., Hallam, A. & Walton, E. K. 1967 *Cyclic sedimentation.* Amsterdam: Elsevier. (280 pages.)

Dyson, I. A. & von der Borch, C. C. 1986 A field guide to the geology of the late Precambrian Wilpena Group, Hallett Cove, South Australia. In *One day geological excursions of the Adelaide region* (comp. A. J. Parker), pp. 17–40. Adelaide: Geol. Soc. Australia, S. Australia Division.

Evans, J. W. 1972 Tidal growth increments in the cockle *Clinocardium nuttalli. Science, Wash.* **176**, 416–417.

FitzGerald, D. M. & Nummedal, D. 1983 Response characteristics of an ebb-dominated tidal inlet channel. *J. Sediment. Petrol.* **53**, 833–845.

Fralick, P. 1987 Depositional environment of Archean iron formation: inferences from layering in sediment and volcanic hosted end members. In *Precambrian iron-formations* (ed. P. W. U. Appel & G. L. LaBerge), pp. 251–266. Athens: Theophrastus.

Garrels, R. M. 1987 A model for the deposition of the microbanded Precambrian iron formations. *Am. J. Sci.* **287**, 81–106.

Gérard, J.-C. & François, L. M. 1987 A model of solar-cycle effects on palaeoclimate and its implications for the Elatina Formation. *Nature, Lond.* **326**, 577–580.

Godin, G. 1972 *The analysis of tides.* Liverpool: Liverpool University Press. (264 pages.)

Jackson, M. J. 1985 Mid-Proterozoic dolomitic varves and microcycles from the McArthur Basin, northern Australia. *Sediment. Geol.* **44**, 301–326.

Lambeck, K. 1978 The Earth's palaeorotation. In *Tidal friction and the Earth's rotation* (ed. P. Brosche & J. Sündermann), pp. 145–153. Berlin: Springer.

Lambeck, K. 1980 *The Earth's variable rotation: geophysical causes and consequences.* Cambridge University Press. (449 pages.)

Lambeck, K. 1988 *Geophysical geodesy: the slow deformations of the Earth.* Oxford University Press. (718 pages.)

Lisitzin, E. 1974 *Sea-level changes.* Amsterdam: Elsevier. (286 pages.)

Özsoy, E. 1986 Ebb-tidal jets: a model of suspended sediment and mass transport at tidal inlets. *Estuarine, coastal shelf Sci.* **22**, 45–62.

Pariwono, J. I., Bye, J. A. T. & Lennon, G. W. 1986 Long-period variations of sea-level in Australasia. *Geophys. Jl R. astr. Soc.* **87**, 43–54.

Preiss, W. V. (comp.) 1987 The Adelaide Geosyncline. *S Australian Dept. Mines and Energy Bull.* no. 53. (438 pages.)

Rees, A. I. 1966 Some flume experiments with a fine silt. *Sedimentology* **6**, 209–240.

Richter-Bernburg, G. 1964 Solar cycle and other climatic periods in varvitic evaporites. In *Problems in palaeoclimatology* (ed. A. E. M. Nairn), pp. 510–521, 532. London: Interscience.

Rinehart, J. S. 1972*a* Fluctuations in geyser activity caused by variations in earth tidal forces, barometric pressure, and tectonic stresses. *J. geophys. Res.* **77**, 342–350.

Rinehart, J. S. 1972*b* 18.6-year earth tide regulates geyser activity. *Science, Wash.* **177**, 346–347.

Rinehart, J. S. 1974 Geysers. *Eos, Wash.* **56**, 1052–1062.

Schove, D. J. (ed.) 1983 *Sunspot cycles.* Benchmark Papers in Geology no. 68. Stroudsburg: Hutchinson Ross. (397 pages.)

Simonson, B. M. 1985 Sedimentological constraints on the origins of Precambrian iron-formations. *Bull. geol. Soc. Am.* **96**, 244–252.

Smith, N. D. 1978 Sedimentation processes and patterns in a glacier-fed lake with low sediment input. *Can. J. Earth Sci.* **15**, 741–756.

Sonett, C. P. & Trebiski, T. J. 1986 Secular change in solar activity derived from ancient varves and the sunspot index. *Nature, Lond.* **322**, 615–617.

Sonett, C. P. & Williams, G. E. 1985 Solar periodicities expressed in varves from glacial Skilak Lake, southern Alaska. *J. geophys. Res.* **90**, 12,019–12,026.

Sugisaki, R. 1981 Deep-seated gas emission induced by the earth tide: a basic observation for geochemical earthquake prediction. *Science, Wash.* **212**, 1264–1266.

Trendall, A. F. 1972 Revolution in Earth history. *J. geol. Soc. Australia* **19**, 287–311.

Trendall, A. F. 1973 Varve cycles in the Weeli Wolli Formation of the Precambrian Hamersley Group, Western Australia. *Econ. Geol.* **68**, 1089–1097.

Trendall, A. F. & Blockley, J. G. 1970 The iron formations of the Precambrian Hamersley Group, Western Australia. *Geol. Surv. W. Australia Bull.* no. 119. (366 pages.)

Udden, J. A. 1928 Study of the laminated structure of certain drill cores obtained from Permian rocks of Texas. *Carnegie Institution of Washington Year Book* no. 27, p. 363.

Visser, M. J. 1980 Neap-spring cycles reflected in Holocene subtidal large-scale bedform deposits: a preliminary note. *Geology* **8**, 543–546.

Walter, M. R. 1972 A hot spring analog for the depositional environment of Precambrian iron formations of the Lake Superior region. *Econ. Geol.* **7**, 965–972.

Williams, G. E. 1981 Sunspot periods in the late Precambrian glacial climate and solar-planetary relations. *Nature, Lond.* **291**, 624–628.

Williams, G. E. 1985 Solar affinity of sedimentary cycles in the late Precambrian Elatina Formation. *Aust. J. Phys.* **38**, 1027–1043.

Williams, G. E. 1986 Precambrian permaforst horizons as indicators of palaeoclimate. *Precambrian Res.* **32**, 233–242.

Williams, G. E. 1987 Cosmic signals laid down in stone. *New Scient.* **114**, (1566), 63–66.

Williams, G. E. 1988 Cyclicity in the late Precambrian Elatina Formation, South Australia: solar or tidal signature? *Climatic Change* **13**, 117–128.

Williams, G. E. 1989*a* Late Precambrian tidal rhythmites in South Australia and the history of the Earth's rotation. *J. geol. Soc. Lond.* **146**, 97–111.

Williams, G. E. 1989*b* Precambrian tidal sedimentary cycles and Earth's paleorotation. *Eos, Wash.* **70**, 33, 40–41.

Williams, G. E. & Sonett, C. P. 1985 Solar signature in sedimentary cycles from the late Precambrian Elatina Formation, Australia. *Nature, Lond.* **318**, 523–527.

Williams, G. E. & Tonkin, D. G. 1985 Periglacial structures and palaeoclimatic significance of a late Precambrian block field in the Cattle Grid copper mine, Mount Gunson, South Australia. *Aust. J. Earth Sci.* **32**, 287–300.

Discussion

J.-C. Gérard (*Institut d'Astrophysique, Université de Liège, Belgium*). Are the deposition rates derived from the thickness of the Elatina Formation laminae compatible with the tidal interpretation of the periodicities?

G. E. Williams. Yes, they are. By the tidal interpretation, the 10 m thick rhythmite member of the Elatina Formation at Pichi Richi Pass was deposited in about 60 years, giving a mean rate of deposition of around 17 cm a^{-1}. This rate is compatible with deposition rates of up to 1 m or more per year determined for estuarine and shallow-water clastic tidal deposits; the comparatively low rate of deposition for the Elatina rhythmite member accords with its envisaged distal, offshore setting. Of course, such rates of deposition are not maintained continuously in the same area over very long intervals of time. Many sedimentary sequences, including tidal deposits, comprise beds or sedimentary packages deposited relatively rapidly and bounded by bedding planes representing long intervals of erosion, reworking or non-deposition.

Phil. Trans. R. Soc. Lond. A **330**, 459–462 (1990)

Printed in Great Britain

Environmental records from polar ice cores

By C. Lorius

*Laboratoire de Glaciologie et Géophysique de l'Environnement, 54 Rue Molière, Domaine Universitaire,
BP 96, 38402 St-Martin-d'Hères Cedex, France*

Polar ice cores provide a wide range of information on past atmospheric climate (temperature, precipitation) and environment (gas and aerosol concentrations). The dating can be very accurate for the more recent part of the records but accuracy decreases with depth and time. Measurements of cosmogenic isotope concentrations (such as ^{10}Be) provide information on palaeo-precipitation rates and particular events can be used to correlate ice core records. Besides these climatic applications, ^{10}Be concentration records in ice cores also contain information on solar activity changes.

INTRODUCTION

The recovery of records of the environmental history is one of the keys to a better understanding of climatic changes. In this respect polar ice can provide a wide range of information, as it records atmospheric climatic conditions and samples aerosols and gases in consecutive datable layers over timescales extending to 150000 years before present (BP). In particular, changes in cosmogenic isotope concentration can potentially provide information on production rates that are influenced, among other factors, by solar modulation. This latter aspect is discussed in this Symposium by Dr G. M. Raisbeck and Professor H. Oeschger.

THE ATMOSPHERIC RECORD

Although we cannot expect to find an ideal climatic recorder in nature, glacial ice is proving to be a rather remarkably close approximation to it (Lorius 1990; Dansgaard & Oeschger 1989).

The basis for palaeotemperature reconstruction is the existence of current correlations between the proportion of deuterium to hydrogen and ^{18}O to ^{16}O atoms measured in deposited snow and temperature conditions at the site, resulting from fractionation processes that take place in the atmospheric water cycle. Although the isotopic composition of polar snow depends on several parameters, there exists a linear relation in both of the polar ice sheets, between the mean annual surface temperature and the mean $\delta^{18}O$ of δD value of deposited snow (Lorius & Merlivat 1977; Johnsen *et al.* 1988) and the large isotopic changes (i.e. glacial–interglacial) obtained from ice cores can be interpreted, as a first-order approximation, in terms of local atmospheric temperature changes. However, the climatic signal:noise ratio leads to more unfavourable conditions for the quantitative reconstruction of climatic changes of smaller magnitude such as that which prevailed over the past centuries.

Relating gas concentrations obtained from ice-core extraction to atmospheric values that are of global significance is much more straightforward; although precipitation mechanisms may have secondary effects, there is for instance, within the precision of the measurements, no

460   C. LORIUS

differences between the CO_2 and CH_4 greenhouse gas data monitored by modern observations
and those obtained from corresponding polar ice layers (Siegenthaler & Oeschger 1987;
Stauffer *et al.* 1985).

Such precipitation mechanisms are of importance to establish a quantitative link between
the concentration of aerosols and impurities in surface snow (Davidson 1989). Although there
is no doubt that the concentration of a species in the air is reflected in snow deposits at the site,
there is a lack both of direct observations and in understanding aerosol deposition processes.
The relative contribution of 'wet' (including precipitation scavenging) and 'dry' deposition is
poorly known and may vary with location and possibly with time. Concentration of impurities
in ice, including cosmogenic isotopes may then depend on several factors such as source
intensity or production rate but also on atmospheric circulation and rate of snow precipitation.

An overview summarizing the climate-related information which can be obtained from ice
cores is given in table 1.

TABLE 1. SHORT SUMMARY OF ENVIRONMENT INFORMATION AND CORRESPONDING
ICE-CORE SIGNAL

atmosphere	ice core
temperature	D/H, $^{18}O/^{16}O$
precipitation	D/H, $^{18}O/^{16}O$, ^{10}Be
humidity	D/H, $^{18}O/^{16}O$
aerosols	
natural (continents, sea	chemicals
volcanoes, biosphere)	(Al, Ca^{2+}, Na^+, H^+, SO_4^{2-}, NO_3^-)
man made	SO_4^{2-}, NO_3^-, Pb, radioactive fallout
cosmogenic	^{10}Be, ^{26}Al, ^{36}Cl
circulation	particles
gases: natural and man made	O_2, N_2, CO_2, CH_4, N_2O

TIMESCALES

For the upper part of the ice sheets the accuracy of the chronology can be very high. Annual
layers can be counted from seasonal variations of various parameters such as visual
stratigraphy, physical properties, isotopic composition, electrical conductivity, chemicals, etc.,
but at great depth they are becoming indiscernible. Prominent features found in ice cores can
be used as reference horizons. They can provide ages when causal events are documented
(radioactive fallout from nuclear tests over the past decades, ash or aerosol deposits from
volcanoes); more generally, large-scale atmospheric events can be used for relative
intercomparison between ice cores (see for instance below the use of ^{10}Be peaks to correlate
antarctic ice records), or with other palaeodata such as those from sea, land or lake sediments.
Although there are promising possibilities to obtain absolute ages for ice-core records from
radioactive and other dating techniques long-term timescales have been so far based on
numerical modelling of the age distribution through ice sheets by ice dynamics (Reeh 1989).

The accuracy of the method depends on a number of factors, but change in accumulation
rate is the most important parameter to consider.

Cosmogenic isotopes: the ^{10}Be records

A number of long-lived radioactive nuclides are formed by cosmic-ray-induced nuclear reactions in the atmosphere. These 'cosmogenic' isotopes are then either transported to ice surfaces in precipitation or dry fallout (for non-volatile species such as ^{10}Be, ^{26}Al, ^{36}Al) or occluded, along with trapped air in bubbles (^{14}C). These isotopes have two potential applications in ice core studies: (i) they might possibly be used to 'date' (either in an absolute, or relative way) ice cores, (ii) they might give a time record of the cosmogenic production rate.

The first application may be illustrated by evaluation of past accumulation (precipitation) changes in Antarctica on a glacial–interglacial timescale. One approach is based on the fact that current accumulation rates in Antarctica are governed by the amount of water vapour circulating above the inversion layer. This amount is itself controlled by temperature via the saturation vapour pressure and the isotope-based temperature record can thus be used to estimate past accumulation. Results give quite similar accumulation values during interglacials while precipitation appears to have been reduced to about 50 % the current value during the coldest periods of the ice age. A second approach (Yiou *et al* 1985) is based on ^{10}Be measurements performed along an ice core. Concentrations of this long-lived cosmogenic radioisotope are quite similar during the two interglacials; they increase during colder periods, by up to a factor of 2 during full glacial conditions. Assuming a constant ^{10}Be deposition flux, the measured concentrations are reflecting the amount of snow precipitation. Indeed there is a very good agreement between accumulation rates derived independently from the isotope temperature and ^{10}Be records. The coherence of these results lends some support to this estimate of past precipitation and also indirectly further supports possible reasons why ^{10}Be concentration might be inversely correlated with palaeoprecipitation rate.

Furthermore ^{10}Be peaks have been found around 60000 and 35000 years BP in Antarctica ice (Raisbeck *et al.* 1987) not correlated with any obvious climatic feature and it was suggested that these peaks may be due to increased production in the atmosphere. There are three possible causes for such production changes: variations in primary cosmic ray flux, changes in solar modulation or changes in geomagnetic field intensity. But regardless of the final explanation these peaks have been used as stratigraphic markers to correlate various climatic records from Antarctic ice (Jouzel *et al.* 1990).

Over shorter timescales (centuries) ice core ^{10}Be and isotope temperature records can be used to test the influence of solar variability on climate (G. M. Raisbeck *et al.*, this Symposium) and ^{10}Be concentrations in ice cores can be compared with the ^{14}C variations in tree rings (H. Oeschger, this Symposium). As already pointed out there may be some limitations in the quantitative reconstruction of relatively small climatic changes from the ice-core isotopic record while on the other hand ^{10}Be concentrations may then more favourably reveal solar variability features than in long-term glacial–interglacial records.

References

Dansgaard, W. & Oeschger, H. 1989 Past environmental long term records from the Arctic. In *The environmental record in glacier and ice sheet* (ed. H. Oeschger & C. C. Langway Jr), pp. 287–317. Chichester: Wiley.

Davidson, C. I. 1989 Mechanisms of wet and dry deposition of atmospheric contaminants to snow surfaces. In *The environmental record in glacier and ice sheet* (ed. H. Oeschger & C. C. Langway Jr), pp. 29–51. Chichester: Wiley.

Johnsen, S. J., Dansgaard, W. & White, J. 1989 The origin of arctic precipitation under glacial and interglacial conditions. *Tellus* B**41**, 452–468.

Jouzel, J., Raisbeck, G., Benoist, J. P., Yiou, F., Lorius, C., Raynaud, D., Petit, J. R., Barkov, N. I., Korotkevitch, Y. S. & Kotlyakov, V. M. 1989 The Antarctic climate over the late glacial period from ice cores. *Quaternary Res.* **31** (2), 135–150.

Lorius, C. 1989 Polar ice cores and climate. In *Understanding climate change* (ed. A. Berger, R. E. Dickinson & J. W. Kidsson), *Geophysical monograph* **52**, 11–16.

Lorius, C. & Merlivat, L. 1977 Distribution of mean surface stable isotope values in East Antarctica: observed changes with depth in a coastal area. In *Isotopes and impurities in snow and ice, Proc. Grenoble Symp., August and September 1975*, pp. 127–137. Int. Ass. Sc. Hyd., 118.

Raisbeck, G. M., Yiou F., Bourles D., Lorius C., Jouzel J. & Barkov, N. I. 1987 Evidence for two intervals of enhanced ^{10}Be deposition in Antarctic ice during the last glacial period. *Nature, Lond.* **326**, 273–277.

Reeh, N. 1989 Dating by ice flow modelling: a useful tool or an exercise in applied mathematics. In *The environmental record in glacier and ice sheet* (ed. H. Oeschger & C. C. Langway Jr), pp. 141–159. Chichester: Wiley.

Siegenthaler, U. & Oeschger, H. 1987 Biospheric CO_2 emissions during the past 200 yrs reconstructed by deconvolution in ice core data. *Tellus* B**39**, 140–154.

Stauffer, B., Fischer, G., Neftel, A. & Oeschger, H. 1985 Increase of atmospheric methane recovered in antarctic ice. *Science, Wash.* **229**, 1386–1388.

Yiou, F., Raisbeck, G. M., Bourles, D., Lorius, C. & Barkov, N. I. 1985 ^{10}Be at Vostok Antarctica during the last climatic cycle. *Nature, Lond.* **316**, 616–617.

Phil. Trans. R. Soc. Lond. A **330**, 463–470 (1990)

Printed in Great Britain

^{10}Be and δ^2H in polar ice cores as a probe of the solar variability's influence on climate

By G. M. Raisbeck[1], F. Yiou[1], J. Jouzel[2,3] and J. R. Petit[3]

[1] *Centre de Spectrométrie Nucléaire et de Spectrométrie de Masse, IN2P3/CNRS, Bât. 108, 91405 Campus Orsay, France*

[2] *Laboratoire de Géochemie Isotopique, CEA/CEN Saclay, 91191 Gif sur Yvette, France*

[3] *Laboratoire de Glaciologie et Géophysique de l'Environnement, 54 Rue Molière, Domaine Universitaire, BP 96, 38402 St-Martin-d'Heres Cedex, France*

By using the technique of accelerator mass spectrometry, it is now possible to measure detailed profiles of cosmogenic (cosmic ray produced) ^{10}Be in polar ice cores. Recent work has demonstrated that these profiles contain information on solar activity, via its influence on the intensity of galactic cosmic rays arriving in the Earth's atmosphere. It has been known for some time that, as a result of temperature-dependent fractionation effects, the stable isotope profiles δ^2O and δ^2H in polar ice cores contain palaeoclimate information. Thus by comparing the ^{10}Be and stable isotope profiles in the same ice core, one can test the influence of solar variability on climate, and this independent of possible uncertainties in the absolute chronology of the records. We present here the results of such a comparison for two Antarctic ice cores; one from the South Pole, covering the past *ca.* 1000 years, and one from Dome C, covering the past *ca.* 3000 years.

Introduction

One of the most fundamental, and controversial, aspects of solar–terrestrial relations is the extent to which solar activity – as reflected by sunspot number (or their more modern analogues such as the geomagnetic 'aa' index or 10.7 cm and He spectral lines) – is related to climate. Therefore, any system that can potentially address this question, especially in an objective and quantitative fashion, deserves careful attention. It is one such system, ^{10}Be and stable isotope ratios in polar ice cores, that we consider here.

It has been established for some time now that the flux of galactic cosmic rays arriving at the Earth is inversely correlated with the 11-year solar activity cycle. Although the mechanism for this relation is still not understood in detail, it basically appears to arise mainly because the interplanetary magnetic fields embedded in the outflowing solar wind are disturbed by shock waves associated with energetic solar flare events. Because the production rate of cosmogenic (i.e. cosmic ray produced) isotopes in the atmosphere is proportional to the galactic cosmic ray flux, their abundance is also inversely correlated with solar activity. (Although energetic solar flare particles can also produce cosmogenic isotopes, averaged over time their contribution compared with the galactic cosmic ray production is estimated to be quite small.)

The best-known cosmogenic isotope is ^{14}C. However, because it forms a gaseous molecule (CO_2) and is in dynamic equilibrium with large reservoirs of carbon in the biosphere and oceans, short-term changes in the ^{14}C production rate lead to strongly damped changes in the atmospheric ^{14}C:^{12}C ratio. Thus for example, the *ca.* 30% production rate changes in the 11-

[65]

year solar cycle are barely discernable in $^{14}C:^{12}C$ ratios from dendrochronologically dated tree-ring sequences. However, for longer time periods, high-precision ^{14}C measurements in these tree-ring sequences do show significant variations, at least some of which are believed to be caused by changes in solar activity. Indeed, the Maunder Minimum type 'wiggles' in the tree-ring ^{14}C record constitute probably one of our strongest evidences that solar activity has varied on timescales longer than the 11-year cycle (see H. E. Suess & T. W. Linick and C. P. Sonett & S. A. Finney, this Symposium).

The only other long-lived isotope that is formed by cosmic-ray-induced nuclear reactions with nitrogen and oxygen, the principal components of the atmosphere, is ^{10}Be (half-life 1.5 Ma). Because beryllium does not form a gaseous species in the atmosphere it quickly attaches itself to aerosols, and is transported to the Earth's surface by precipitation and dry deposition in *ca.* 1 year. Changes in the production rate of this isotope are thus rapidly and strongly reflected in its deposition rate. Unfortunately, the short atmospheric residence time and meteorological influence on ^{10}Be deposition result in changes in the ^{10}Be concentration in geological reservoirs that are not as directly related to the global production rate as is $^{14}C:^{12}C$. However, recent work – including results presented here – demonstrate that ^{10}Be profiles in polar ice cores do appear to give a record of solar activity comparable with that of $^{14}C:^{12}C$ in tree rings.

It has been known for some time that the stable isotope ratios (generally presented as $\delta^{18}O$ or δ^2H, the deviation from standard ocean water) in polar ice cores are related to the climate at the time the precipitation occurred. The best illustration of this (see, for example, figure 1) are the observed changes in these isotopic ratios during the last glacial to interglacial transition *ca.* 10000 years ago.

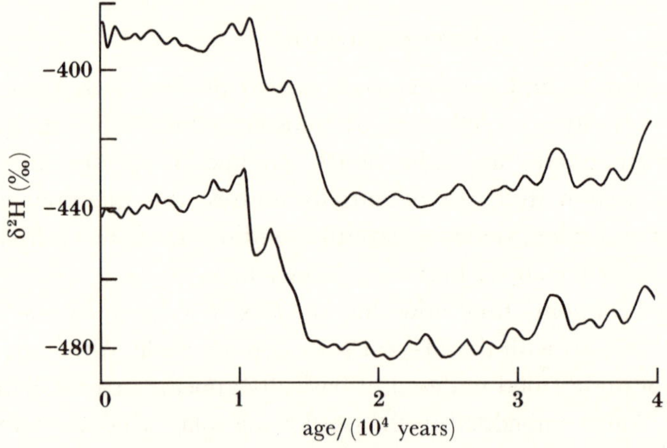

FIGURE 1. Smoothed profiles of δ^2H as a function of time in ice from Dome C (upper curve) and Vostok (lower curve), Antarctica (see Jouzel *et al.* 1989).

We thus see that we have a system that fulfils the conditions mentioned earlier, namely, a geological reservoir containing an objective, quantitative record of both solar activity and climate. By comparing these two records, one can thus estimate the degree of influence of solar activity on the climate. It is important to point out that, because both records can be obtained from the same core, such a comparison can be made *independently* of the absolute timescale for the samples.

RESULTS

South Pole

We begin by considering a 127 m ice core drilled at the South Pole in 1984. The ^{10}Be profile from this core, which has been obtained recently (Raisbeck & Yiou 1990), is shown in figure 2. The core has been dated by identifying previously dated volcanic signals in the ice (Kirchner 1988). To average out some of the meteorological 'noise' in the ^{10}Be signal, and make the time resolution more comparable with that observed in the ^{14}C record, this raw ^{10}Be signal can be subjected to various degrees of smoothing. An example of this, using a spline function, is shown in figure 2 as the dark solid curve.

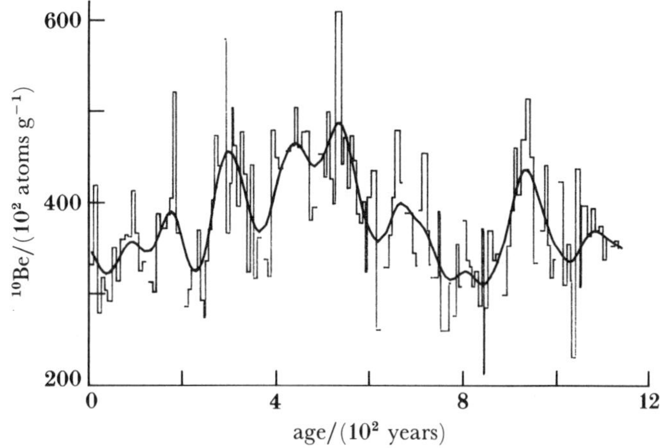

FIGURE 2. Concentration of ^{10}Be as a function of time in a 127 m ice core from the South Pole station, Antarctica. The histogram represents the raw data and the smoothed curve has been obtained by using a spline function.

In figure 3 we show a comparison of the smoothed ^{10}Be record with that of the bidecadal tree ring δ^{14}C record of Stuiver & Pearson (1986) and Pearson & Stuiver (1986). Taking into account an expected lag in the ^{14}C response (because of longer atmosphere residence time and exchange with the biological and oceanic reservoirs) compared with that of ^{10}Be, there is a strong resemblance of these two completely independent curves. This resemblance is, in our

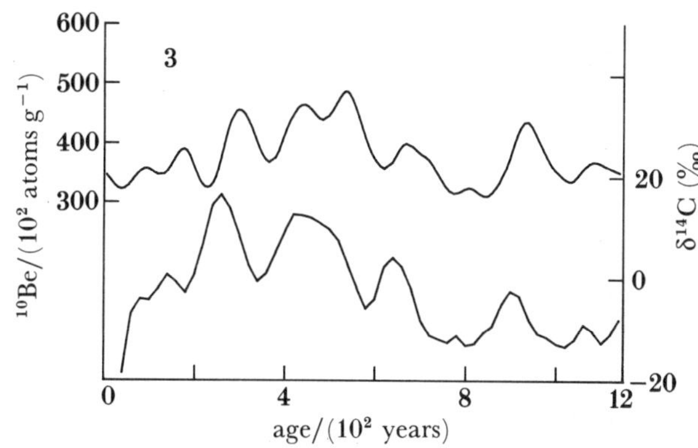

FIGURE 3. Comparison of smoothed ^{10}Be concentration in ice at South Pole (from figure 2) with δ^{14}C in dendrochronologically dated tree rings (see text).

opinion, one of the strongest pieces of evidence to date that the ^{10}Be ice core data are indeed recording solar modulation induced production rate changes.

Because there has been considerable discussion at this meeting on possible cyclicity of solar activity, we show in figure 4 the results of a preliminary spectral analysis of the ^{10}Be profile from the South Pole core, using a multitaper method (Yiou *et al.* 1990), based on a technique pioneered by D. J. Thompson (this Symposium). The results suggest statistically significant cyclicity at frequencies of 92 and 202 years.

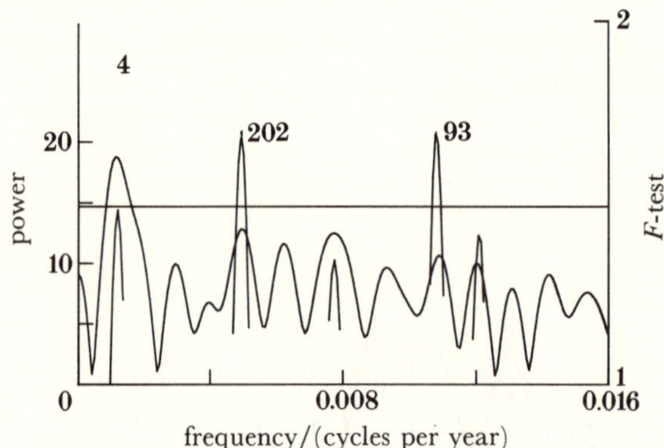

FIGURE 4. Multitaper spectral analysis of ^{10}Be concentration in South Pole ice over the past *ca.* 1100 years, showing power (left scale) and significance according to *F*-test (right scale). The horizontal line represents an *F*-test confidence level of 90%.

In figure 5 we show the recently obtained δ^2H profile for the South Pole core (Petit *et al.* 1990). The same depth-to-age conversion as for the ^{10}Be curve has been used here. It can be observed that the raw δ^2H data contains even more high-frequency 'noise' than the ^{10}Be signal. Once again the spline smoothed data is shown as a heavy solid curve.

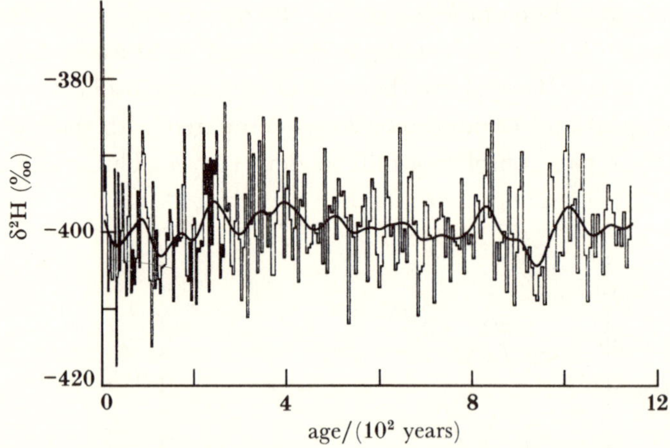

FIGURE 5. δ^2H as a function of time in the same South Pole ice core used for ^{10}Be measurements shown in figure 2. Histogram represents raw data and the smoothed curve has been obtained by using a spline function.

We can now look for a possible correlation in the ^{10}Be and δ^2H data. To make a visual comparison, it is obviously preferable to show the smoothed curves (figure 6). However, the data are then no longer independent, making it impossible to obtain a quantitative estimate of the degree of significance from a cross-correlation analysis. We have therefore carried out the correlation analysis on the raw ^{10}Be and δ^2H data, after resampling to have the same sampling intervals. Depending on the averaging interval used, the correlation coefficient varies from 0.007 (127 intervals of *ca.* 8 years) to 0.205 (32 intervals of *ca.* 35 years). None of these is statistically significant at the 10% level.

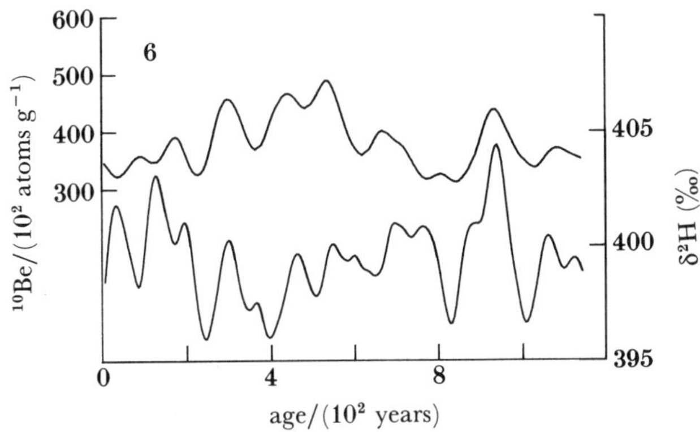

FIGURE 6. Comparison of smoothed ^{10}Be concentration and δ^2H in South Pole ice core.

Dome C

To extend the above type of analysis to a longer timescale, and another site, we now consider results from another Antarctic ice core at Dome C. Both the ^{10}Be (Raisbeck & Yiou 1988) and δ^2H (Benoit *et al.* 1982) data from this core have been previously published. We thus show directly in figure 7 the smoothed profiles derived from these measurements, adopting the timescale of Raisbeck & Yiou (1988). We stress that, although the absolute timescale in this case is considerably more uncertain than for the South Pole core, the use of the same depth-to-age conversion for both profiles means that our correlation analysis is independent of the timescale adopted. Once again the correlation has been carried out on the raw data after resampling. The result is a correlation coefficient of 0.136 (160 intervals of *ca.* 20 years) or 0.150 (80 intervals of *ca.* 40 years) that again is not significant.

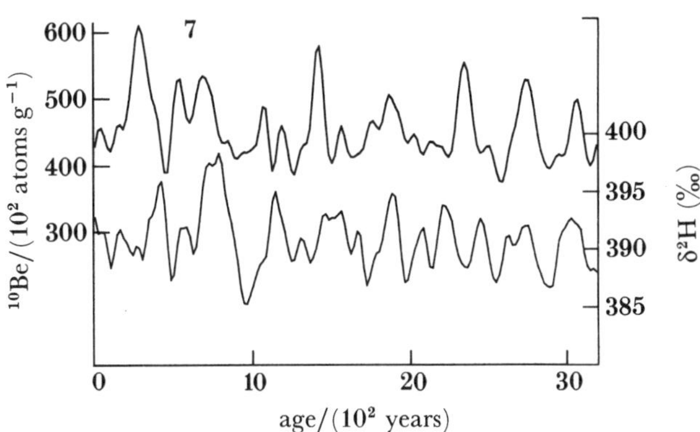

FIGURE 7. Comparison of smoothed ^{10}Be concentration and δ^2H in Dome C ice core.

Discussion and conclusions

We have shown above that there is no obvious correlation between solar activity (as indicated by ^{10}Be) and climate (as indicated by δ^2H) in two ice cores from Antarctica representing approximately the past 1000 and 3000 years. There are at least three possible explanations for this.

1. ^{10}Be is not a reliable indicator of solar activity. Although the ^{10}Be concentration in polar ice obviously contains a certain amount of meteorologically induced 'noise', we believe the remarkable correlation with ^{14}C seen in figure 3, as well as a recent comparison of ^{10}Be from Arctic and Antarctic ice cores (Beer *et al.* 1990) argue strongly in favour of a production rate origin for the main 100–200-year structure in the ^{10}Be record. Moreover, the magnitude of these variations is quite consistent with that predicted between periods of solar maximum and minimum. We thus consider it highly likely that the ^{10}Be record in polar ice is a good proxy indicator of solar activity for the time periods considered here.

2. δ^2H is not a reliable indicator of climate. As was mentioned earlier, and illustrated in figure 1, there seems to be no doubt that large changes in global climate, such as the glacial to interglacial transition occurring *ca.* 10–15 thousand years ago, are clearly recorded in the δ^2H (or δ^{18}O) signal of Antarctic ice. The change in surface temperature corresponding to this transition has been estimated as *ca.* 9 °C (Jouzel *et al.* 1989). The similarity in fine structure of the two curves in figure 1, from ice cores separated by *ca.* 500 km, suggests that, provided they last *ca.* 10^3 years, regional temperature changes as small as 1–2 °C can also be inferred from such records, although the global significance of these has not yet been demonstrated. For temperature changes on shorter timescales (*ca.* 10^2 years) characteristic of the solar variability being studied here, the situation is less clear. For example, even δ^2H records from two adjacent cores drilled at Dome C showed insignificant correlation, except possibly when smoothed over fairly long (*ca.* 500 year) time intervals (Benoit *et al.* 1982). This can probably be attributed to the low (*ca.* 3 g cm^{-2} a^{-1}) and irregular precipitation rate at this site, which can introduce significant noise into the stable isotope records over shorter time periods. On the other hand, a comparison of the δ^2H signal and annual mean temperature over a 20-year period at the South Pole, where the precipitation rate is significantly larger (*ca.* 8 g cm^{-2} a^{-1}), showed a high degree of correlation (Jouzel *et al.* 1983). Thus it seems reasonable to assume that this site should be recording temperature variability on the timescale being examined here, although once again the global significance of such changes remains uncertain. In summary then, although stable isotope ratios in polar ice cores have demonstrated their validity as indicators of large global climate change, their sensitivity for discerning modest global temperature changes on timescales (*ca.* 10^2 years) relevant to the solar activity variations under consideration here requires further studies.

3. There is no significant correlation between solar activity and climate. Such a conclusion would undoubtedly be unpalatable to many of the participants at this meeting, especially in the light of recent results showing an apparent correlation of solar irradiance with the activity level (P. Foukal, this Symposium). Because of the uncertainties in the climate signal mentioned above, we do not believe the data presented here exclude some level of solar activity–climate correlation. However, at the very least we have shown that there have been century-long periods of substantial change in solar activity during the past few thousand years, without

correspondingly large changes in Antarctic temperatures. Thus if a solar activity–climate relation does exist, it apparently is rather subtle, and will not be readily resolvable by the approach considered here.

REFERENCES

Beer, J., Raisbeck, G. M. & Yiou, F. 1990 In *Sun in time* (ed. C. P. Sonett, M. S. Giampapa & M. S. Mathews). University of Arizona Press. (Submitted.)
Benoit, J. P., Jouzel, J., Lorius, C., Merlivat, L. & Pourchet, M. 1982 *Ann. Glaciology* **3**, 17–22.
Jouzel, J. *et al.* 1989 *Quaternary Res.* **31**, 135–150.
Jouzel, J., Merlivat, L., Petit, J. R. & Lorius, C. 1983 *J. geophys. Res.* C **88**, 2693–2703.
Kirchner, S. 1988 Ph.D. thesis, University of Grenoble, France.
Pearson, G. W. & Stuiver, M. 1986 *Radiocarbon* **28**, 839–862.
Petit, J. R. *et al.* 1990 (In preparation.)
Raisbeck, G. M. & Yiou, F. 1988 In *Secular solar and geomagnetic variations in the last 10 000 years* (ed. F. R. Stephenson & A. W. Wolfendale), pp. 287–296. Kluwer Academic Publishers.
Raisbeck, G. M. & Yiou, F. 1990 (In preparation.)
Stuiver, M. & Pearson, G. W. 1986 *Radiocarbon* **28**, 805–838.
Yiou, P., Genthon, C., Jouzel, J., Ghil, M., Le Treut, H., Barnola, J. M., Lorius, C. & Korotkevitch, Y. N. 1990 In *Interactions of the global carbon and climate system* (ed. R. Keir). Electric Power Res. Inst. Report. (In the press.)

Discussion

N. O. Weiss (*University of Cambridge, U.K.*). I know that the 11-year activity cycle can be followed in the Milcent ice core from Greenland. Is it possible to detect this cycle in the ice core from the South Pole as well? At present the [10]Be record offers the only means of following the solar cycle back through the Maunder Minimum and over the past 1000 years. From the theoretical point of view it is extremely important to establish whether phase was maintained through the Maunder and Spörer Minima or not. This would allow us to distinguish between nonlinear dynamos and oscillators with stochastic perturbations. I hope therefore that the groups involved in analysing ice cores will succeed in providing an annual record of [10]Be abundance that covers the last millennium.

G. M. Raisbeck. The low accumulation rate on the Antarctic plateau, compared with the Arctic, makes it more difficult to record short-term variations such as the 11-year solar cycle. In the South Pole core, however, annual cycles of stable isotopes are apparently retained, so there is some hope that an 11-year cycle of [10]Be could be extracted. Although the time interval of the samples used in the present work were too long for such a study, we plan to examine this problem in the future.

Jenny Allsop (*Deep Geology Research Group, British Geological Survey, Nottingham, U.K.*). The slide, comparing the profile for ^{14}C with those for [10]Be and ²H from polar ice cores, provided a good comparative index between ^{14}C and [10]Be indicating a relation between solar activity and variations in the Earth's climate. However, the use of a single comparative index for the whole of the ²H profile may be misleading. The ²H profile showed a good optical match on either side of a large, central area of divergence between ²H and [10]Be that appeared to be almost a mirror image. A 'running' statistical comparative index may have illustrated that one large inconsistency between these profiles had caused the low index value.

If the reason for the low index is purely because of the large, single divergence, then ²H may, in fact, reflect the same variations seen in the ^{10}Be and ^{14}C profiles under 'normal' conditions. Therefore, if a reasonable explanation for this divergence can be found, this may provide additional data on a particular climatic situation that has not as yet been considered.

G. M. RAISBECK. It is true that there are some portions of the ^{10}Be and δ^2H curves that seem to be correlated, and may indicate a causal relation. However, even two random curves will show similar behaviour over limited intervals and, in the absence of an *a priori* reason for choosing some intervals and not others, it is statistically dangerous to draw any conclusions from such *a posteriori* observations.

G. DE Q. ROBIN (*Scott Polar Research Institute, Cambridge, U.K.*). Are the ^{10}Be data expressed in terms of ^{10}Be per unit mass corrected for changes in the rate of accumulation on ice sheets? The data shows that this was approximately halved during the last ice age.

I also comment that the noise level in δ-records in ice cores is highly dependent on the accumulation rate. Data from Dome C indicate that we need a core length between 100 and 150 years to estimate the mean temperature change of precipitation to within 1 °C from the δD record. We are not likely to see an 11- or 22-year periodicity in the δD record from Dome C, although we may just see this periodicity at the South Pole, where precipitation is at least double that at Dome C.

G. M. RAISBECK. Our data are expressed as ^{10}Be concentration per unit mass of ice. As Dr Grove points out, we do find an approximately factor of two increase in this concentration during ice age periods, and we have interpreted this as being as a result of a corresponding reduction in accumulation rate of ice during those periods. It is in fact this complication that prevents us from estimating possible differences in the average solar activity between these ice ages and interglacials. However, using the same temperature-accumulation relations, one would predict a maximum effect on the ^{10}Be concentration during the Holocene of only *ca.* 10%, i.e. considerably smaller than the effects attributed to solar modulation.

With regard to your remarks on extracting short-term temperature estimates from the δ^2H data in low accumulation rate areas, we are in complete agreement, as I hope the written text makes clear.

Phil. Trans. R. Soc. Lond. A **330**, 471–480 (1990)

Printed in Great Britain

The past 5000 years history of solar modulation of cosmic radiation from ^{10}Be and ^{14}C studies

By H. Oeschger[1] and J. Beer[2]

[1] *Physics Institute, University of Bern, CH-3012 Bern, Sidlerstrasse 5, Switzerland*
[2] *Environmental Physics, ETHZ, c/o EAWAG, CH-8600 Dübendorf, Switzerland*

^{10}Be is produced in a similar way as ^{14}C by the interaction of cosmic radiation with the nuclei in the atmosphere. Assuming that the ^{10}Be and ^{14}C variation are proportional and considering the different behaviour in the Earth system, the ^{10}Be concentrations in ice cores can be compared with the ^{14}C variations in tree rings. A high correlation is found for the short-term variations (^{14}C-Suess-wiggles). They reflect with a high probability production rate variations. More problematic is the interpretation of the long-term trends of ^{14}C and ^{10}Be. Several explanations are discussed.

The reconstructed CO_2 concentrations in ice cores indicate a rather constant value (280 ± 10 p.p.m. by volume) during the past few millenia. Measurements on the ice core from Byrd Station, Antarctica, during the period 9000 to 6000 years BP indicate a decrease that might be explained by the extraction of CO_2 from the atmosphere–ocean system to build the terrestrial biomass pool during the climatic optimum.

Introduction

After W. F. Libby introduced the radiocarbon dating method around 1950, by studying ^{14}C:C ratios in absolutely dated samples like tree-rings several authors investigated the constancy of the atmospheric ^{14}C:C ratio.

H. E. Suess pointed both at the existence of a long-term trend and short-term wiggles in the ^{14}C-bristlecone tree-ring record (Suess 1971, this Symposium). Whereas the existence of the long-term trend was confirmed in a number of laboratories, it was roughly a decade before general agreement regarding the existence of the short-term wiggles was reached. Today a very precisely measured record of the ^{14}C trend over the past 9000 years exists (Stuiver & Kra 1986). With the development of accelerator mass spectrometry about a decade ago, it became possible to measure additional long-lived radioisotopes (^{10}Be, ^{26}Al, ^{36}Cl) in natural archives.

In this paper we compare the ^{14}C record measured in tree-rings with ^{10}Be records measured on ice cores and deduce the present knowledge of short-term and long-term variations of cosmic radiation.

The ^{14}C-variations

Figure 1 shows the present continuous record on ^{14}C:C variations, expressed in per mille deviations from a standard, determined on dendrochronologically dated tree-rings (Stuiver & Kra 1986).

Additional information on the ^{14}C:C behaviour during the glacial–post-glacial transition exists from peat bog studies (Oeschger *et al.* 1980) and plant macrofossils deposited in lake sediments (Zbinden *et al.* 1990), which indicated variations of the order of 50–100‰ during

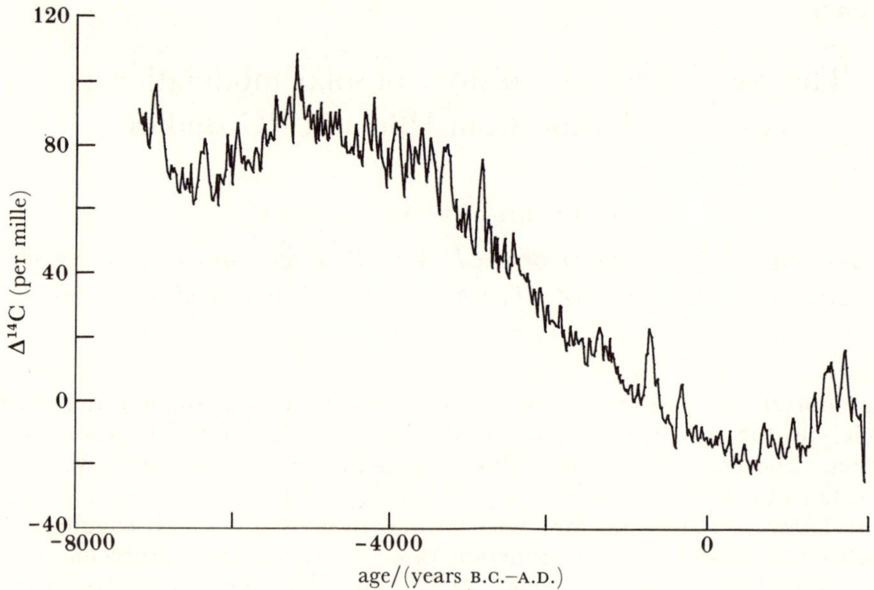

FIGURE 1. Atmospheric ^{14}C:C variations expressed in per mille deviation Δ^{14}C from a standard
(Stuiver & Kra 1986).

this period of major climate change. It is likely that these ^{14}C variations from 13000 to 10000
years before present (BP) mainly reflect the effects of carbon system changes as a result of
repartitioning of carbon between atmosphere, ocean, biosphere, sediments, etc., as well as
possible changes in the system dynamics, like the mixing of ocean surface and deep water. The
variations during the Holocene probably reflect mainly changes in the ^{14}C production rate. In
the following we draw some independent conclusions on the stability of the carbon system
during the past *ca.* 20000 years to investigate the potential influence of carbon system
variations on the atmospheric ^{14}C:C ratio.

HOW STABLE IS THE ATMOSPHERIC CO_2 CONCENTRATION AND THE CARBON SYSTEM DYNAMICS IN GENERAL?

Information on the stability of the carbon system during the past 30 years can be drawn from
the direct CO_2 concentration observations at Mauna Loa, Hawaii, and at the South Pole,
which started in 1985. In spite of the very large yearly exchange fluxes between atmosphere
and ocean and atmosphere and biomass, which together correspond to 25% of the atmospheric
CO_2 per year, the atmospheric CO_2 concentration reacted relatively monotonously to the
anthropogenic input (figure 2). The annual increase of CO_2 was *ca.* 1 ± 1 p.p.m. CO_2 per year,
varying not more than *ca.* 3% of the atmospheric CO_2 content. Figure 3 shows the CO_2
increase reconstructed from measurements of the air occluded in an ice core from Siple Station,
Antarctica, over the past 200 years (Neftel *et al.* 1985; Siegenthaler & Oeschger 1987). Again
a very homogeneous increase of atmospheric CO_2 is observed and deconvolution of the increase
by using a carbon cycle model resulted in a total CO_2 production from fossil fuel combustion
as well as from deforestation and land managing that might be very close to the actual ones
(Siegenthaler & Oeschger 1987). Based on this work we conclude that the carbon cycle, in

† p.p.m. is parts per million by volume.

[74]

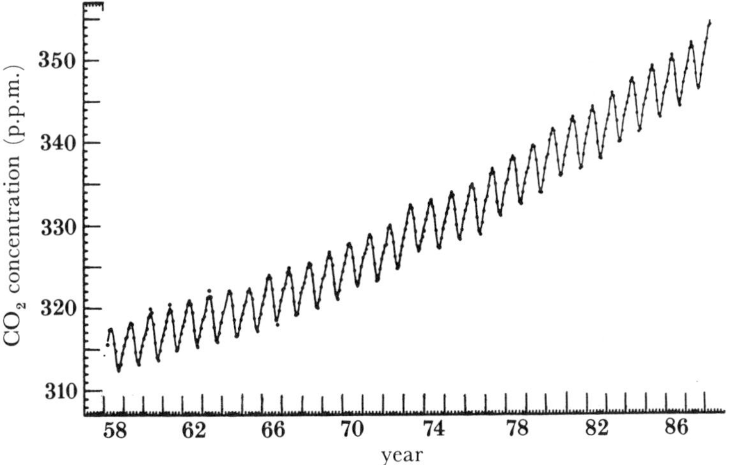

FIGURE 2. Concentrations of atmospheric CO_2 (p.p.m.) at Mauna Loa Observatory, Hawaii, from 1958 to 1988.

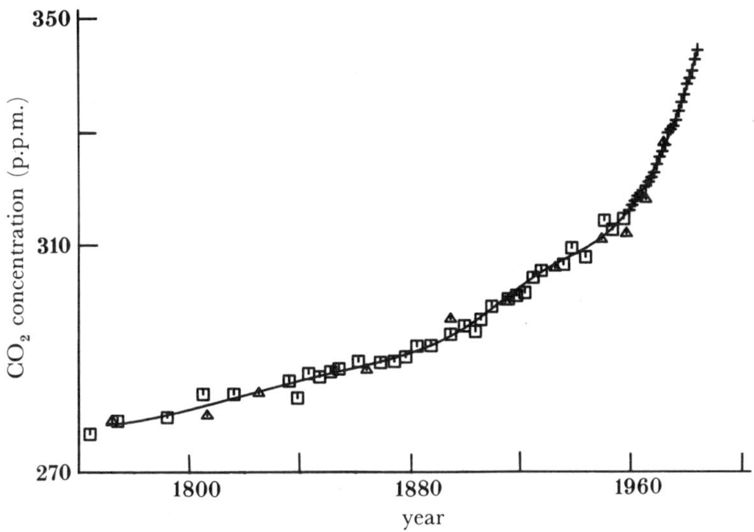

FIGURE 3. Atmospheric CO_2 increase in the past 200 years as indicated by measurements on air trapped in old ice from Siple Station, Antarctica, (Neftel *et al.* 1985) and by the annual mean values from Mauna Loa. The full line is a spline fit through all data (Siegenthaler & Oeschger 1987).

particular the CO_2 exchange atmosphere–ocean, during the past 200 years functioned essentially as it does at present.

Going further back in time there are indications of minor variations in atmospheric CO_2, e.g. an increase by *ca.* 5 p.p.m. between 1200 and 1300 A.D.

Estimates using carbon cycle models indicate that such small variations probably had little influence on the atmospheric $^{14}C:C$ ratio. A somewhat different situation might have existed at the beginning of the Holocene when the CO_2 system was adjusting to the new conditions. CO_2 measurements on the Byrd Station core, Antarctica (figure 4) showed a decrease from *ca.* 280 p.p.m. to 255 p.p.m. during a period of *ca.* 4000 years. An obvious explanation for this change is the regrowth of soils and plants on the continents in areas that during the glacial had been covered by ice.

The decrease in atmospheric CO_2 was about 10 % and, based on the assumption of an ocean behaving, regarding its chemistry, like the present one, the 10% CO_2 decrease in the atmosphere and the CO_2 partial pressure in the surface ocean correspond according to a rough estimate to an extraction of CO_2 from the atmosphere–ocean system of *ca.* 400 Gt C.

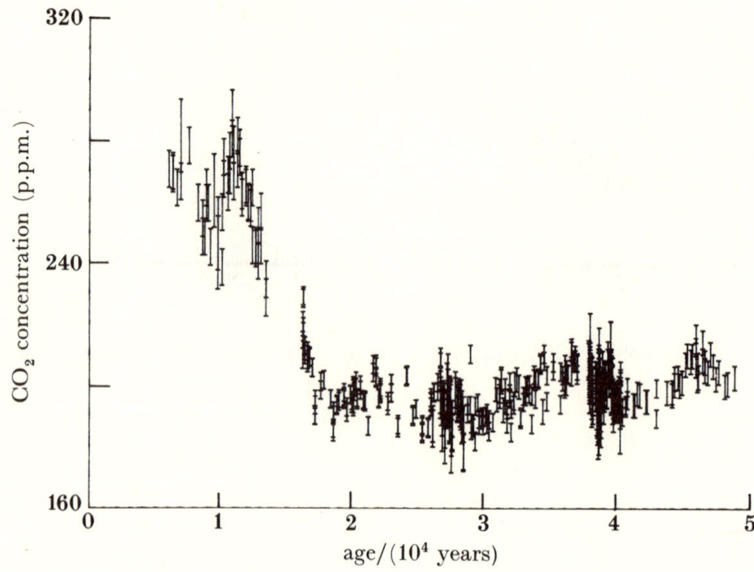

FIGURE 4. CO_2 concentrations of the past 5000–50 000 years, measured on air trapped in an ice core drilled at Byrd Station, Antarctica (B. Stauffer & J. Schwander, personal communication).

It is not impossible that this phenomenon had some impact on the long-term $^{14}C:C$ trend. In fact 7000 years BP the ^{14}C record (figure 1) shows a decline of the $^{14}C:C$ ratio by *ca.* $30\%_0$. For final conclusions the ^{14}C record should be extended further back in time and model estimates should be made.

Figure 4 shows the CO_2 concentrations during the glacial–post-glacial transition 12 000–10 000 years BP measured on the ice core from Byrd Station, Antarctica. Evident are the relatively low CO_2 concentrations during glacial period and the parallel increase with the rising $\delta^{18}O$ values that mark the climatic transition. The data are not dense enough to exclude a CO_2 increase in two steps as suggested from studies of ice cores from Greenland (Dansgaard & Oeschger 1989).

There is little doubt that the CO_2 concentration change from 180 to 280 p.p.m. during this period had an influence on the ratio $^{14}C:C$ in atmospheric CO_2, because on the one side the increase probably was due to CO_2 with a lower $^{14}C:C$ ratio escaping from the ocean. In addition, with the new partitioning of CO_2 also the CO_2 system dynamics changed. Indicators for the $^{14}C:C$ variation during this period have been obtained from plant macrofossils in Swiss lakes (Zbinden *et al.* 1990).

THE ^{10}Be VARIATIONS

Both ^{14}C and ^{10}Be are produced by the interaction of cosmic rays and their secondaries with atomic nuclei of air gases. Though the detailed production mechanisms for the two isotopes are different, it is a valid assumption to consider as a first approximation the relative ^{14}C and ^{10}Be production variations to be equal.

Their geochemical pathways are very different: whereas ^{14}C is oxidized to $^{14}CO_2$, ^{10}Be is attached to aerosol particles. $^{14}CO_2$ mixes rapidly with atmospheric CO_2, but is also in exchange with the carbon in the ocean and in the biomass and soils. ^{10}Be is precipitated after a residence time of 1–2 years and can be found in polar ice and in sediments. Thus short-term production variations in principle are recorded in polar ice and can be studied in annual layers.

In the case of ^{14}C, production fluctuations are considerably damped and delayed. Attenuation factors and delay times are given for different models in figure 5 (Siegenthaler *et al.* 1980). Typical examples of attenuation factors for the periods 11 years, 200 years and 2000 years are *ca.* 100, *ca.* 20 and *ca.* 8, respectively.

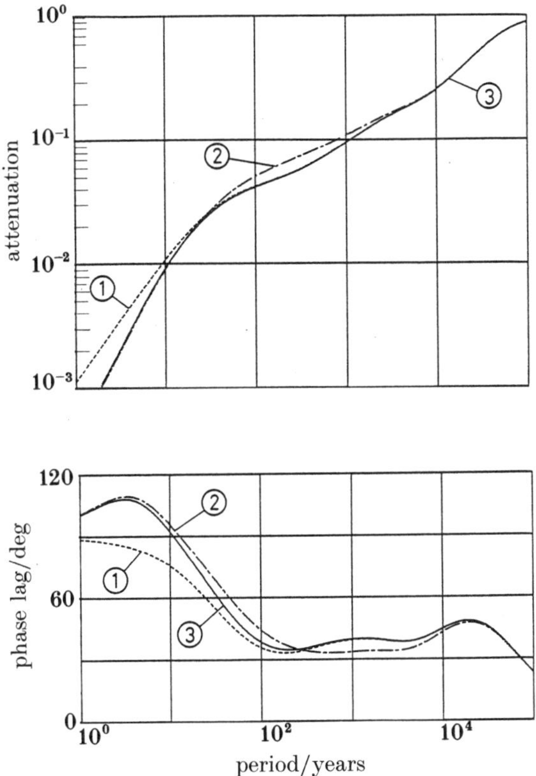

FIGURE 5. Attenuation and phase-shift of tropospheric ^{14}C concentration for a sinusoidal variation of the ^{14}C production rate. (1) Well-mixed atmosphere, model includes biosphere; (2) troposphere separated from stratosphere, no biosphere; (3) troposphere and stratosphere separated, model includes biosphere (Siegenthaler *et al.* 1980).

This large damping of ^{14}C production variations, which renders their detection difficult, is partly compensated by the fact that ^{14}C can be measured relative to a tracer, i.e. ^{14}C:C in CO_2, which is globally well mixed within a few years in the atmosphere. Because of its short atmospheric residence time ^{10}Be is not well mixed in the atmosphere and is therefore influenced by local parameters. Also a valid carrier is missing.

The ^{10}Be concentrations are expressed in atoms per gram of water. Both the atmospheric transport of the aerosols and the dilution with water may vary from year to year.

Because the damping of ^{10}Be-production due to the short residence time is essentially negligible and the production variations are relatively large (20–30 % for the 11-year cycle) the terrestrial noise does not prevent the observation of the so-called 11-year cycle of the ^{10}Be production due to solar modulation of the cosmic-ray flux.

COMPARISONS BETWEEN ^{10}Be AND ^{14}C

The first attempts to investigate if ^{10}Be in ice cores shows variations comparable with those of ^{14}C in tree rings were performed on an ice core from station Milcent in central Greenland (Beer *et al.* 1983; Oeschger *et al.* 1987). The ice core covers the past 800 years. Large variations in the ^{10}Be concentrations have been found. For a comparison with the measured ^{14}C variations, a carbon cycle model was used to calculate the ^{14}C variations assuming that the ^{10}Be record directly reflects the ^{14}C production variations. The agreement is satisfactory. The three periods of high ^{14}C:C (corresponding to the Maunder, Spörer and Wolf solar minima) also appear in the calculated ^{14}C:C record. However, whereas the ^{10}Be based ^{14}C maximum agrees during the Wolf in amplitude with the measured one, during the Spörer Minimum it underestimates and during the Maunder Minimum it overestimates it.

^{10}Be IN THE CAMP CENTURY ICE CORE, NORTH GREENLAND

Thanks to the collaboration with Dr C. C. Langway, SUNY, Buffalo, U.S.A., it was possible to measure the ^{10}Be concentrations in the ice core from Camp Century, North Greenland, which covers the past *ca.* 100000 years. The data on both ^{10}Be and δ^{18}O are given in figure 6.

The most striking feature is that after a flat part with minor fluctuations, both records show a rapid increase at 250 m above bedrock. This height above bedrock corresponds to the transition from Pleistocene to Holocene.

This parallel change in ^{10}Be and δ^{18}O can be explained by the assumption of a roughly constant ^{10}Be flux and a precipitation rate that was smaller during the cold period (δ^{18}O values more negative) than during the warm post-glacial period. During the glacial period, therefore, ^{10}Be was considerably less diluted by water vapour than during the post-glacial one. In fact the variations of the ^{10}Be concentrations have been used to estimate accumulation rate changes during the past 160000 years in the Vostok core (Raisbeck *et al.* 1987).

There are also some interesting differences in the ^{10}Be and δ^{18}O profiles: below 100 m above bedrock ^{10}Be concentrations are low, comparable with the Holocene level, with the exception of one value, whereas the δ^{18}O still indicates glacial conditions. At present we cannot explain this phenomenon.

The small variations in δ^{18}O and ^{10}Be during the Holocene show no correlation, except for the range 1200–1250 m above bedrock, which corresponds to the Maunder Minimum period of sunspots (1645–1715 A.D.) known as a cold period in Europe. Most of the ^{10}Be variations therefore have to be explained by another phenomenon than terrestrial modulation.

For a further analysis the ^{10}Be variations were again converted in apparent ^{14}C variations, based on the assumption that the production variations of ^{10}Be and ^{14}C had been proportional and that the carbon cycle model also holds for the entire post-glacial period. Because both the ^{10}Be and the ^{14}C records are strongly affected by terrestrial phenomena in the glacial and the glacial–post-glacial transition period this comparison was only performed for the period 7000 B.C. till present (figure 7). The choice of this boundary for the assumption of more or less

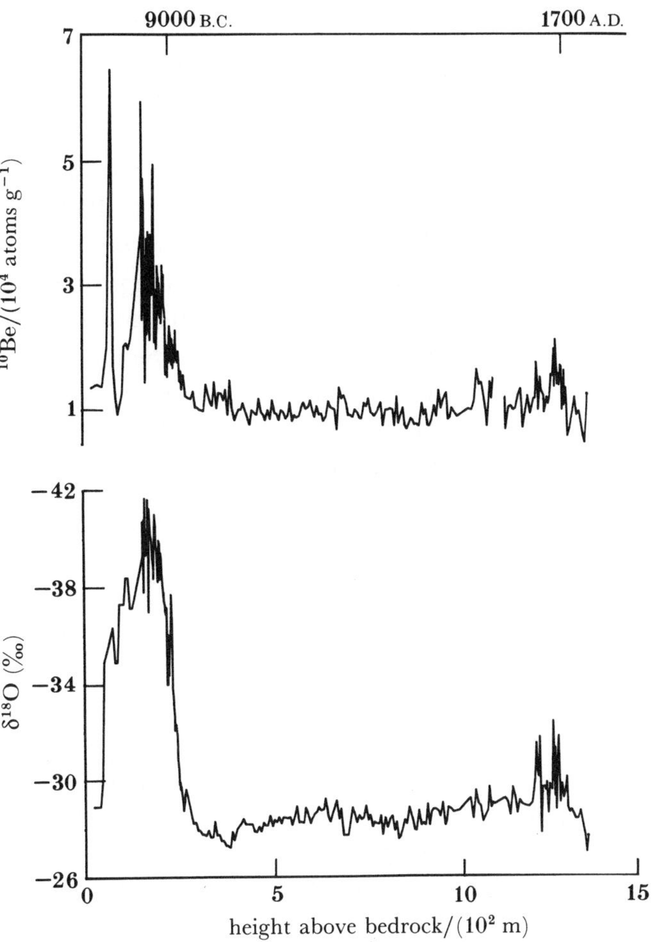

FIGURE 6. ^{10}Be concentrations and δ^{18}O from the Camp Century ice core, Greenland (Beer *et al.* 1988). The strong changes at 250 m correspond to the transition into the last glaciation (*ca.* 9000 B.C.).

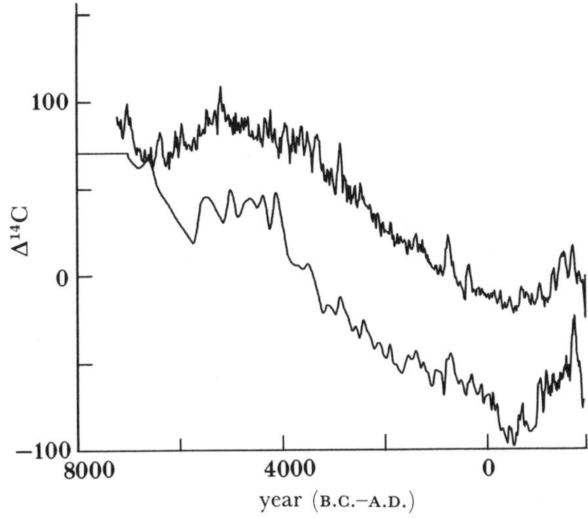

FIGURE 7. Comparison of measured Δ^{14}C (figure 1) with calculated Δ^{14}C based on the ^{10}Be concentrations of figure 6*a*. The calculated curve is shifted by 50 per mille for easier comparison.

stable conditions both for the ^{10}Be and the CO_2 system is somewhat arbitrary and it cannot be excluded that the long-term trends are influenced by after effects following the transition to the Holocene.

Such effects, however, probably would have little influence on the short-term variations that we discuss next. To compare the short-term ^{10}Be and ^{14}C variations from both the model-generated ^{14}C data and the tree ring data of the period 3000 B.C. to 1800 A.D., the long-term trend was removed by applying a binomial high-pass filter. The resulting curves are shown in figure 8 (Beer *et al.* 1988). They are very similar; for example, the prominent maxima of the tree-ring record at *ca.* 2800 B.C., 900 B.C., 300 B.C., 800 A.D., 1100 A.D. and 1700 A.D. (Maunder Minimum) are also found in the ^{10}Be-based model curve. Also the cross-correlation function was calculated. The correlation coefficient has a clear maximum for a phase shift near zero years; its value of 0.58 is highly significant. Although the amplitudes of the ^{10}Be-based ^{14}C variations are slightly higher (4.4% compared with 3.9%) the general agreement is satisfactory.

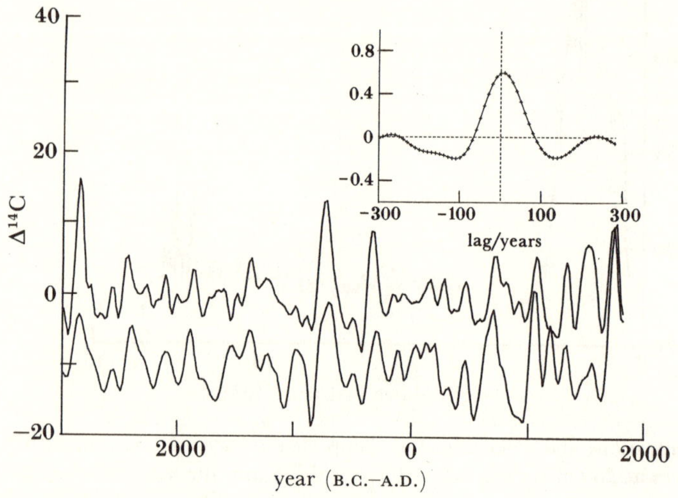

FIGURE 8. (*a*) Expansion of the period 3000 B.C. to A.D. 2000 of figure 7 after removing the long-term trend (Beer *et al.* 1988). The upper curve represents the measured, the lower one the calculated Δ^{14}C values. The lower curve is shifted by 10 per mille. (*b*) Cross-correlation between the two curves as a function of lag time.

From this comparison we conclude with two points.

1. Both the ^{10}Be and the ^{14}C variations have a common cause, i.e. variations in the production rates.

2. The good agreement between the direct ^{14}C- and the ^{10}Be-deduced short-term variations indicates that for the period investigated the carbon cycle model (Siegenthaler & Oeschger 1987) describes very well the response of the carbon system to an atmospheric disturbance with a characteristic time of *ca.* 200 years.

THE LONG-TERM EFFECTS IN ^{10}Be AND ^{14}C

The long-term effects in the cosmogenic isotope records are of lesser relevance to the topic of this Symposium and we summarize the major results and conclusions briefly. Beer *et al.* (1988) studied how far the ^{10}Be long-term record of the past 9000 years can be explained by the hitherto determined long-term variation of the geomagnetic dipole moment. They came to the conclusion that the observed ^{10}Be date and those derived from the geomagnetic modulation are not in good agreement. Possible explanations for the poor agreement might be (1) the fact that a high-latitude record reflects to some degree regional production of ^{10}Be and might not therefore be ideal for the study of the geomagnetic effects that are relatively small in the polar atmosphere; (2) the reconstruction of the geomagnetic dipole moment from palaeomagnetic intensity data is hampered by serious uncertainties.

When comparing the measured ^{14}C record with the ^{10}Be-derived record one can draw the following conclusions regarding the long-term behaviour (figure 7).

The ^{14}C decrease from 5000 B.C. to 500 A.D. is not reflected in the ^{10}Be concentration record. It can, however, be simulated by assuming that the production rate of cosmogenic radionuclides during the last *ca.* 10000 years of the Pleistocene was higher by 20 % than during the Holocene leading to higher ^{14}C levels in all the carbon reservoirs. The decrease during the Holocene can be understood as an exponential adjustment to the new steady state corresponding to the lower production rate. Such an adjustment would also occur for a constant ^{14}C production if after the last glaciation the CO_2 system had not been in equilibrium regarding ^{14}C production and decay. Astonishingly high are the ^{10}Be concentrations during the past 1000 years, partly as a result of the three periods of a quiet Sun, partly as a result of a decrease of the geomagnetic field intensity. The increase of the ^{14}C : C ratio during this period is also reflected in the direct ^{14}C measurements.

To obtain a perfect picture of the ^{10}Be production one should in principal measure its precipitation history over the entire globe. Samples suited for this purpose are hard to get from mid-latitude glaciers, but correspondingly high resolved information from Antarctica is available (G. M. Raisbeck, this Symposium). The Wolf, Spörer and Maunder Minima are also visible in these records. The Maunder Minimum is less pronounced than in the Greenland cores.

This raises the question whether transport and deposition, or polarity effects caused the different distribution of cosmic-ray produced ^{10}Be in the two polar area during the periods of the quiet Sun.

CONCLUSIONS

The ^{10}Be record in ice cores and the ^{14}C record in tree rings give insights into Earth and planetary system processes that hitherto seemed beyond the potential of the natural sciences. In the following some of the major results are summarized.

1. During quiet periods of the Earth system the records essentially give information on the production rate of ^{10}Be and ^{14}C, whereas in periods of major climatic change, like the glacial–post-glacial transition, *ca.* 15000–10000 years BP, the fluctuations in the records are dominated by Earth system changes.

2. A major result is the widely consistent information on the periods of the quiet Sun with high cosmogenic isotope production; during the past 5000 years seven periods of high ^{10}Be and ^{14}C concentrations have been identified. They provide a basis to study the possible effects of variations of solar properties on climate.

3. The good agreement between the short-term ^{14}C variations measured on tree rings and those models calculated from the short-term ^{10}Be variations supports the validity of the carbon cycle model used, at least regarding variations with characteristic times of *ca.* 200 years.

However, with potential effects of global change this model may lose part of its predictability.

4. More difficult is the explanation of the long-term ^{14}C change. There are two possible reasons.

(i) The decrease of ^{14}C after last glaciation until 500 A.D. may reflect a carbon system not in equilibrium regarding production and decay after the drastic effects during the glacial–post-glacial transition. At maximum such a disequilibrium was of the order of Δ^{14}C $= 100\permil$, which seems at the low side of expectations.

(ii) There is geomagnetic modulation of production as partly suggested from the ^{10}Be record.

5. Astonishing is the increase in ^{14}C and ^{10}Be after 500 A.D. Possible reasons are a decrease in the Earth's magnetic field; an unusual summation of periods of a quiet Sun; meteorological effects in the ^{10}Be deposition process; some unknown carbon cycle effects regarding the ^{14}C increase.

To answer some of the questions raised here, there is an urgent need for additional ^{10}Be measurements on ice cores in both hemispheres, but also attempts to resolve the 11-year cycle in the ^{14}C record, especially during, for example, the solar Maunder Minimum.

References

Beer, J., Siegenthaler, U., Bonani, G., Finkel, R. C., Oeschger, H., Suter, M. & Wölfli, W. 1988 *Nature, Lond.* **331**, 675–679.

Beer, J., Siegenthaler, U., Oeschger, H., Andrée, M., Bonani, G., Hofmann, H. J., Nessi, M., Suter, M., Wölfli, W., Finkel, R. C. & Langway, C. C. 1983 In *Proc. 18th Int. Cosmic Ray Conf., Bangalore* (ed. N. Durgaprasad, S. Ramadurai, P. V. Ramana Murthy, M. V. S. Rao & K. Sivaprasad), vol. 9, pp. 317–320.

Daansgard, W. & Oeschger, H. 1989 In *The environmental record in glaciers and ice sheets* (ed. H. Oeschger & C. C. Langway Jr), pp. 285–317. New York: Wiley.

Neftel, A., Moor, E., Oeschger, H. & Stauffer, B. 1985 *Nature, Lond.* **315**, 45–47.

Oeschger, H., Welten, M., Eicher, U., Moell, M., Riesen, T., Siegenthaler, U. & Wegmueller, S. 1980 *Radiocarbon* **22**, 299–310.

Oeschger, H., Beer, J. & Andrée, M. 1987 *Phil. Trans. R. Soc. Lond.* A **323**, 45–56.

Raisbeck, G. M., Yiou, F., Bourles, D., Lorius, C., Jouzel, J. & Barkov, N. I. 1987 *Nature, Lond.* **326**, 273–277.

Siegenthaler, U., Heimann, M. & Oeschger, H. 1980 *Radiocarbon* **22**, 177–191.

Siegenthaler, U. & Oeschger, H. 1987 *Tellus* B **39**, 140–154.

Stuiver, M. & Kra, R. S. 1986 *Radiocarbon* **28**, 969–979.

Suess, H. E. 1971 In *Radiocarbon variations and absolute chronology, Proc. 12th Nobel Symp.* (ed. I. U. Olsson), pp. 303–313. New York: Wiley.

Zbinden, H., Andrée, M., Oeschger, H., Ammann, B., Lotter, A., Bonani, G. & Wölfli, W. 1990 *Radiocarbon* (In the press.)

Phil. Trans. R. Soc. Lond. A **330**, 481–486 (1990)

Printed in Great Britain

481

On the presence of regular periodicities in the thermoluminescence profile of a recent sea sediment core

By Giuliana Cini Castagnoli, G. Bonino, A. Provenzale and M. Serio

Istituto di Cosmogeofisica del C.N.R., Corso Fiume 4, 10133 *Torino, Italy*
Istituto di Fisica Generale dell'Universita', Torino, Italy

We briefly discuss how the thermoluminescence (TL) profile of a young marine sediment provides phenomenological information on the changes in the environmental conditions in the past 18 centuries. The main periodicities present in the TL profile are studied and the similarities between the TL variations and the fluctuations in the contemporary tree-ring Δ^{14}C signal are considered. An interesting result is the presence, in the TL data, of a well-defined 11-year cycle which is stable and 'in phase' for the entire period analysed. We also discuss how four dominant periodicities present in the TL data may be rewritten as the sum of an 11.4-year and of an 82.6-year cycle (reminiscent respectively of the Schwabe and of the Gleissberg cycles of solar activity), which are both amplitude modulated by a 206-year wave. This last periodicity has already been shown to play a dominant role in the Δ^{14}C record. These results suggest that the TL profiles of recent marine sediments may be successfully used as a new line of evidence for solar variability in the past centuries.

1. Introduction

In previous studies (Cini Castagnoli *et al.* 1984*a, b*, 1988*a, b*; Cini Castagnoli & Bonino 1985, 1988) we analysed the variations of the thermoluminescence (TL) level accumulated in the fine polymineral material (grains less than 44 µm in size) of young marine sediments. In recent cores (about 2000 years before present (BP)), the predepositional exposure of the crystals is in general not completely altered by the irradiation produced *in situ* by the local radioactivity. Thus the analysis of the TL fluctuations in these cores indicate, it is hoped, how the environmental conditions, which determine the TL level accumulated by the crystalline minerals before deposition, change with time. In the above papers the principal periodicities present in the TL profiles have been determined and the affinity of the TL signal to the solar variability has been explored. In this paper we further discuss the TL periodicities and consider the possible correlations between the TL signal and tree-ring radiocarbon data. These results may thus provide new phenomenological information on the characteristic temporal variations of the solar–terrestrial system and on the mechanisms that drive the terrestrial rhythms.

Let us now briefly consider how TL time series are measured. When the material of a core is heated some of the charges that are trapped in the polymineral crystals of the sediment are released, generating a TL signal. If the sedimentation rate is measured by independent methods (such as measurement of the ^{210}Pb concentration) then the TL depth profile (i.e. the succession of the TL relative intensities measured in contiguous layers of the sediment) may be transformed in a TL time series. In past years we analysed the TL profiles of several cores from different coastal sites, each core having a different sedimentation rate and being sampled with a different

[83]

sampling interval. Common features were detected in the TL profiles of the different cores, and consistent periodicities on both decennial and secular timescales have been determined.

In the case of the GT14 core, drilled in the Ionian Sea in 1979, it was possible to conduct a detailed analysis on a TL profile that spans approximately 18 centuries (Cini Castagnoli *et al.* 1988*a*, *b*). The GT14 core is 1.17 m long and has been drilled in the Gulf of Taranto near Gallipoli (Italy), at 39°45′55″ N, 17°53′30″ E. The drilling site is located on the continental shelf at 18 km from the coast, in a water depth of 166 m. The core has been sampled at equal intervals $\Delta d = 2.5$ mm and well-defined, regular oscillations with periods of 89 mm, 7.8 mm and 6.97 mm have been detected. The use of the ^{210}Pb method and a careful tephroanalysis of known volcanic events, together with the information deduced from the presence of a ^{137}Cs peak at the top of the sediment, allowed for the precise determination of the sedimentation rate. The sampling interval Δd has been found to correspond approximately to a time interval $\Delta t = 3.87$ years and the TL depth profile was consequently transformed in a TL time series. The temporal periods of the main regular oscillations detected in the TL profile have correspondingly been determined. The 11-year solar cycle in particular was detected with a high confidence level in the resulting TL signal (Cini Castagnoli *et al.* 1988*b*). Here we (1) discuss the principal features present in the TL profile of the GT14 core and (2) we analyse, using the method of cyclograms (Galli 1988), the stability of the phase of the 11-year cycle during the time interval spanned by the core.

2. TL FEATURES DURING THE SOLAR EVENTS OF MAUNDER, SPÖRER AND WOLF

To compare the TL signal with other proxy data for solar activity in figure 1 we show the radiocarbon tree-ring time series (*a*) and the TL time series from the GT14 core (*b*) for the period 1100–1900 A.D. The Δ^{14}C data are in the decadal form given by Stuiver & Braziunas (1988) and are plotted on an inverted scale. We can see that the TL data show the expected Maunder Minimum during the period 1645–1715 A.D. In this period the Δ^{14}C signal has a

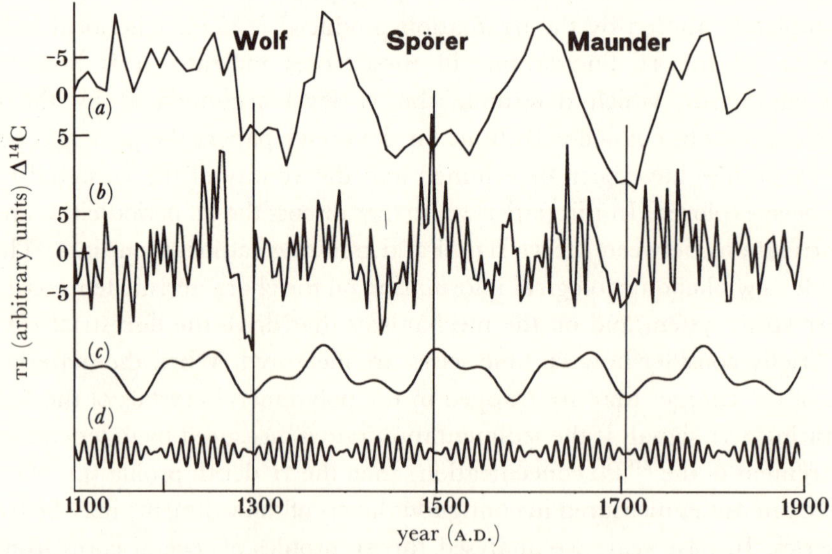

FIGURE 1. (*a*) The decadal Δ^{14}C record as given by Stuiver & Braziunas (1988). (*b*) The detrended TL signal from the GT14 core. (*c*) The sum of the two TL waves with periods of 137.7 and 59 years. (*d*) The sum of the two TL waves with periods of 12.06 and 10.8 years.

well-known maximum. From the TL signal one also sees that the Spörer Minimum starts at the end of the fourteenth century, in agreement with the Δ^{14}C data. In the TL data, however, this event seems to have recovered around 1500 A.D., showing a double feature not present in Δ^{14}C signal. The end of the Spörer event is vice versa at about the same time for both series (around 1600 A.D.). The Wolf Minimum is finally observed in the TL series at the expected time (around 1300 A.D.), in agreement with the Δ^{14}C data.

3. THE TL PERIODICITIES

The power spectral density of the TL data (Cini Castagloni *et al.* 1988*b*) was computed by standard Fourier techniques from the detrended 427 points time series and is shown in figure 2. The spectrum shows sharp significant peaks among which the well-known 11-year cycle appears. The 95 % significativity implies a significant interval $\frac{1}{2}P(f) \leqslant \pi(f) \leqslant 4\,P(f)$ where $\pi(f)$ is the 'true' spectrum and $P(f)$ is our spectral estimate. This because we have four independent glow curves for each layer and therefore four replicas of the 427 points time series. In figure 2 a second, low-level dashed curve is also reported. This curve is the spectrum computed for a fifth series, obtained from 'bleached' samples. One sample for each layer was in fact exposed to the action of a standard sunlamp for 420 min before measuring the glow curves. The bleached spectrum does not show any evident periodicity and provides a background noise level for the original TL spectra. In this respect we stress the fact that the secular and the decennial peaks of 137.7, 59, 12.06 and 10.8 years are all significant at the 95 % significance level.

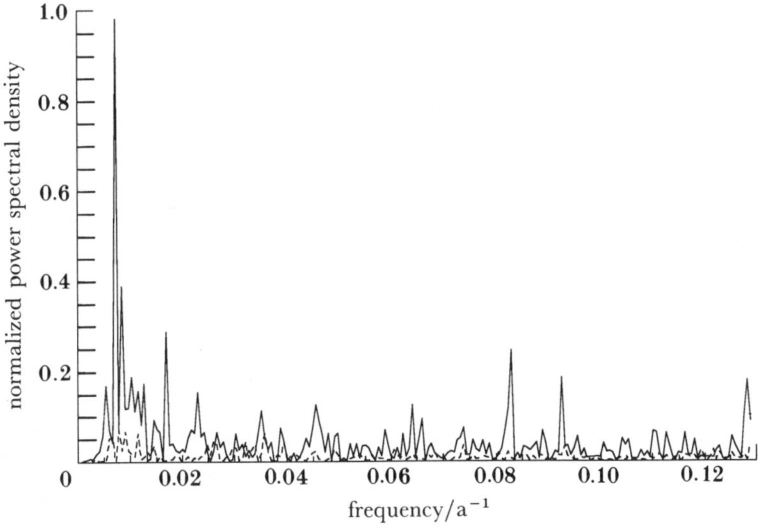

FIGURE 2. Power spectrum of the TL signal.

To further explore whether the 11-year cycle is present over the whole time interval spanned by the TL data we consider the cyclogram analysis for the 10.8-year and 12.06-year waves. The cyclogram method allows a time series to be analysed interval-wise by plotting running averages of the complex amplitudes of the Fourier transform for a given periodicity (Galli 1988). A window is moved along the time series, continuously testing for the chosen periodicity. When the chosen periodicity is present and correctly determined, the cyclogram tends to a straight line. For test periods shorter than a 'true' period present in the data the cyclogram tends to turn anticlockwise, while for test periods larger than the true period the cyclogram turns clockwise. If no regular oscillation with periodicity around the chosen test period is present in the data, then the cyclogram has the appearance of a random walk on the plane of the complex Fourier amplitudes. Thus computing several cyclograms for different test periods around a periodicity of interest allows (1) for its careful determination and (2) for testing its stability along the time series.

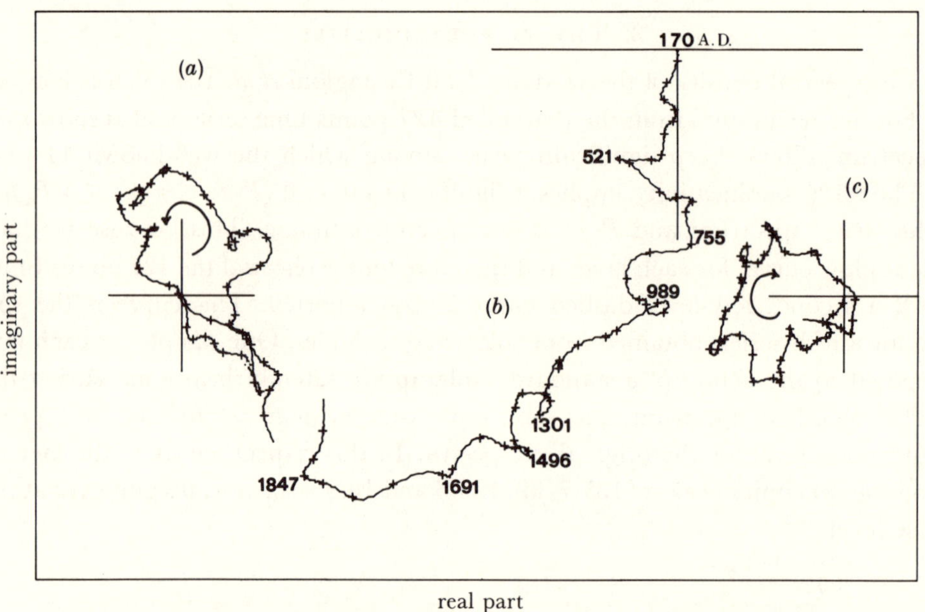

FIGURE 3. Cyclograms obtained from the TL signal using a window of period $T = 39$ years and test periods (a) $\tau = 10.7$ years, (b) $\tau = 10.8$ years, and (c) $\tau = 10.85$ years.

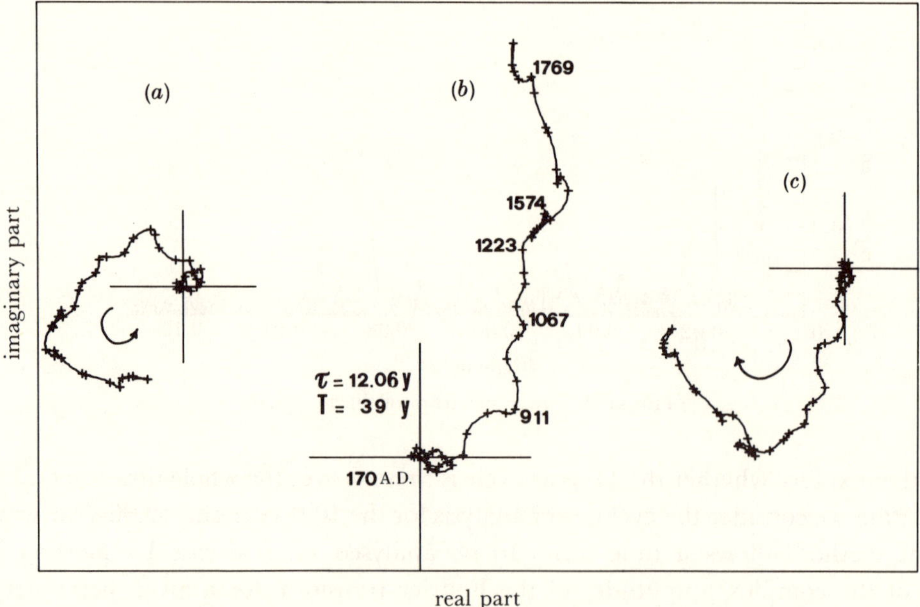

FIGURE 4. Cyclograms obtained from the TL signal using a window of period $T = 39$ years and test periods (a) $\tau = 12.0$ years, (b) $\tau = 12.06$ years, and (c) $\tau = 12.15$ years.

In figure $3b$ we show the cyclogram for the test period $\tau = 10.8$ years with a window of $T = 39$ years (equivalent to 10 data points). The cyclogram is clearly straight and shows that 45 independent vectors (whose extrema are indicated by the crosses) are aligned with the same phase from 170 to 1970 A.D. The cyclograms are anticlockwise when performed on a test period $\tau = 10.7$ years (figure $3a$) and clockwise for $\tau = 10.85$ years (figure $3c$), indicating a very regular, non-random character of the 10.8-year cyclicity. In figure $4b$ we show the analogous cyclogram for $\tau = 12.06$ years. Again the cyclogram is straight. The cyclogram is on the other hand anticlockwise for $\tau = 12.0$ years (figure $4a$) and clockwise for $\tau = 12.15$ years (figure $4c$). The quantitative significance of the observed periodicities is difficult to assess, however the distance between the first and the last points of the cyclograms of figures $3b$ and $4b$ significantly exceeds the distance that would be attained by a random walk with 45 independent vectors (steps) with non-correlated increments.

4. RELATIONS WITH THE SOLAR CYCLES

In a previous paper (Cini Castagnoli *et al.* 1988b) we discussed how the two high-frequency components of 12.06-year and 10.8-year beat producing a wave train whose carrier has a period of 11.4 years. This wave train is amplitude modulated with a period of approximately 206 years, a periodicity already detected in the $\Delta^{14}C$ data (Beer *et al.* 1988; Cini Castagnoli *et al.* 1989). In Cini Castagnoli *et al.*'s paper the phases and amplitudes of the waves were determined by the method of the superposition of epochs. It is interesting to note that the nodes of the modulated wave train coincide with the well-known secular minima in the yearly sunspot number recorded in 1810 A.D. and in 1913 A.D. The modulated wave train is shown in figure $1d$ for the period 1100–1900 A.D. The beats coincide with the centres of the Maunder and Spörer events, as given by the $\Delta^{14}C$ data. The same approximately happens for the Wolf event. This may well indicate that the presence of small sunspot cycles every 103 years may be a regular feature of the solar output.

The oscillation obtained as the sum of the two long waves of 137.7 and 59 years is shown in figure $1c$. We notice that the centres of the Maunder and of the Wolf events lie in minima of this oscillation, while the centre of the Spörer event lies in a maximum. We recall that also the sum of the two low-frequency TL waves of 137.7 and 59 years may be rewritten as the product of two component waves of 82.6 and 206 years, plus a residues of 137.7 years (Cini Castagnoli *et al.* 1988b).

As a conclusion we thus stress that (1) a few well-defined dominant periodicities are evident in the TL signal; (2) among these two decennial oscillations are present, which have been shown to be stable over the entire time interval spanned by the core; and (3) four of the most energetic significant TL peaks may be rewritten as the sum of two oscillations of 11.4 and 82.6 years, which are amplitude modulated with the same period of 206 years. The two 'carrier' oscillations have periods that are similar to those generally attributed to the Sun as the Schwabe and the Gleissberg cycles, while the amplitude modulation has the same period of a very energetic wave present in the $\Delta^{14}C$ signal (Beer *et al.* 1988; Cini Castagnoli *et al.* 1989). These results thus suggest that the TL profiles of recent sea sediment cores may be successfully utilized as a new line of evidence for the solar variability and encourage to further pursue this type of analysis.

We thank Professor C. Castagnoli and Professor L. Bergamasco of the University of Torino (Italy) and Professor S. K. Runcorn, F.R.S., of the University of Newcastle upon Tyne (U.K.) for useful suggestions and valuable encouragement. The technical assistance of Mr P. Cerale and Mr A. Romero is gratefully acknowledged.

References

Beer, J., Siegenthaler, U., Bonani, G., Finkel, R. C., Oeschger, H., Suter, M. & Wolfli, W. 1988 Information on past solar activity and geomagnetism from ^{10}Be in the Camp Century ice core. *Nature, Lond.* **331**, 675.

Cini Castagnoli, G. & Bonino, G. 1985 Comparison of TL profiles in recent sea cores. *Nucl. Tracks* **10**, 759.

Cini Castagnoli, G. & Bonino, G. 1988 Solar imprint in sea sediments: the thermoluminescence profile as a new proxy record. In *Secular solar and geomagnetic variations in the last 10000 years* (ed. F. R. Stephenson & A. W. Wolfendale), p. 341. Amsterdam: Kluwer.

Cini Castagnoli, G., Bonino, G. & Provenzale, A. 1988a On the thermoluminescence profile of an Ionian Sea sediment: evidence of 137, 118, 12.1 and 10.8 y cycles in the last two millennia. *Nuovo Cim.* **C11**, 1.

Cini Castagnoli, G., Bonino, G. & Provenzale, A. 1988b The thermoluminescence profile of a recent sea sediment core and the solar variability. *Sol. Phys.* **117**, 187.

Cini Castagnoli, G., Bonino, G. & Provenzale, A. 1989. The 206 yr cycle in tree-ring radiocarbon data and in the thermoluminescence profile of a recent sea sediment. *J. geophys. Res.* **94**, 11971.

Cini Castagnoli, G., Bonino, G., Attolini, M. R. & Galli, M. 1989a The 11 yr cycle in the thermoluminescence profile of sea sediments. *Nuovo Cim.* **C7**, 69.

Cini Castagnoli, G., Bonino, G., Attolini, M. R., Galli, M. & Beer, J. 1984b Solar cycles in the last centuries in ^{10}Be and δ^{18}O in polar ice and in the thermoluminescence signal of a sea sediment. *Nuovo Cim.* **C7**, 235.

Galli, M. 1988 Time series analysis with power spectrum and cyclograms. In *Solar–terrestrial relationships and the Earth environment in the last millennia* (ed. G. C. Castagnoli), p. 246. Amsterdam: North Holland.

Stuiver, M. & Braziunas, T. F. 1988 The solar component of the atmospheric Δ^{14}C record. In *Secular solar and geomagnetic variations in the last 10000 years* (ed. F. R. Stephenson & A. W. Wolfendale), p. 245. Amsterdam: Kluwer.

Phil. Trans. R. Soc. Lond. A **330**, 487–497 (1990)

Printed in Great Britain

Astronomical determinations of the solar variability

By Elizabeth Ribes

Observatoire de Paris, Section d'Astrophysique, 5 Place Jules Janssen,
92195 Meudon Principal Cedex, Paris, France

Historical evidence of solar variability is based upon sunspot occurrence and rotation rates, and on observations of the Sun's radius. I present here the methods of observation used to this end, and assess their accuracy and limitations. I then discuss and interpret the periodicities I have found. The main conclusions are that, during the past three centuries solar variability has appeared in the form of recurrent expulsions of magnetic activity at the Sun's surface, accompanied by a modulation of its rotation and periodic changes in its radius. Periodicities of from 150 days up to as much as 80 years, including the well-known 11-year cycle, are present in the historical observations and contemporary observations. No exact theoretical model of the change in solar radiation output under the influence of periodic magnetic fields has yet been developed; but the presence of periodicities common to the Sun's activity and to the Earth's environment suggests a causal relation between solar variability and the meteorology of the Earth.

1. Introduction

Systematic observation of the Sun have been going on ever since Galileo first used the telescope for astronomy in 1611. Although two centuries went by before Schwabe, in 1837, formulated the concept of a regular 11-year cycle, observers had already noticed the variability of the Sun's activity (Picard 1671) as well as of its diameter (Lalande 1771) long before. The question of whether the observed variability was real or was only the result of systematic errors has raised passionate debates over all this time (Lalande 1771; Eddy & Boornazian 1979; Parkinson *et al.* 1980; Gilliland 1981), and remains controversial to our day. The grounds for uncertainty stem from the fact that comparing observations made with different techniques used by different observers is subject to criticism, except when observations overlap and can be calibrated (Ribes *et al.* 1987). The problem is even more difficult when dealing with historical observations obtained with optics that are no longer available. Assuming, however, that the periodicities are real, the question that comes immediately to mind is whether the sunspot cycle is a periodic oscillation of the entire Sun or rather the product of a turbulent dynamo in the convective zone alone. In the former case magnetic fields of alternate polarities are released from the solar interior by some unspecified processes and rise to the surface through a turbulent convective zone (Dicke 1978; Bracewell 1988). The phase of the cycle would somehow be locked, in spite of quite irregular cycles due to the turbulent convective zone. In the latter case, the dynamo is located within the convective zone, and is due to the interaction between the rotational and convective motions. Such a turbulent dynamo could not maintain its phase over a period of centuries.

The choice between these two alternatives is important because the timescales of the processes involved and the consequent effects on the solar envelope and on the Earth's environment are quite different. Very long records of solar activity (sunspots) and of its effect on the solar envelope (diameter, luminosity), are needed to answer this very fundamental question (D. O. Gough, this Symposium). As reviews are available on the subject of the astronomical means of detecting solar variability (Wittman & Débarbat 1990), I discuss them only briefly here, and emphasize instead the results obtained from homogeneous time series, so that the periodicities have some significance. Special attention is devoted to seventeenth-century observations of reduced magnetic activity (Maunder 1894) coincident with a severe cold on Earth, 'the Little Ice Age' (Le Roy Ladurie 1967). A connection between solar variability and the Earth's meteorology, if proved, would be of invaluable importance.

2. Historical methods for detecting solar variability

2.1 Sunspot observations

(i) Sunspot occurrence

Naked-eye observations of sunspots have been reported as far back as 2000 years ago. Because there is to be a paper on this subject (F. R. Stephenson, this Symposium), I concentrate here on telescopic observations. In the early seventeenth century, sunspots were observed on images projected by an objective lens of 2 cm aperture (an aperture ratio of 22). Larger, though strongly chromatic, lenses were available at the time; however, modest as they were, these primitive lenses enabled observers to detect most of the sunspots we moderns would have found with our powerful instruments. Only pores, about 10% of the total sunspot number, were out of reach. Furthermore, the number of observations made each year by our Paris counterparts during the second half of the seventeenth century was substantial (from 150 to 230 days) and, knowing that an average sunspot lasts almost as long as one 27-day rotation of the Sun, we may rest assured that our skilful observers did not miss many sunspots, and that their count provides a reliable index of the activity of the entire Sun. Their data are shown in figure 1.

The relative sunspot number is defined as $R = K(10g+s)$, where g is the number of sunspot groups present on the solar disk and s is the total number of sunspots for a given day. The parameter K is a scaling factor depending upon the observer, and is used to convert the scale to the original one established by Wolf (1856). The monthly sunspot counts are then smoothed by a six-month running mean to appear in figure 1. The activity index represented by the smoothed relative number is biased, as individual sunspots and pores are given greater comparative weight than clusters of spots. The physical reason of this is to emphasize the occurrence of magnetic fields rather than their size.

It is clear from figure 1 that the sunspot distribution is cyclic in time, although the periodicity is far from constant. Also, the peak of the sunspot number and the duration of the 11-year cycle both vary considerably from one cycle to the next, and the sunspot maxima are clearly modulated, with a periodicity of roughly 80 years. Finally, there is a period of 50 years or so when sunspot activity was practically non-existant, as the Astronomer Royal R. Maunder pointed out in 1894. A close inspection of the historical records of the seventeenth-century observations (Ribes *et al.* 1987) confirmed the existence of the magnetic anomaly of the Sun reported by the Paris observers (Picard 1671, among others). The 11-year signal as well as long-term modulation of the solar activity are clearly present in auroras (Gleissberg 1966; Schröder 1988) and other proxy records, showing that the cyclic character of solar activity is a fundamental feature of the Sun.

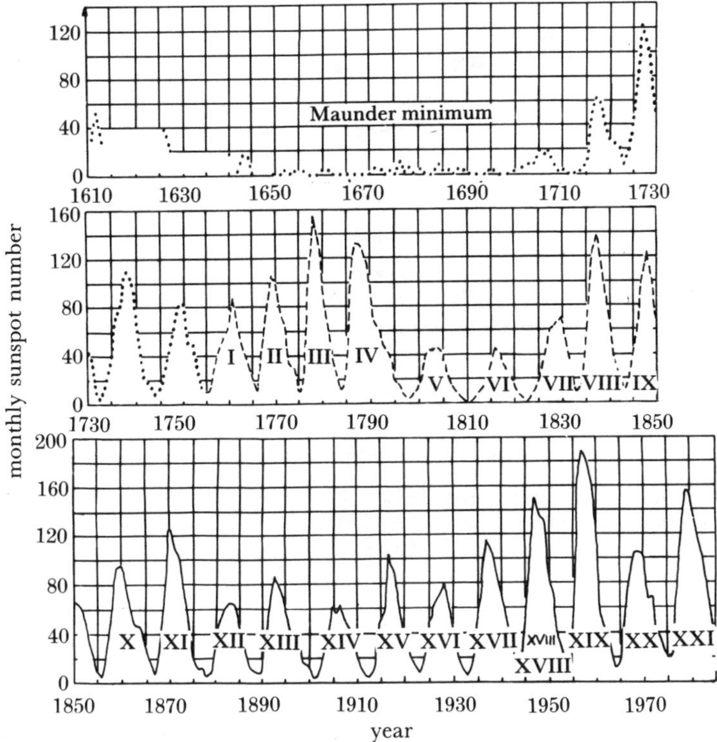

FIGURE 1. Monthly sunspot number against time, as determined by Wolf (1856) and pursued by the National Oceanic and Atmospheric Administration Space environment Services Center.

(ii) *Latitudinal distribution of sunspots*

Sunspots appear at preferential latitudes during the 11-year solar cycle. Classically, sunspot zones appear at the mid-latitudes (about $\pm 45°$) at the beginning of the cycle and this point of origin descends closer to the equator as the cycle goes on, following a 'butterfly wing pattern'. The amplitude of the sunspot maximum and the duration of the cycle are both related to the latitude where the activity starts: for low-activity cycles, for example, the activity starts closer to the equator. This was particularly clear during the Maunder Minimum, throughout which the meagre sunspot activity was concentrated at low latitudes (below 20°).

(iii) *Rotation of sunspots*

Sunspots have been used in measuring the rotation of the Sun. Apart from the well-known fact that the Sun's surface rotates differentially, with the equator rotating faster than the mid-latitudes, the rate of rotation varies from one cycle to the next and fluctuates through a given solar cycle. Again, I should stress that this phenomenon was particularly enhanced at the time of the Maunder Minimum (figure 2), when the whole surface seemed to rotate more slowly than now (Ribes *et al.* 1987). Early in the present century too, the rotation rate decreased by a few per cent (Balthasar *et al.* 1986), and this decrease corresponds to two cycles of low amplitude (cycles 12 and 13, see figure 1). The modulation of the rotation of the solar envelope results from recurrent emergences of magnetic activity.

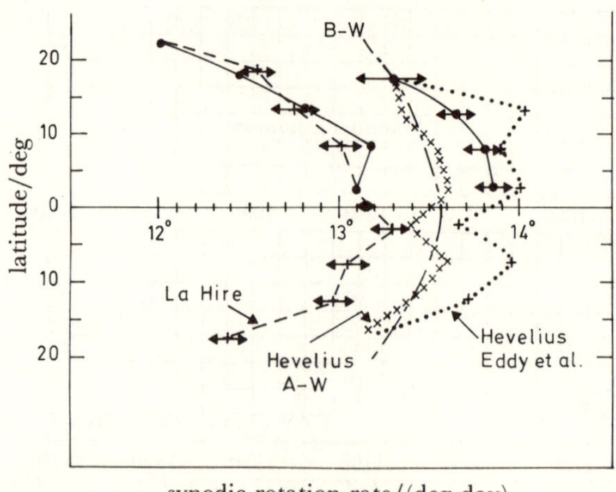

FIGURE 2. Rotation of sunspots against latitude from 1672 to 1719. The dashed line represents the modern rotation of sunspots (Balthasar & Wohl 1980). I have also plotted some controversial results of Hevelius drawings (1642–44) published by Eddy *et al.* (1976) and Abarbanell & Wohl (1981).

2.2. *Solar diameter measurements*

(i) *Filar micrometer method*

This method consists of measuring the linear breadth of the solar image as formed in the focal plane of a lens. The micrometer was invented by Gascoygne in 1640 and was considerably improved by Auzout (1729). It has given accurate readings of the Sun's edges. There are two problems connected with micrometer measurements: the first is determining the focal plane accurately, and the second is the calibration of the micrometer screw. Auzout claimed that an accuracy of ± 1 arcsec could be achieved by Picard when he measured the diameter of the Sun or any planet. There are two indications that such an accuracy was in fact achieved with Auzout's micrometer. First, Picard was able to detect a reduction of 0.5 arcsec in the equatorial diameter of Jupiter, when the planet was in quadrature with the Sun (Ribes *et al.* 1988*a*), the true effect being 0.4 arcsec. Secondly, the statistical error found for the 60 diameter observations performed by Picard each year does not exceed ± 1.5 arcsec (Ribes *et al.* 1987), despite the fact that day-observations may not be as accurate as those made at night since solar heating induces local atmospheric turbulence. However, if the amplitude of this statistical error contains a solar signal, then, Auzout's claim would be confirmed.

(ii) *Meridian timings*

The Sun's diameter can be estimated from records of the time it takes its image to travel through the meridian plane of a refractor (or reflector). In the seventeenth century, the pendulum clock was already very accurate (to within one or two seconds per day). The main limitation was the 'ear and eye' method of quantifying the timing. As the pendulum beat was one second, the uncertainty in each radius determination was about ± 3.5 arcsec. One immediately sees that a great many meridian timings must be made in a year to reduce the statistical noise just a little. A proper analysis of meridian timings should take into account a large number of effects carefully listed by Wittman & Débarbat (1990). For most meridian timings, the determination of the solar limb was, and often still is, visual; so one of the most serious errors affecting the measurements was the observer bias. According to Parkinson *et al.* (1980), the horizontal diameter can vary by as much as 2.3 arcsec among observers. For this reason, detecting periodicities on the basis of observations made by different people is questionable, unless the observations overlap significantly to permit scaling.

Another very serious error is the image blurring due to the Earth's atmosphere. This affects all Earth-based observations. The definition of the solar limb has been improved by photoelectric methods, which are, indeed more objective than the visual method. None the less, the time variation of the solar limb has not yet been properly modelled, so one should keep in mind that any periodicity that is found may be contaminated systematically or randomly by the observation method.

(iii) *Astrolabe*

This method consists of timing the edges of the solar limb, when the trajectory of the Sun crosses the almucantar defined by the instrument (Laclare *et al.* 1983). The Danjon-type astrolabe offers several advantages over meridian timings. The reference system (the mercury surface plane defining the horizon, and the time-stable prism angle defining the zenith distance) is not subject to the distortions that affect all transit methods. Nor is the method sensitive to errors in the refraction modelling, except if the reflection properties change during a given timing. On the other hand, the method is sensitive to the personal estimate of the coincidence of the two images, which explains why the diameter may vary from one observer to another. Also, in the same way as for the other methods (filar micrometer methods of meridian timings), there is the blurring due to atmospheric turbulence. In our day, an accuracy of ± 0.15 arcsec is possible on each radius determination (Laclare 1983).

(iv) *Solar eclipse timings*

Timing the duration of a solar eclipse is one very good way of measuring the solar diameter, as it avoids all of the usual atmospheric problems. To succeed in using it, however, we must know the surface topology of the moon around its entire limb; and this is not available, at least not with the accuracy needed. This is why timings made near border of totality are preferred: the contacts there correspond to the polar region of the Moon where the topology is better known. The accuracy of the solar radius obtained this way in the past could be of the order of ± 0.2 arcsec (Dunham *et al.* 1980). But unfortunately, solar eclipses are neither regular nor frequent and cannot provide any reliable short-term periodicities.

(v) *Timings of planet transits across the solar disk*

Somewhat like solar eclipse timings, the timing of the contacts of a planet with the solar limb can provide an estimate of the length of the cord joining the contact points, and hence of the solar diameter. However, contact determinations suffer from various difficulties and cannot be used to estimate of the solar diameter to within better than ± 0.5 arcsec (Parkinson *et al.* 1980; Shapiro 1980).

3. PERIODICITIES PRESENT IN THE DIAMETER SERIES

As shown in the recent summary of historical observations published by Wittman & Débarbat (1990), the Sun's diameter exhibits a very large dispersion because all of the different techniques used. The fluctuations are much larger than those expected from real solar variability. As systematic errors are difficult to extract, the problem of absolute calibration is a very delicate one. So I consider only long series of observations made by a single observer using a given technique. This way, we can be sure that any periodicities we find from Fourier analysis will not be affected by personal bias or any systematic error of calibration.

Such long time series are available. For the seventeenth century, timings by P. La Hire (1718) using a $9\frac{1}{2}$ ft (*ca.* 3 m) quadrant, cover a large period during the Maunder Minimum and extend a little beyond it, when the solar activity has resumed (from 1705 onwards). Two independent time series (astrolabe observations at the Cerga observatory and meridian timings at the Belgrade observatory) covering the last solar cycle (1978–87) will enable us to detect periodicities of the present time and to compare them with the past Sun.

Finally, the eclipse and planet timings will be used to detect any possible changes in the Sun's diameter. The results are shown in figures 3 and 4.

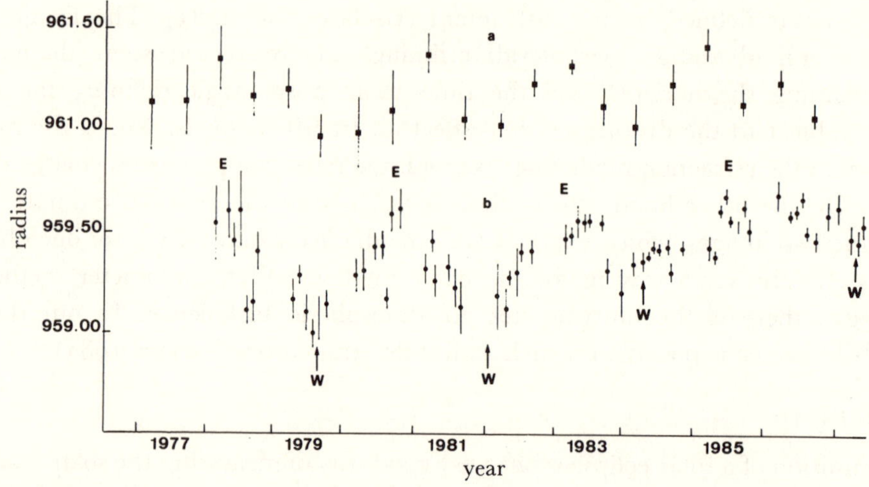

Figure 3. Diameter observations with the CERGA astrolabe against time provided by F. Laclare. The onset of the stratospheric winds at the 10 mbar† pressure (Naujokat 1986) are indicated by E (easterlies) and W (westerlies).

In the contemporary data, the remarkable result is a quasi-biennial oscillation of the solar diameter (figure 3). The fact this is present in two independent modern time series (Laclare 1987; Ribes *et al.* 1988*b*) shows that such a pseudo-periodic phenomenon is probably not the result of the technique used nor of the personal bias of the observer. A third independent time series of solar diameter has been obtained with an astrolabe (Leister, personal communication) and exhibits the 1000-day periodicity. As this series comes from the Southern Hemisphere (Abrahao de Moraes, in Brazil), no doubt that a comparative analysis will be of great importance for understanding the solar (or atmospheric) nature of the oscillation.

Other periodicities are present in these series, and are also found in the historical timings (figure 4). Their characteristics (amplitude) are somewhat different, which might be due to the secular variability of the Sun over the centuries. In particular, the amplitudes of the peaks present during the Maunder Minimum are much larger (by a factor of four to five) than those found in the modern time series. This can be understood as a superoscillation of the solar envelope, associated with the dearth of magnetic activity (Ribes *et al.* 1989).

† 1 mbar = 10^2 Pa.

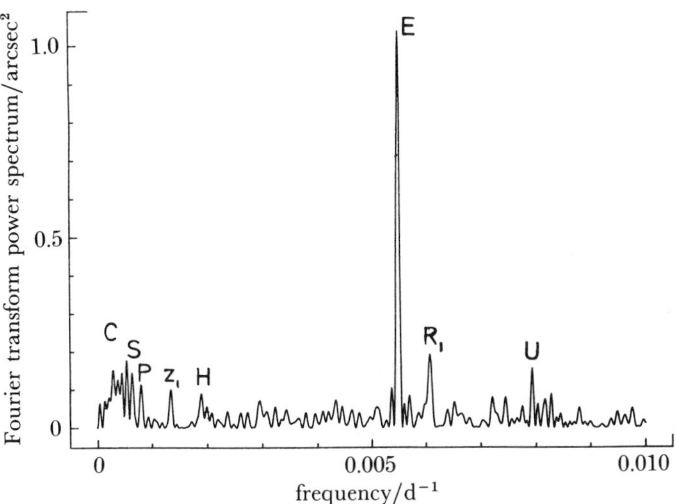

FIGURE 4. Fourier analysis (modulus) of the historical timings performed by P. La Hire from 1683 to 1718 against frequency. The most prominent features are the signature of the 11-year cycle (denoted as 'C') and the semi-annual signal indicating the change in the Sun's shape (denoted as 'E').

In the present day, there is also a particular difference between the characteristics of the quiet and active phases of the 11-year solar cycle: The apparent Sun looks smaller when its surface activity is large, and larger when its activity is reduced. The characteristics of the 11-year cycle during the seventeenth century are similar, and more pronounced at the time of the dearth. This supports the conclusion that the periodicities found originate from the Sun itself rather than from error or bias. The solar nature of the periodicities is further confirmed by the study of other solar and geomagnetic indicators (§4).

There are also a few singularities in both the modern and historical spectra which deserve some attention. The most striking feature is a line that rises far above the 15σ level in the Fourier analysis of historical time series (figure 4), but is weak in the Fourier analysis of modern time series. Before considering a solar origin for this semi-annual periodicity, one wonders whether some special meteorological conditions might be responsible for it. A seasonal nebulosity occurring in winter could change the apparent diameter and would yield an annual signal. A spurious effect of this kind has been investigated by Wittman (1980), and can be partly explained by a second-order effect of atmospheric refraction. After correction for the Wittman effect, the annual peak is eliminated but the diameter time series, still exhibits the 182.5-day peak. One remaining possibility is that the Sun is elliptical in shape, which would give rise to a semi-annual modulation of the horizontal radius, because the horizontal diameter approximates the equatorial diameter around the soltices. The observed diameter is smaller near the equinoxes and larger near solstices. To help settle this question, a simulation of solar oblateness has been performed: an ellipticity of 10^{-2} would be needed to explain the 182.5-day signal present in the historical time series (Ribes *et al.* 1989). This is three orders of magnitude larger than the oblateness expected from the dynamics of the interior. For the 182.5-day signal present in the modern timings of Belgrade, the ellipticity required is still 10 times larger than the assumed oblateness (10^{-5}). Alternatively, a recent study by Kuhn *et al.* (1988) has shown that the shape of the Sun is not at all regular: the Sun is hotter at latitudes of solar activity. As a consequence, the solar disk may be oblate with a solar-cycle-dependent bump at the active latitudes that moves toward the equator during the solar cycle. Such an effect could lead to a change of 10^{-4} in the diameter, which is precisely the value found in the Belgrade timings. Along this line of thought, the signal found in the historical timings can be explained by a drastic apparent oblateness due to the lack of surface activity at mid-latitudes. The resulting equatorial bulge would have reached about 3 arcsec on the diameter.

I should stress that my conclusion is completely independent of any calibration problems, as I have used data obtained by a single observer using the same equipment. However, it is interesting the compare this finding to other measurements performed at the time, and at the end of the Maunder Minimum. Picard, after measuring the horizontal diameter of the Sun, by the filar micrometer method, through the period from 1666 to 1673, came to the conclusion that the horizontal radius was about 965 arcsec. This is in agreement with timings performed by P. La Hire, whenever the solar surface was empty of sunspots. There is also the radius determination from the solar eclipse observed in 1715 by Halley, which gives a value of 960.11 arcsec (Dunham *et al.* 1980). Using the eclipse value of 1715 to calibrate the timings, a correction of 3 arcsec on the radius should be applied to the La Hire timings, reducing this highest record of the radius to 962 arcsec. Our conclusion remains unchallenged, i.e. that the Sun's apparent radius was larger in the deep of the Maunder Minimum (where practically no sunspots were seen) than it was at the end, when solar activity has resumed (from 1705 onwards). Although the change in the disk shape provides a plausible explanation of radius variability, one cannot exclude *a priori* a real radius change. An expansion of the solar envelope of about 3 arcsec on the diameter would be consistent with the cooling of the envelope and with a slowing down of the surface rotation.

4. Periodicities in atmospheric and climatic indicators

The variability of the Sun's diameter is interesting not only *per se*, for the understanding of the internal properties of the Sun. A variable Sun might also effect the Earth's environment, and hence the climatic conditions. Although the Sun–climate connection is the focus of this meeting and is approached from many angles, I simply report the apparent relation between the solar periodicities and those found in the Earth's stratospheric circulation. The most interesting feature might be the relation between the extrema of the solar diameter at the 1000-day period and the Earth's stratospheric winds (Ribes *et al.* 1988 *b*): the diameter maxima coincide with the onset of the Easterlies, at 10 mbar pressure, while the diameter minima coincide with the onset of the Westerlies (figure 4). Although a causal relation still remains to be established, there is some indication that solar variability is the force sustaining this quasi-biennal oscillation of the stratosphere. The pseudo-periodic magnetic flux expulsion observed at the solar surface seems to induce a large-scale convective pattern, the azimuthal rolls. As a result, a modulation of the solar envelope would lead to concomitant variations of the radius and luminosity (Ribes & Laclare 1988). The observed changes of radius and luminosity are of the order of 10^{-4} over the 11-year cycle (Willson & Hudson 1988) and do not induce drastic changes in the climate. This is different for the Maunder Minimum, where the apparent ellipticity found was two orders of magnitude larger with a probable reduced luminosity of the order of 1 %.

Shorter solar periodicities are also present in various modern geomagnetic, atmospheric and meteorological records (Delache 1989). In particular, periods around 50–100 days in the solar diameter data correspond to analogous recurrent phenomena in the geomagnetic activity, the

Earth's rotation, the angular momentum of the Earth's atmosphere (Feyssel & Nitschelm 1985; Djurovič & Paquet 1987). It is not firmly established, however, that there is a causal relation between solar variability and the meteorological effects.

5. Tentative interpretation

Although the periodicities present in the diameter data seem to be real, it is still premature to conclude that they correspond to a real change of the solar radius. An alternative could be a change in the thermal structure of the Sun, under the action of the magnetic fields as suggested by Kuhn *et al.* (1988). As the limb brightness changes with latitude and is time dependent over a large range of scales (from days to the 11-year cycle), these changes can induce peaks in the diameter data of power spectra. The limb brightness contributes about several tenths of an arcsecond to the radius, which is the order of magnitude of the radius changes found in the modern time series (Delache *et al.* 1985; Ribes *et al.* 1988*b*).

There are some theoretical arguments indicating that magnetic fields located within the convective zone could not produce detectable radius changes, thus lending support to the thermal structure as the main cause of the apparent radius changes. On the other hand, the numerical simulations predict some significant radius change if the magnetic fields are concentrated beneath the convective zone (Dappen 1983). The crucial point, then, is the location of the dynamo (within or below the convective zone). Again, I refer to modern observations of the solar cycle as well as recent to helioseismological data. The convective zone does not exhibit the radial velocity gradient required to drive the dynamo (Libbrecht 1990). Moreover, convective motions in a fluid dominated by rotation would have the form of meridional cells. No such cells have yet been observed on the Sun. Instead, the large-scale circulation detected by means of the magnetic tracers are azimuthal rolls, i.e. orthogonal to the expected meridional cells (Ribes *et al.* 1985; Ribes 1986). These observational facts seem to question the convective dynamo. Moreover, the recent findings that the interior of the Sun rotates rigidly (that is, without the differential rotation characteristics of the surface (Brown & Morrow 1987)), and that solar activity rotates rigidly when emerging, all support the idea of a deep-seated magnetic field source. Although a core-driven solar cycle is completely beyond the scope of this paper, such speculations would shed light on the variability of solar neutrinos detected by Davies.

It is clear that the Sun pulsates magnetically with different periods, and with concomitant modulations of the solar envelope. Whether or not this pulsation leads to any significant change of the solar output cannot be answered yet, since long and accurate time series of the solar constant are not available. Deeply confined magnetic fields could also be preferable for causing significant changes in the solar output, although the thermal timescale is a constraint on short-term variations.

It remains that the Sun's shape changes in the course of its cycle as well as its rotation, as the result of a strong coupling between the magnetic fields and the solar envelope. The relation between solar variability and stratospheric winds is a very promising investigation, even though we are as yet far from modelling it.

I am grateful to Dr F. Laclare, Dr N. Leister and Dr S. Sadsakov, who generously provided their diameter data.

References

Abarbanell, C. & Wohl, H. 1981 Solar rotation velocity as determined from sunspot drawings of J. Hevelius in the 17th century. *Sol. Phys.* **70**, 197–203.

Auzout, A. 1729 Du micrometre ou Maniere exacte. *Mém. Acad. Sci., Paris* **7**, 118–130.

Balthasar, H. & Wohl, H. 1980 Differential rotation and meridional circulation of sunspots in the year 1940–1968. *Astron. Astrophys.* **92**, 111–116.

Bracewell, R. N. 1988 Varves and solar physics. *Q. Jl R. astr. Soc.* **29**, 110–128.

Brown, T. M. & Morrow, C. A. 1987 Depth and latitude dependence of solar rotation. *Astrophys. J.* **314**, L21–L26.

Dappen, W. 1983 Hydrostatic reaction of the Sun to local disturbances. *Astron. Astrophys* **124**, 11–22.

Delache, P. 1989 Variability of the solar diameter. *Adv. Space. Res.* **8**, 119–128.

Delache, P., Laclare, F. & Sadsaoud, H. 1985 Long period oscillations in solar diameter measurements. *Nature, Lond.* **317**, 416–418.

Dicke, R. H. 1978 Is there a chronometer hidden deep in the Sun? *Nature, Lond.* **276**, 676–680.

Djurovič, D. & Paquet, P. 1988 The solar origin of the 50-day fluctuation of the Earth rotation and atmospheric circulation. *Astron. Astrophys.* **204**, 306–312.

Dunham, D. W., Sofia, S., Fiala, A. D. & Muller, P. M. 1980 Observations of a change in the solar radius between 1717 and 1979. *Science, Wash.* **210**, 1243–1244.

Eddy, J. A. & Boornazian, A. A. 1979 Secular decrease in the solar diameter, 1863–1953. *Bull. Am. astr. Soc.* **437**.

Eddy, J. A., Gilman, P. A. & Trotter, D. E. 1976 Solar rotation during the Maunder minimum. *Sol. Phys.* **46**, 3–14.

Feyssel, M. & Nitschelm, N. 1985 Time-dependent aspects of the atmospheric driven fluctuations in the duration of the day. *Annls Geofisicae* **3** (2), 181–186.

Gilliland, R. L. 1981 Solar radius variations over the past 265 years. *Astrophys. J.* **248**, 1144–1155.

Gleissberg, W. 1966 Ascent and descent in the eighty-year cycles of solar activity. *J. Br. astr. Assoc.* **76**, 265–270.

Kuhn, J., Libbrecht, K. G. & Dicke, R. H. 1988 The surface temperature of the Sun and changes in the solar constant. *Science, Wash.* **242**, 908–911.

Laclare, F. 1983 Mesures du diametre solaire a l'astrolabe. *Astron. Astrophys.* **125**, 200–203.

Laclare, F. 1987 Sur les variations du diametre du Soleil observees a l'astrolabe solaire du CERGA. *C. r. Acad. Sci, Paris* II **305**, 451–454.

Laclare, F., Demarcq, J. & Chollet, F. 1983 Sur la realisation d'un astrolabe solaire au Centre d'Etudes de Recherches Geodynamiques et Astronomiques. *C. r. Séanc. Acad. Sci., Paris* **291**, 189–192.

La Hire, P. 1683–1718 Archives Observatoire de Paris. Manuscripts D2, 1–10.

Lalande, J. 1771 *Astronomie* (ed. J. Desaimp), 2nd edn, p. 1771. Paris.

Le Roy Ladurie, E. 1967 *Histoire du climat depuis l'an mil* (ed. C. Flammarion). Paris. (Republication 1983.)

Libbrecht, K. G. 1990 Solar P-mode frequency splittings. *Proc. Symp. Seismology of the Sun and Sun-like stars, Tenerife.* (In the press.)

Maunder, R. W. 1894 A prolonged sunspot minimum. *Knowledge* **17**(106), 173–176.

Naujokat, B. 1986 An update of the observed quasi-biennal oscillation of the stratospheric winds over the tropics. *J. atmos. Sci.* **43**, 1873–1877.

Parkinson, J. H., Morrison, L. V. & Stephenson, F. R. 1980 The constancy of the solar diameter over the past 250 years. *Nature, Lond.* **288**, 548–551.

Picard, J. 1671 Archives Observatoire de Paris. Manuscripts D1, 14–16.

Ribes, E. 1986 Étude de la dynamique de la zone convective solaire et ses consequences sur le cycle d'activité. *C. r. Acad. Sci., Paris* II **302**, 871–876.

Ribes, E. & Laclare, F. 1988 Toroidal convection rolls in the Sun: a challenge to Theory. *Geophys. Astrophys. Fluid Dynam.* **41**, 171–176.

Ribes, E., Mein, P. & Mangeney, A. 1985 A large-scale meridional circulation in the convective zone. *Nature, Lond.* **318**, 170–171.

Ribes, E., Merlin, P., Ribes, J. C. & Barthalot, R., 1989 Absolute periodicities of the solar diameter, derived from historical and modern time-series. *Annls Geofisicae* **7**(4), 321–329.

Ribes, E., Ribes, J. C. & Barthalot, R. 1987 Evidence for a larger Sun with a slower rotation during the seventeenth century. *Nature, Lond.* **326**, 52–55.

Ribes, E., Ribes, J. C. & Barthalot, R. 1988*a* Size of the Sun in the seventeenth century. *Nature, Lond.* **332**, 689–690.

Ribes, E., Ribes, J. C., Vince, I. & Merlin, P. 1988*b* Sur l'oscillation du diametre solaire. *C. r. Acad. Sci., Paris* II **307**, 1195–1201.

Schroder, W. 1988 Aurorae during the Maunder Minimum. *Meteorology Atmos. Phys.* **38**, 246–251.

Shapiro, I. I. 1980 Is the Sun shrinking. *Science, Wash.* **208**, 51–53.

Willson, R. C. & Hudson, H. S. 1988 Solar luminosity variations in solar cycle 21. *Nature, Lond.* **332**, 810–812.

Wittman, A. D. 1980 Tobias Mayer's observations of the Sun: evidence against a secular decrease of the solar diameter. *Sol. Phys.* **66**, 223–229.

Wittman, A. D. & Débarbat, S. 1990 The solar diameter and its variability. *Sterne Weltraum.* (Submitted.)
Wolf, R. 1856 Die Sonnenflecken. *Astron. Mitt., Zürich* **61**, 1856.

Discussion

J. S. STANFORD (*Department of Atmospheric, Oceanic and Planetary Physics, University of Oxford, U.K.*). I am interested in the solar diameter oscillations Dr Ribes mentioned, with periods of seasonal length (150–180 days). Can she comment further on these?

ELIZABETH RIBES. The strongest periodicity occurring at the time of the Maunder Minimum is a semi-annual one (at 182.5 days). Such a periodicity would show up if the solar shape were elliptical. The reason for this is that meridian timings measure the horizontal diameter in a plane that wobbles 26° to either side of the Sun's equator due to the difference between the Sun's and the Earth's axes of rotation. An apparent oblateness of 10^{-2} is required to account for the 182.5-day peak amplitude in the Fourier spectrum. Moreover, there is a phase shift of 70 days between a constant oblateness and an apparent oblateness due to the concentration of solar activity near the equator. Therefore, I surmise that the semi-annual peak during the Maunder Minimum was the consequence of the concentration of sunspot activity in a very narrow range of latitudes (0–18°)

Other periodicities (from 50 to 155 days) seem to be common to solar (irradiance, radius, sunspot number) geomagnetic and Earth atmosphere indicators. The causal relation is not well understood at present.

P. FOUKAL (*CRI, Cambridge, Massachusetts, U.S.A.*). 1. Dr Ribes mentioned that during the Maunder Minimum the butterfly diagram lost one of its wings, yet she showed rotation measurements by La Hire from that epoch which refer to both hemispheres. Am I misunderstanding something?

2. Is she concerned about the fact that the most accurate diameter data – those of Brown at HAO – show no variation down to about 0.03 arcsec per year precision?

ELIZABETH RIBES. 1. From 1666 to 1712 the sunspot activity was concentrated in the Southern Hemisphere only, giving the corresponding rotation curve shown in figure 2. From 1712 onwards, the sunspot activity resumed in the Northern Hemisphere, leading to the corresponding rotation rate in figure 2.

2. The diameter data obtained at the high-altitude observatory (HAO) rely on an objective definition of the solar limb. There is, however, a fundamental problem which affects ground-based observations of the solar diameter: the solar limb definition is distorted by the Earth's atmosphere. The photoelectric method aims to decouple the atmospheric effects from the solar signal. This is a difficult problem to solve, however. The HAO data and solar activity indicators have been subjected to a multi-correlation analysis: the atmospheric correction in the HAO data is found to be significantly correlated with solar activity. This means that the residual HAO diameters are likely to have been overcorrected. At this point, it is preferable to be cautious about drawing any strong conclusions on the basis of the HAO data.

Phil. Trans. R. Soc. Lond. A **330**, 499–512 (1990)

Printed in Great Britain

Historical evidence concerning the Sun: interpretation of sunspot records during the telescopic and pretelescopic eras

By F. R. Stephenson

Department of Physics, University of Durham, South Road, Durham DH1 3LE, U.K.

The value of sunspot observations in investigating solar activity trends – mainly on the centennial to millennial timescale – is considered in some detail. It is shown that although observations made since the mid-eighteenth century are in general very reliable indicators of solar activity, older data are of dubious quality and utility. The sunspot record in both the pretelescopic and early telescopic periods appears to be confused by serious data artefacts.

1. Introduction

The main theme of this paper is the investigation of long-term trends in solar activity on the centennial to millennial timescale. In principle, these trends may be inferred from observations of both sunspots and auroras, reports of which extend back more than two millennia. However, for reasons given immediately below, only sunspot data is considered in detail here.

Unlike auroras, sunspots provide a *direct* indication of solar activity. Almost daily reports of the number of spots visible on the Sun's disc have been compiled since early in the nineteenth century, whereas even as far back as A.D. 1750 fairly consistent observations are still preserved (McKinnon 1987).

By using this material, solar activity can be studied in detail over the last 240 years. Before 1750, the extant sunspot record becomes progressively poorer and, in particular, relatively few observations survive from the seventeenth century. Not surprisingly, pretelescopic data are even more incomplete. Hence it is in these earlier times that alternative information would be most valuable.

Although numerous reports of the aurora borealis are available, especially in recent centuries, the solar signal is only weakly preserved in the auroral record. Since about A.D. 1700, this is largely a consequence of the biased distribution over the Earth's surface of the majority of observers. Most accounts of auroras after this date originate from regions in fairly high geomagnetic latitude (especially Northern Europe and, latterly, North America); there are relatively few from more southerly zones. Auroras visible at high latitudes are mainly produced by charged particles of rather low energy; these particles are expelled from the Sun in solar wind streams that originate in coronal holes. Because coronal holes may develop at all phases of the solar cycle with comparable frequency, the more recent auroral record is not expected to show a good correlation with solar activity.

The limited utility of even modern auroral observations in studying solar activity is well exemplified by the work of Siscoe (1980). He investigated the frequency with which auroras were observed during the past 100 years at three separate sites. These locations, two in the U.S.A. and one in Germany, were all in much the same geomagnetic latitude. Siscoe found only dubious evidence for the existence of the solar cycle from the observations reported at any one station; in order to reveal the cycle clearly it was necessary to combine data from all three places. In view of the poorer consistency of the older records, there seems little hope of using auroral observations to trace the solar cycle further into the past in any detail.

In lower geomagnetic latitudes (e.g. China), most displays of the aurora borealis are produced by energetic particles that are expelled from the Sun during solar flares. Hence, had auroral data from these regions been more complete, they might have been expected to show a useful correlation with the solar cycle. As it happens, the auroral record from China and elsewhere in East Asia (mainly Japan and Korea) during the past few centuries is far from systematic. Hence its value as an index of solar activity is probably minimal.

In principle, auroral data should have greater value during the pretelescopic period for in general they are considerably more numerous than sunspots sightings. The higher incidence of observations of auroras in these early times is hardly surprising because these phenomena are often rather spectacular; on the contrary, the detection of sunspots with the unaided eye is a relatively esoteric pursuit. Some 2000 accounts of auroras are extant from before A.D. 1600, fully an order of magnitude greater than the surviving number of sunspot reports. Of these auroral records, the vast majority originate from only two areas of the globe: Europe and East Asia. The contribution from each source will be discussed in turn.

Roughly half of the auroral data before A.D. 1600 are from Europe. These have been carefully catalogued by Link (1962). The early European record of auroras appears likely to show little indication of variation in solar activity both because of (i) the relatively high geomagnetic latitude of the places of observation, as already discussed, and (ii) the sporadic nature of most auroral sightings. Few astronomical records of any kind are preserved from ancient Europe. Although the degree of preservation from medieval times is much higher, most observations of celestial phenomena were then made on a casual basis. Observers tended to have no more than a superficial interest in astronomy and noted only the more spectacular occurrences (e.g. comets, eclipses and meteor showers, as well as auroras). Reports of these events are usually cited in chronicles of towns and monasteries, rather than in works devoted to astronomy.

During ancient and medieval times, the more consistent oriental auroral record might be expected to prove superior to that from Europe for investigating past trends in solar variability. As well as the advantage of lower geomagnetic latitudes for the places of observation, most sightings in East Asia were made by official astronomers who tended to keep a fairly regular watch of the sky. However, there are still interpretational problems. Whereas historical accounts of sunspots are usually readily recognizable, the unambiguous identification of many of the descriptions of possible auroras is difficult. For example, numerous instances of reddening of the night sky are reported in sixteenth- (and also early seventeenth-) century Korean history. Although these reports would seem to relate to the aurora borealis, the frequency of occurrence is sometimes as much as two orders of magnitude greater than the expected auroral frequency (Stephenson 1988). Difficulties such as these need to be satisfactorily resolved before an objective investigation of the relevance of East Asian auroral data to the question of solar variability can be attempted.

Even if the early oriental record of auroras could be adequately purged of false sightings, it is doubtful whether at present useful corrections could be made to the observed frequency of auroral displays to allow for long-term changes in the terrestrial magnetic field. The rate of occurrence of auroras at any particular location is strongly dependent both on the intensity of the Earth's magnetic field and the position of the geomagnetic pole of the time. Although variations of the terrestrial magnetic field in recent centuries are fairly well mapped, changes on the millennial timescale are currently only poorly determined (Tarling 1988).

In the study of past solar variability, observations of sunspots thus possess a number of distinct advantages over auroral sightings. Nevertheless, the incompleteness of the sunspot record before A.D. 1750 has already been emphasized. Because regular monitoring of the Sun for spots was not maintained before 1750, any evidence for long-term trends in solar activity is likely to be contaminated by data artefacts. No more than about 160 reports of sunspots are known to have survived from the entire pretelescopic period. As several careful literature searches for ancient and medieval sunspot observations have been made in recent years (see §2), it seems unlikely that this number will be improved upon significantly by future work. With only about one reported sighting per decade, the value of pretelescopic sunspot data is obviously severely limited. Nevertheless, it is important to try to make an objective assessment of the utility of this unique record.

2. SOLAR ACTIVITY DURING THE TELESCOPIC PERIOD AS DEDUCED FROM SUNSPOT RECORDS

As discussed most recently by McKinnon (1987), solar activity during the past 170 years can be studied in detail by using the almost daily record of sunspot numbers that is available throughout that time. A systematic watch of the Sun for spots was initiated in 1849 by the Swiss astronomer Rudolph Wolf and the scheme that he devised for estimating sunspot numbers is still use in today. Much of what is known regarding solar activity in the earlier part of the telescopic period is also a consequence of the pioneering researches of Wolf. As the result of an extensive search of European literature (scientific journals and observatory archives), Wolf was able to construct a fairly complete sequence of daily sunspot numbers as far back as 1818. These data, edited by Waldmeier (1961), are fully tabulated by McKinnon (1987). There are only a few gaps lasting for more than a week in the whole of the sunspot record between 1818 and 1848, except for the years 1820, 1825 and 1835, which are rather less well represented. Between 1749 and 1817 the data were of inferior quality, but Wolf was still able to derive monthly means.

2.1. *Solar activity since* A.D. 1715

Figure 1, which shows the smoothed monthly sunspot numbers between A.D. 1749 and 1985, is taken from McKinnon (1987). Over this period of rather less than 250 years, both the length of any particular cycle and the sunspot number at different maxima were very variable.

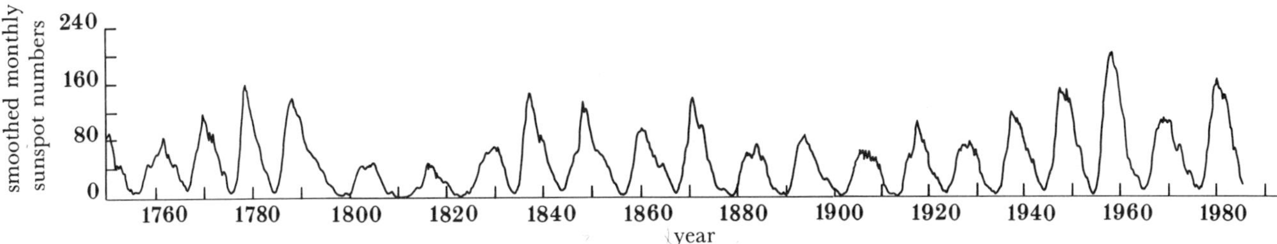

FIGURE 1. Smoothed monthly sunspot numbers: July 1749 to June 1985. (After McKinnon 1987.)

[103]

Although the mean interval between successive maxima was 10.9 years, the actual interval could be as short as 7.3 years (1829–37) or as long as 17.1 years (1788–1805). The peak sunspot number in any cycle ranged over a factor of 4: from 49 in 1805 and 1816 to 201 in 1957 (overall mean 116.1). Hence despite the fact that the sunspot cycle is a remarkably persistent feature of solar behaviour, it is obviously no more than quasi-regular. Recently, Weiss (1988) has argued that the aperiodic behaviour of the solar cycle is an example of deterministic chaos.

Although solar activity in the period since A.D. 1749 is fairly well determined, the situation before that date is less encouraging. Using historical records, Wolf deduced annual mean sunspot numbers between 1700 and 1749, but previous to 1700 he made no attempt to deduce even annual means. Since the efforts of Wolf roughly 150 years ago, no systematic literature search has been made to improve on his results for the period between about 1715 and 1817, an undertaking that would certainly seem desirable today.

2.2. *The Maunder Minimum*

For the period between A.D. 1610 and 1715, Eddy (1976) made a revised compilation of European sunspot records from a variety of sources. He also incorporated Wolf's original data. Eddy used these observations to estimate annual mean sunspot numbers during the period in question, but especially before 1650 his information was very fragmentary.

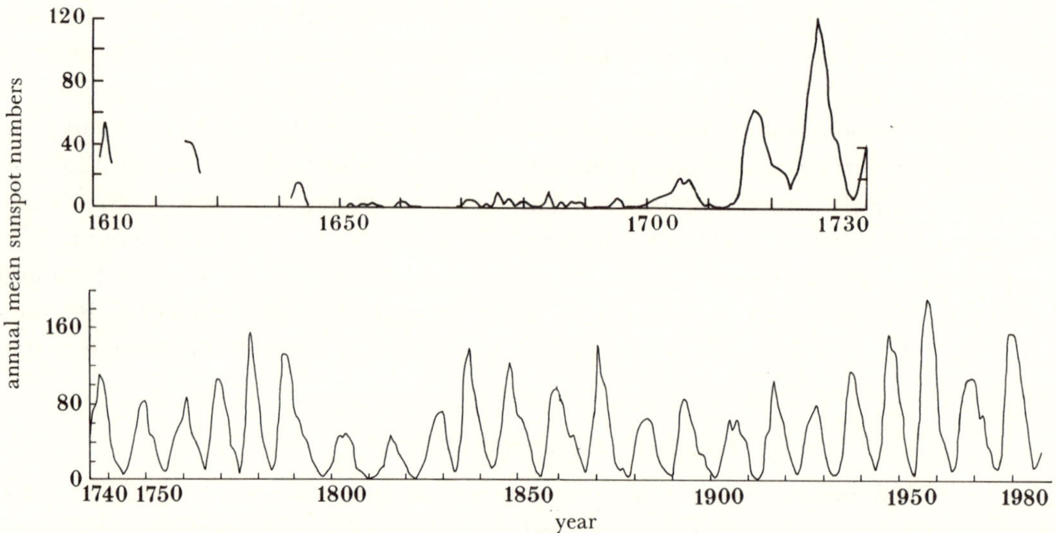

FIGURE 2. Annual mean sunspot numbers: A.D. 1610–1988 (J. A. Eddy, personal communication).

Figure 2, which shows derived annual mean sunspot numbers from A.D. 1610 to 1988, is due to Eddy (personal communication). It is apparent from this diagram that although the sunspot cycle has been clearly evident since A.D. 1715, it cannot be traced in the previous century. In particular, the level of solar activity over the entire interval between about A.D. 1645 and 1715 (the 'Maunder Minimum') would appear to have been considerably depressed relative to that in more recent centuries. There can be little doubt that during much of the Maunder Minimum the Sun was unusually quiet. Thus Eddy (1976) noted that astronomers of the time, such as Cassini and Flamsteed, had commented on the scarcity of sunspots over many years.

However, for a variety of reasons, it is very difficult to quantify the level of solar activity with any confidence during the Maunder Minimum. Factors that might be mentioned include the following: (i) the poor quality of telescope optics at this early period, so that only larger sunspots would be noticed; (ii) the lack of systematic observations of the Sun until much later; and (iii) the relative inaccessibility of astronomical records of any kind from most of the seventeenth century compared with in more recent times. The establishment of the official journals of the Royal Societies of London and Paris in the late seventeenth century did much to encourage the dissemination of reports of celestial phenomena.

2.3. *Comparison data: occultations of stars by the Moon*

To illustrate the difficulties in interpreting the early telescopic sunspot record, it is helpful to investigate the changing annual frequency with which some other kind of astronomical phenomenon is reported during the same period. A useful comparison is provided by occultations of stars by the Moon; detailed catalogues of these events have been compiled in recent years. Although occultations represent a completely independent phenomenon to sunspots, they were often observed by the same astronomers and are recorded in much the same literature. To some extent, the frequency with which early observations of occultations are still preserved today thus seems likely to be affected by similar factors to those pertaining to sunspot records. However, although the observed occultation frequency might be expected to show evidence of short-term periodicity (related to the 18-year regression of the lunar nodes), no significant long-term cyclical behaviour is expected, in marked contrast to sunspots.

Since the early seventeenth century, astronomers have measured the time of occurrence of occultations on account of the value of such data in improving knowledge of the lunar motion. The most recent and extensive catalogue of these data is that of Morrison *et al.* (1981). Covering the period from A.D. 1623 to 1942, this compilation contains some 40000 individual timings. Details for nearly 3000 separate observations between 1623 and 1860, virtually all from Europe, are printed in the above work. However, the quantity of later measurements is so vast that these are listed in microfiche. Here I concentrate on the earlier set of data.

The number of separate occultations recorded in the selected period of 238 years, although large, is substantially less than 3000. Many astronomers timed both the beginning and end of an occultation of the same star and Morrison *et al.* listed these as distinct events. Further, the same occultation would frequently be reported independently by several observers, particularly if the star were bright or a member of a well-known group such as the Pleiades. In enumerating the annual frequency of recorded occultations between 1623 and 1860 I have thus counted only one occultation of any given star per night. In this way the occultation record will have a closer parallel to that for sunspots. The total number of individual occultations reported up to 1860 is about 1300.

In figure 3 is plotted the annual frequency of occultation observations (with the restrictions discussed above) between 1623 and 1860. The diagram reveals clear indication of cyclical behaviour of approximate period 9 years (half the nodal regression) in the later record; this arises from the very uneven distribution of stars in the ecliptic zone. Before about 1800, evidence for such periodicity is confused by other factors: largely data artefacts due to the lack of any regular observing programme and relative inaccessibility of observational reports.

The main longer-term features of figure 3 are (i) very few preserved data before 1670 (a total of no more than 30 observations since the beginning of the telescopic era); (ii) a significant

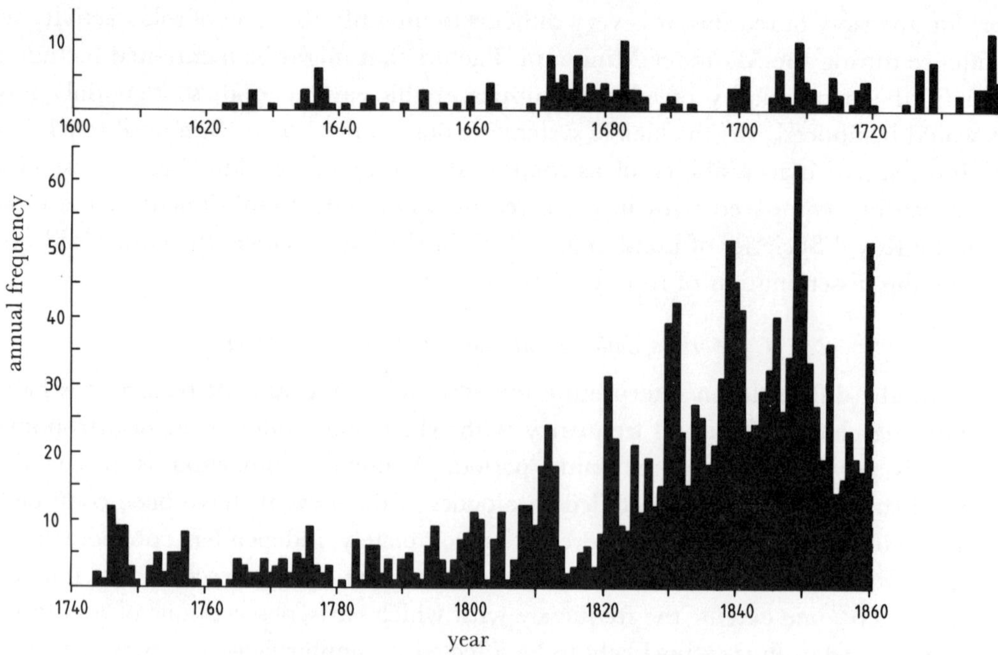

FIGURE 3. Annual frequency of reports of occultations of stars by the Moon
observed telescopically: A.D. 1623–1860.

increase in the mean annual frequency of recorded observations around 1670 with little further
change until 1810; (iii) a sharp rise in annual frequency at 1810 followed by still further growth
in the remaining 50 years of the selected interval. Investigation of the typical brightness of the
occulted stars at different epochs between 1623 and 1860 does not reveal a marked trend
towards observation of fainter stars in more recent times. Hence improvement in telescope
optics was only partly responsible for the observed long-term features in figure 3. Evidently
only a small and quite variable proportion of the observable events was actually witnessed or
recorded or both during the eighteenth century, whereas in the seventeenth century the
situation was even worse.

2.4. *Interpretation of the early telescopic sunspot record*

The results of the above investigation can be applied in a general way to the interpretation
of early sunspot data. Comparing figure 2 with figure 3, it seems very likely that the preserved
sunspot record from the earlier telescopic period may also be affected by significant trends of
a spurious nature. In particular, observations of sunspots made during the seventeenth century
were possibly too infrequent to reveal evidence of the solar cycle. It is also probable that recent
estimates of the mean level of solar activity during the Maunder Minimum, e.g. as displayed
in figure 2, may be too low by a large (and appreciably time-variant) factor. Summarizing,
although the sunspot record since 1818, and probably since 1749, is a reliable indicator of solar
activity, earlier observations are of very limited value. Especially during the seventeenth
century, it seems more reasonable to infer long-term changes in solar activity from the available
proxy records such as ^{14}C and ^{10}Be (Eddy 1988).

3. The pretelescopic sunspot record (*ca.* 150 B.C. to A.D. 1610)

Pretelescopic records of sunspots are preserved from only three regions of the world: (i) East Asia (almost exclusively China and Korea); (ii) Europe; and (iii) the Arab dominions. The contribution from sources (ii) and (iii) is virtually negligible compared with that from (i). For instance, the detailed catalogue of unaided-eye sunspot observations by Wittmann & Xu (1987), which contains fairly complete entries from both oriental and occidental history, cites less than 10 individual observations from outside East Asia during the entire period anterior to the invention of the telescope.

3.1. *Babylonian texts*

Although the surviving Late Babylonian astronomical diaries are extremely detailed (Sachs & Hunger 1988), they contain no recognizable allusions to sunspots. This circumstance may be at least partly a result of the Babylonians' concern with observing regular and thus predictable phenomena. Eclipses or lunar and planetary movements are recorded in abundance, but references to sporadic events (e.g. auroras and meteor showers) are rare. Nevertheless, it should be emphasized that no more than about 5% of the original texts have survived, with possible loss of key records.

3.2. *European and Arab records*

In ancient and medieval Europe, the aristotelian notion of a faultless Sun probably contributed to what appears to be an almost complete lack of interest in observing sunspots. Nevertheless, it is likely there are additional explanations for the small number of sunspots preserved in European history during this period. Because very few writings of an astronomical nature have survived from before the Renaissance, some sunspot records may well have been lost. Ptolemy's *Almagest* is unique among surviving ancient European works in quoting a variety of celestial observations: notably, eclipses and lunar and planetary phenomena. However, in this rather specialized treatise, sightings of sunspots and other seemingly random occurrences (e.g. new stars or meteor showers) would have no place.

As noted in the Introduction to this paper, most allusions to celestial phenomena from Europe in ancient and medieval times are recorded in chronicles and other historical works. These compilations often make reference to the more spectacular celestial events. However, because the observers had, in general, little interest in astronomy, they may have hardly ever noticed minor phenomena such as sunspots. In this context, it is perhaps significant that two of the very few sunspot observations made in Europe during medieval times (in A.D. 1365 and 1371) took place when smoke from forest fires dimmed the Sun sufficiently for it to be viewed directly (Vyssotsky 1949).

Medieval Arab astronomers tended to explain sunspots in terms of transits of Venus or Mercury across the solar disc. Several such instances are on record: in A.D. 840, *ca.* 1030, 1068 and *ca.* 1130. These observations have been discussed in detail by (Goldstein 1969). In practice, Venus (unlike Mercury) is quite visible to the unaided eye when silhouetted against the Sun's disc, as was true at the last transit in 1882. However, by reference to the calculations of Meeus (1958), Goldstein was able to show that no suitable transits occurred in or near the stated years; hence the observers must have witnessed sunspots instead. Attempts by the Arab astronomers to interpret sunspots as planetary transits may have been partly because of their familiarity with Aristotle's hypothesis of a blemish-free Sun. Nevertheless, they were obviously fully aware that the inner planets (like the Moon) could pass directly in front of the Sun. The appearance of sunspots would thus seem to the observers to provide 'proof' that these planets could actually be seen in silhouette on the solar disc.

There is, in fact no evidence that a Venus transit was ever observed before the introduction of the telescope. Recently (Stephenson 1990), I compared Meeus's (1958) list of computed dates for Venus transits with the dates of pretelescopic sunspot sightings (mainly from East Asia) compiled by Wittmann & Xu (1987) and Yau & Stephenson (1988). There proved to be no coincidences or near-coincidences. Hence it may be concluded that Jeremiah Horrocks still holds the distinction of making the first known observation of a transit of Venus. This observation took place in 1639, following a successful prediction by Horrocks of the event.

3.3. *Oriental observations*

In the remainder of this section, I concentrate on East Asian sunspot sightings. These comprise some 95% of all pretelescopic sunspot reports. A pioneering study of Far Eastern observations of sunspots was made by Kanda more than 50 years ago (Kanda 1933). More recently (some ten years ago), several further contributions to the subject were published; among the more detailed investigations were those by the Yunnan Observatory (1976), Clark & Stephenson (1978) and Chen & Dai (1982). Following the publication during the last two years of revised catalogues of unaided-eye sunspot sightings by Wittmann & Xu (1987) and Yau & Stephenson (1988), it is appropriate to reconsider just what the early observations are capable of revealing with regard to past trends in solar activity. The latter paper, referred to subsequently as Y.S., forms the basis of much of the succeeding discussion.

Apart from slight differences in translation, most of the entries in the catalogue of Wittmann & Xu are identical with those listed by Y.S. However, the former paper is concerned with both oriental and occidental sightings whereas the latter work is restricted to observations from East Asia. In addition, Y.S. correct a few minor errors in the compilation of Wittmann & Xu and also include several additional observations in the pretelescopic period. Both catalogues contain oriental sightings made after the invention of the telescope, the compilation of Y.S. extending down to as late as 1918. Nevertheless, it is probable that all of these more recent observations were made with the unaided eye; as an astronomical device, the telescope made little impact throughout East Asia until the present century. The number of reported sightings of sunspots in the Orient since A.D. 1610 (65) is minute compared with the number of telescopic observations made in the West. For example, as many as 750 reports of sunspots are preserved from Europe during the Maunder Minimum alone. Hence the value of these more recent naked-eye data is severely limited in the study of past solar variability. In this section I thus restrict attention to those observations made before the telescopic era.

The entries in the catalogues of Wittmann & Xu and Y.S. are based entirely on *written* records. It is regrettable that apart from a few highly schematic illustrations, drawings of sunspots from the pretelescopic era are extremely rare. Accounts of sunspots are mainly cited in astronomical treatises; these form important sections of the official dynastic histories of China and Korea. Numerous celestial observations of all kinds (eclipses, lunar and planetary movements, comets, novae and supernovae, meteor showers and auroras, as well as sunspots) are found in these treatises. Such data are largely summaries of the reports of the court astronomers, whose task it was (at least in principle) to maintain a regular watch of both the day and night sky. Throughout East Asia the prime motive for celestial observation was astrological. Even in recent centuries, astronomers still closely followed the traditions established in ancient China. Thus the terminology used to describe celestial phenomena remained almost unchanged over two millennia.

Although the earliest reliable report of a sunspot from the Orient dates from 165 B.C., relatively few accounts have survived before A.D. 300. With a single exception (an observation made in Japan in A.D. 851), all Far Eastern sunspot records earlier than A.D. 1150 are from China. After that date there is a sizeable contribution from Korea, but Chinese observations still tend to dominate. Very occasionally a report from Vietnam is also encountered (in A.D. 1276, 1593 and 1603).

Most oriental sunspot records follow only two basic forms: (i) 'Within the Sun there was a black spot' (*hei-tzu*); and (ii) 'Within the Sun there was a black vapour' (*hei-ch'i*). Entries of the former type are only found after A.D. 300; use of *hei-ch'i*, although generally rarer, occurs at all periods. The word *tzu*, rendered 'spot' in (i), has several more general meanings such as 'seeds' or 'pellets'. Hence when referring to the Sun, objects of small angular size are clearly indicated. Especially as some records state that *hei-tzu* were seen for several days within the Sun, there can be no viable alternative to a sunspot interpretation. The expression *ch'i* ('air', 'vapour', etc.) can have a variety of meteorological as well as astronomical meanings; it is the standard term to indicate auroras, whereas in the appropriate context it may be used to describe cloud formations. However, when a black vapour is specifically described as appearing 'within the Sun', a sunspot identification can be confidently assumed. As in the case of *hei-tzu*, *hei-ch'i* were from time to time seen on the Sun for several days.

Other, comparatively rare descriptions that fairly obviously relate to sunspots are miscellaneous objects appearing within the Sun, e.g. birds (especially crows) or stars. Again, these phenomena were sometimes of several days' duration. Because there is no grammatical plural in Classical Chinese (the language in which virtually all East Asian astronomical observations are recorded), it is best to assume that all sunspot reports refer to a single spot or group unless the number of objects visible is specified. Several sightings of double or multiple spots are on record, the earliest in A.D. 355 when two separate sunspots 'as large as peaches' were seen together.

3.4. *Temporal distribution of the oriental sunspot sightings*

In figure 4 is shown the decade distribution of unaided-eye oriental sunspot observations from earliest times (200 B.C.) until A.D. 1610, as listed in the catalogue of Y.S. The diagram is

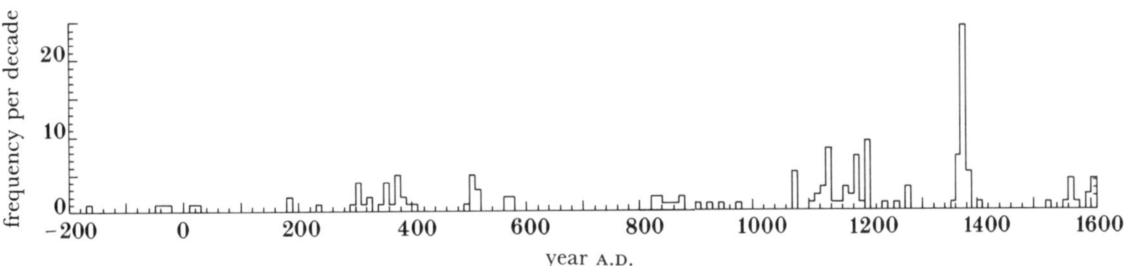

FIGURE 4. Decade distribution of unaided-eye sunspot observations recorded in East Asia: 165 B.C. to A.D. 1610. (After Yau 1988*b*.)

taken from a corresponding figure in the paper by Yau (1988*b*). In all, 157 separate occurrences of sunspots are represented. The individual totals are 122 from China, 34 from Korea, three from Vietnam and one from Japan. Owing to three instances when the same spot (or spot group) was observed in both China and Korea, the sum total is 160 rather than 157.

The main features of figure 4 may be summarized as follows: (i) only sporadic sightings of sunspots before about A.D. 1100; (ii) comparatively frequent observations both in China and Korea during the twelfth century of our era; (iii) very few spots noted between about A.D. 1200 and 1360; (iv) a marked peak around 1370, mainly resulting from an excess of Chinese records; (v) few subsequent references to sunspots in either Chinese or Korean history until the end of the period covered by the diagram. If interpreted literally, the oriental sunspot record would thus imply a number of lengthy intervals when the Sun was quiet, interspersed with active periods lasting up to about a century. However, although the existence of the sunspot cycle can be detected from pretelescopic observations (Wittmann & Xu 1988; Yau 1988*b*), a variety of evidence suggests that the existing record is contaminated by significant data artefacts. This evidence will be discussed below.

3.5. *Efficiency of the oriental observers*

Judging from the fact that almost every apparition of Halley's Comet since 240 B.C. is recorded in oriental history (Stephenson & Yau 1986), it might be inferred that during ancient and medieval times the astronomers of East Asia were remarkably efficient observers. (Compared with some long-period comets, Halley's Comet is not particularly brilliant even when passing fairly close to the Earth.) However, as will be discussed below, there is good evidence that the astronomers were far from systematic observers of most phenomena. This remark is especially true of sunspots. Reference to figure 4 shows that with a single notable exception, the peak number of sunspots (or spot groups) reported in any one decade during the whole of the pretelescopic period was less than 10. (Between A.D. 1367 and 1376, as many as 28 sunspots were sighted, half of them within a period of two years.) Obviously, only large sunspots are visible to the unaided eye, but until recently it has not been possible to make an objective assessment of the efficiency of the oriental observers in noting spots. Fortunately, a systematic survey of the Sun made without a telescope by Mossmann (1989) around the time of the last solar maximum now provides valuable comparison data.

Over an interval of 13 months between 1981 and 1982, Mossmann detected sunspots without any optical aid on 50 separate occasions; using a filter this number increased to 170 observations. In reviewing the results of this project, Eddy *et al.* (1989) estimated that Mossman's rate of sighting sunspots with the unaided eye, when adjusted for cycle phase, is as much as 200 times the long-term rate from the Orient. Even when compared with the unusual achievement of the Chinese and Korean astronomers around A.D. 1370, Mossman was still an order of magnitude more successful. Overall, in reporting sunspots visible to the unaided eye, the astronomers of East Asia evidently achieved an efficiency of less than 1 %.

Mossman's remarkable success at detecting sunspots without any optical aid suggests that there is no need to assume that the early oriental astronomers were in the habit of using some kind of optical filter to dim the Sun when looking for spots. This conclusion is confirmed by the East Asian records themselves. Thus several accounts mention that the Sun was unusually dull, or appeared yellow or red, when sunspots were sighted; other reports imply observation close to sunrise or sunset. Willis *et al.* (1988) showed that fully 40 % of all pretelescopic sightings of sunspots from the Orient were made in the two months March and April. At these times, dust storms are prevalent in East Asia. Airborne dust would act as a natural filter, reducing the solar glare and enabling sunspots to be more readily perceived.

The figure of 157 pretelescopic sunspot sightings in East Asian history, rather than the many thousands that might be expected, may in part be explained by emphasis on reporting only the very largest visible sunspots (or spot groups) by the oriental astronomers. For example, some 10 sunspot groups were recorded as having an obviously non-circular shape. However, large spot groups tend to be common around a solar maximum and even during the short time-span of his survey Mossman was able to detect several elementary shapes. In general it would seem that the Sun was only infrequently scrutinized for spots by the East Asian astronomers. Furthermore, of the observations actually made, few were presumably considered to be worth citing in the official histories (Eddy *et al.* 1989).

3.6. *Uniformity of the oriental record*

Of the various gaps evident in figure 4, there can be little doubt that the lacuna between about A.D. 600 and 800 is largely artificial in origin. Large numbers of records, both astronomical and otherwise, were destroyed when the rebel Chinese general An Lu-shan sacked the T'ang Dynasty capital of Ch'ang-an in A.D. 755. Much important material was probably lost at other periods; most of the various capitals that have served China throughout its long history have been extensively looted at one time or another, e.g. during the violence which usually accompanied the fall of a dynasty.

Changing attitudes towards celestial portents was probably a major factor in complicating the existing record of sunspots as well as of other celestial phenomena. Thus it is interesting to note that more than half of all Chinese sightings of sunspots during the pretelescopic period are reported during the reigns of only six emperors. (In all, some 150 emperors ruled China in the 1800 years covered by figure 4.) In this connection, Park (1977) had made a detailed study of the various portants (celestial and otherwise) reported in Korean history. He noted that Buddhist rulers were in general more gullible with regard to omens than Confucians; the latter tended to adopt a more rational view of portents. One king (Yonsan, who reigned between A.D. 1495 and 1506) was so hostile towards portentology that he actually abolished the time-honoured Office of Astronomy. However, this was soon reinstated by his successor.

Park cites an interesting case from A.D. 1204. In February of that year a sunspot was observed for three days in Korea. The Astronomer Royal of the time tried to suppress the report because he knew from Chinese history that a similar occurrence had presaged an emperor's death. Nevertheless, the portent was duly registered. A month later the Korean king died. After this event, there are no further mentions of sunspots in Korean history for 54 years, although several sightings were made in China during the intervening time. Only when a subsequent king (Kojong) had been on the throne for 45 years was it apparently considered safe for the Korean astronomers to report another sunspot. Just how much the long-term sunspot record from both Korea and China is similarly affected is impossible to judge, but it seems likely that the influence of astrology was far from negligible.

Park found marked changes in the relative importance of different omens at various periods. For example, there was a notable surge in recording solar haloes and sightings of Venus in daylight around A.D. 1400 because at the time they seemed to be threatening the intrinsic brightness of the Sun (the symbol of the king). By contrast, interest in reporting meteors and

510   F. R. STEPHENSON

night-time observations of the planets waned simultaneously. No similarly detailed study has
been made of portentology throughout Chinese history, but it seems likely that imperial
attitudes to omens were also extremely fluid.

Figure 5 illustrates just how variable the record of astronomical phenomena during even a
single dynasty can be. This diagram, produced at my suggestion by my colleague K. K. C. Yau

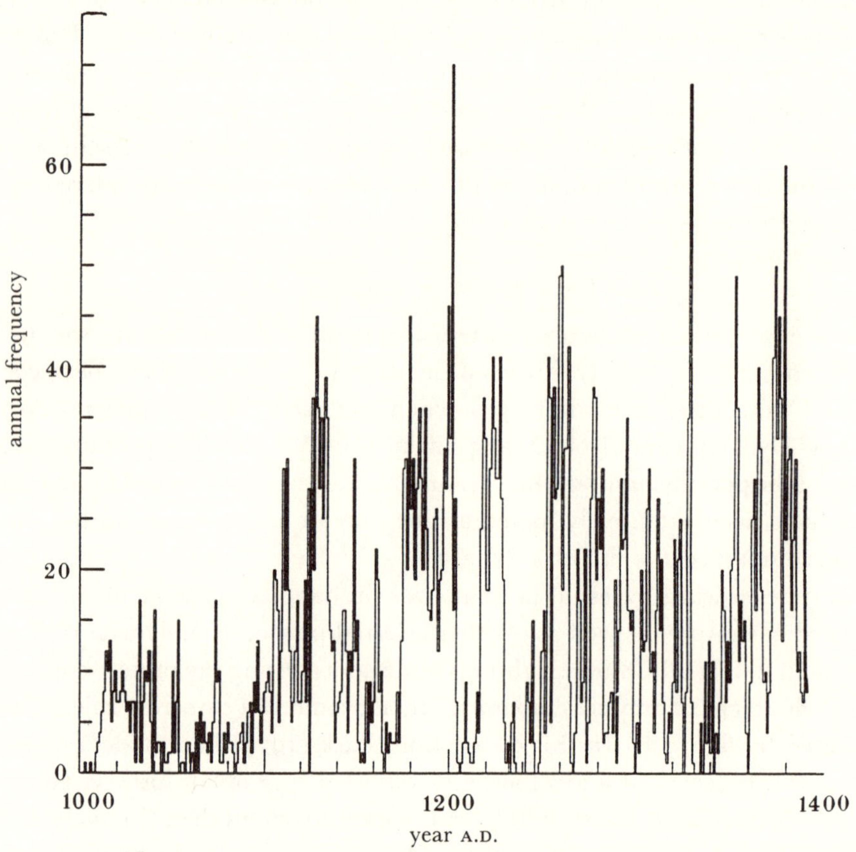

FIGURE 5. Annual frequency of astronomical observations of all kinds recorded in Korea
during the Koryo Dynasty (A.D. 918–1392). (After Yau 1988a.)

(Yau 1988a), is a plot of the annual frequency of astronomical events of all kinds (comets,
eclipses, lunar and planetary conjunctions, sightings of Venus in daylight, meteor showers,
auroras, etc.) recorded in Korea during the Koryo Dynasty (A.D. 918–1392). Most of the
events noted by the Korean astronomers would be expected to show no obvious trends in
frequency on timescales much longer than a decade. Hence the extremely complex pattern
visible in figure 5 must be largely of artificial origin. Figure 6 represents the results of a similar
investigation for China, but this covers many dynasties. This diagram is based on a decade-by-
decade count of celestial phenomena of all types recorded in Chinese history between 200 B.C.
and A.D. 1610 (Yau 1988a). Once again, most of the highly irregular distribution must be of
historical rather than astronomical origin.

3.7. *Interpretation of the oriental sunspot record*

In attempting to assess the relevance of the oriental record of sunspot sightings to the study
of long-term trends in solar variability it is helpful to consider the changing frequency with
which some other solar phenomenon is reported in East Asian history. Solar haloes, although

[112]

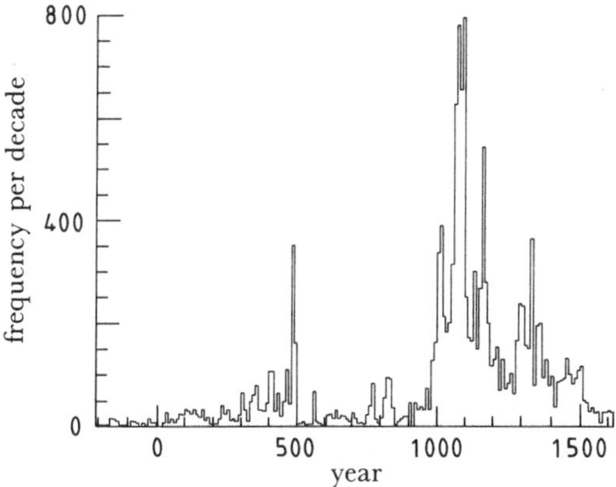

FIGURE 6. Decade distribution of astronomical observations of all kinds recorded
in China: 200 B.C. to A.D. 1610. (After Yau 1988a.)

purely atmospheric phenomena, are often recorded in the same sections of dynastic histories as sunspots. Neither phenomenon was, of course, understood by the oriental astronomers and both are often listed under the heading of 'solar changes'. Haloes provide a particularly useful comparison because they are reported with comparable frequency to sunspots.

Figure 7 is a plot of the decade-by-decade frequency of solar haloes recorded in Chinese history from the beginning of the Christian era to A.D. 1610. This is taken from the paper by Yau (1988b). The diagram reveals marked variation in the numbers of haloes noted per

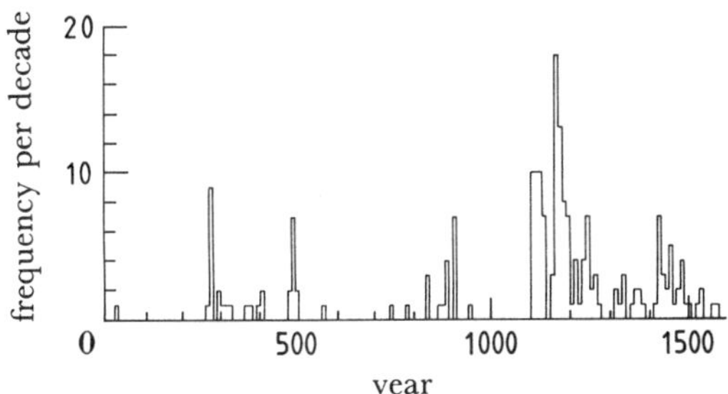

FIGURE 7. Decade distribution of solar haloes recorded in Chinese history from the beginning
of the Christian era to A.D. 1649. (After Yau 1988b.)

decade, ranging from zero for lengthy periods to almost 20. Systematic observation of the sky would be expected to reveal many more haloes and once again we have a very incomplete record. Much of the pattern visible in figure 7 is thus likely to be of artificial rather than atmospheric origin, reflecting both changes in attitudes to celestial portents and varying degrees of preservation of observer's reports. Many of the main features of figure 7 show a fairly good correlation with those of figure 4, peaks occurring around A.D. 400, 1200 and 1400, and troughs around A.D. 700, 1000 and 1300. This suggests that many of the long-term variations present in the sunspot record are spurious, as implied by much of the discussion in the earlier part of this section.

4. CONCLUSIONS

Two main conclusions may be deduced from the foregoing investigation of sunspot observations. (i) In the case of the more recent data (from A.D. 1750 and especially since 1818) important results can be obtained regarding past solar variability. The roughly decadal cycle of solar activity can be mapped on a fairly systematic basis throughout the whole of the last 240 years. During this same interval, long-term trends on the centennial timescale (the so-called Gleissberg cycle) are also apparent. (ii) The solar cycle can in fact be traced as far back as about A.D. 1715 and it is possible to detect its existence even during the pretelescopic period. However, at any time anterior to about 1715, evidence for the existence of long-term trends in solar activity must be regarded as ambiguous. Studies of solar activity before about A.D. 1715 are thus best based on proxy records such as ^{14}C in tree-rings or ^{10}Be in ice cores.

Although many of the arguments presented in this paper might appear to be negative, the aim throughout has been to attempt an objective interpretation of the available sunspot data. It is unfortunate that the excellent series of observations available in more recent centuries is not preceded by data of comparable quality.

The financial support of the SERC and the University of Durham is gratefully acknowledged.

REFERENCES

Chen Mei-dong & Dai Nian-zu 1982 *Stud. Hist. nat. Sci.* **1**, 227.

Clark, D. H. & Stephenson, F. R. 1978 *Q. Jl R. astr. Soc.* **19**, 387.

Eddy, J. A. 1976 *Science, Wash.* **192**, 1189.

Eddy, J. A. 1988 In *Secular solar and geomagnetic variations in the last 10 000 years* (ed. F. R. Stephenson & A. W. Wolfendale), pp. 1–23. Dordrecht: Kluwer.

Eddy, J. A., Stephenson, F. R. & Yau, K. K. C. 1989 *Q. Jl R. astr. Soc.* **30**, 65.

Goldstein, B. R. 1969 *Centaurus* **14**, 49.

Kanda Shigeru 1933 *Proc. imp. Acad. Japan* **9**, 23.

Link, F. 1962 *Geofyz. Sb.* **10**, 297.

McKinnon, J. 1987 *World Data Center A for Solar–Terrestrial Physics, Boulder, Colorado. Report UAG-95.*

Meeus, J. 1958 *J. Br. astr. Ass.* **68**, 98.

Morrison, L. V., Lukac, M. R. & Stephenson, F. R. 1981 *Royal Greenwich Observatory Bull.* 186.

Mossman, J. E. 1989 *Q. Jl R. astr. Soc.* **30**, 59.

Park Seong Rae 1977 Portents and politics in early Yi Korea: 1392–1519. Ph.D. thesis, University of Hawaii, U.S.A.

Sachs, A. J. & Hunger, H. 1988 *Astronomical diaries and related texts from Babylonia*, vol. I. Wien: Osterreichischen Akademie der Wissenschaften.

Siscoe, G. L. 1980 *Rev. Geophys. Space Phys.* **18**, 647.

Stephenson, F. R. 1988 In *Solar–terrestrial relationships and the Earth environment in the last millennia* (ed. G. Cini Castagnoli), pp. 133–150. Amsterdam: North-Holland.

Stephenson, F. R. 1989 (In preparation.)

Stephenson, F. R. & Yau, K. K. C. 1986 *J. Br. interplanet. Soc.* **38**, 195.

Tarling, D. H. 1988 In *Secular solar and geomagnetic variations in the last 10,000 years* (ed. F. R. Stephenson & A. W. Wolfendale), pp. 349–365. Dordrecht: Kluwer.

Vyssotsky, A. N. 1949 *Meddn Lunds astr. Obs.* Historical Papers, no. 22.

Waldmeier, M. 1961 *The sunspot-activity in the years 1610–1960.* Zurich: Schultess.

Weiss, N. O. 1988 In *Secular solar and geomagnetic variations in the last 10 000 years* (ed. F. R. Stephenson & A. W. Wolfendale), pp. 69–78. Dordrecht: Kluwer.

Willis, D. M., Doidge, C. M., Hapgood, M. A., Yau, K. K. C. & Stephenson, F. R. 1988 In *Secular solar and geomagnetic variations in the last 10 000 years* (ed. F. R. Stephenson & A. W. Wolfendale), pp. 187–202.

Wittmann, A. D. & Xu Zhentao 1987 *Astron. Astrophys. Suppl. Ser.* **70**, 83.

Wittmann, A. D. & Xu Zhentao 1988 In *Secular solar and geomagnetic variations in the last 10 000 years* (ed. F. R. Stephenson & A. W. Wolfendale), pp. 131–139. Dordrecht: Kluwer.

Yau, K. K. C. 1988*a* An investigation of some contemporary problems in astronomy and astrophysics by way of early astronomical records. Ph.D. thesis, University of Durham, U.K.

Yau, K. K. C. 1988*b* In *Secular solar and geomagnetic variations in the last 10 000 years* (ed. F. R. Stephenson & A. W. Wolfendale), pp. 161–185. Dordrecht: Kluwer.

Yau, K. K. C. & Stephenson, F. R. 1988 *Q. Jl R. astr. Soc.* **29**, 175.

Yunnan Observatory 1976 *Acta astr. sin.* **17**, 217. (English trans. *Chinese Astr.* **1**, 347.).

Phil. Trans. R. Soc. Lond. A **330**, 513–515 (1990)

Printed in Great Britain

513

Solar observations in ancient China and solar variability

By Xu Zhentao

Purple Mountain Observatory, Academia Sinica, Nanjing, People's Republic of China

In this paper I review both the history of solar observations in ancient China and recent researches on solar variability. The paper consists of three parts. In the first part I describe Sun worship and the early observations of solar phenomena. In the second part I concentrate on sunspot observations and improving the catalogue of naked-eye sunspot records. In the third part I discuss long-term variations of solar activity by using historical sunspot records over 2000 years. The 210-year cycle, which has the largest significance in the power spectrum, may have an important influence on the forecasting of the next cycle (no. 22).

Solar observations in ancient China

Many archaeological materials have shown that in ancient China Sun worship was in vogue and there were daily sacrificial rites for welcoming and seeing off the Sun (see, for example, Xu *et al.* 1985). Clear evidence is found on bone inscriptions of the Shang Dynasty (*ca.* 1500–1050 B.C.). For example: *Yi Si Bu Wang Bin Ri* (Yi Si day, the King received as guest the rising Sun). *Chu Ru Ri Sui San Niu* (sacrifices were offered to the rising and setting Sun and three oxen were killed).

According to the textual researches of several palaeographers, the words *Bin Ri*, *Chu Ri* and *Ru Ri* were all sacrificial rites to the Sun.

The daily worshipping of the Sun certainly caused spontaneous observations of solar phenomena. In fact, from bone inscriptions it is known that the ancient Chinese had already recorded at least four solar phenomena. They are:

(1) *Ri Hui* (the Sun was dark and gloomy);
(2) *Ri Yun* (solar halo);
(3) *Ri Zhi* (no definite explanation);
(4) *Ri Shi* (solar eclipses).

In addition to this evidence, many ancient works of art related to solar phenomena have recently been unearthed in China (see, for example, Xu 1987). Most of them have shown the same solar phenomenon that is known as *Ri Zhong Wu* (a crow – or a three-legged crow – was within the Sun). In fact, they are artistic representations of sunspot phenomena.

Sunspot observations and systematic records

As far as is known, the earliest sunspot observations should be identified by the expression *Ri Zhi* as found in the above mentioned bone inscriptions. This word has been interpreted in different ways, but an explanation in terms of sunspots seems to be most reasonable. In a paper that will be published elsewhere I shall discuss this problem in detail. Sunspot observations were mentioned in one of the famous Chinese classics, the *Book of changes*. In the hexagram of *Feng* ('abundance') there were two specialized terms: *Ri Zhong Jian Dou* ('the *Dou*, which is a measuring rice tool, was seen within the Sun') and *Ri Zhong Jian Mei* ('a star was seen within the Sun'). It has been established that the two terms essentially mean that sunspots were seen within the Sun (see, for example, Xu 1979). These same expressions have been used for recording sunspot phenomena until the beginning of the Qing dynasty (A.D. 1644).

Systematic recording of sunspot phenomena began from the Han dynasty (206 B.C. to A.D. 220), because just at that time the Chinese official histories started to be compiled. A well-known event is the sunspot record of 10 May 28 B.C. The text states that 'On Ji Wei day the Sun was yellow at its rising and a black vapour as large as a coin was observed at its centre'. After the Han dynasty there were some sunspot records in each dynastic history. In the encyclopaedia *Tu Shu Ji Cheng* (*Imperial Encyclopaedia*) of the Qing dynasty, the previous sunspot records were gathered in a special chapter. To trace sunspot activity back in time, several authors have compiled a limited amount of sunspot records and produced catalogues of sunspot observations. The most famous of these works is Kanda's catalogue. It contains 142 items from Chinese histories with a supplement of some records from Korean and Japanese official histories (see, for example, Kanda 1932). However, there are two shortcomings in these researches. Firstly, it is now known that there are at least seven separate specialized terms for recording sunspot phenomena sighted by the naked eye in ancient times (see, for example, Xu 1990). Because all of these terms were not recognized until recently, some sunspot records were overlooked or misinterpreted. Secondly, apart from the official histories there are also numerous local topographies called *Fang Zhi* ('local gazettes'). These are important supplements to the official histories and equivalent to the latter in the reliability of recording events. The *Fang Zhi* were ignored until Xu & Jiang (1982) explained the importance of seventeenth-century records. Using the above mentioned sources, I have reconstructed a new naked-eye sunspot catalogue that assembled both Western and Oriental records and added a substantial amount of new data (see, for instance, Wittmann & Xu 1987).

LONG-TERM VARIATIONS OF SOLAR ACTIVITY

The variability of solar activity before A.D. 1700 has for some time been an interesting problem. During this historical period, Chinese sunspot records were direct reflections of solar activity. Some authors have therefore used them to analyse the periodicity of solar activity. However, on the one hand, their sunspot catalogues were incomplete, and on the other hand the statistical weight of different sunspot records was not considered.

To overcome these shortcomings, I used our essentially complete sunspot catalogue, mentioned in the above section. Then according to descriptive differences in sunspot size, number and sighting duration I assigned weights for each sunspot record. Thus I transformed descriptive sunspot records into a numerical time series. From the time series I obtained its power spectrum, details of which are summarized in table 1 (see, for example, Xu *et al.* 1988). These results enable the following conclusions to be deduced.

1. The 10.62-year period, which actually is the mean periodicity of Schwabe's cycle, is quite significant during the past two millenia, even though the ancient sunspot record is not continuous. This means that Schwabe's cycle is a basic regularity of solar activity and such activity has never ceased.

2. The 212-year period is most significant. Recent researches have also shown that a 210-year cycle of solar activity can be obtained from analysing some astrophysical and geophysical

[116]

TABLE 1. DETAILS OF THE POWER SPECTRUM

frequency cycles per year	period/years	phase	amplitude
0.0047	212.77	0.0267	0.2555
0.0038	263.16	1.3251	0.2448
0.0159	62.89	−0.2926	0.1352
0.0024	416.67	−0.7763	0.1279
0.0942	10.62	−1.5087	0.1156
0.1506	6.64	−0.8893	0.0858
0.0090	110.10	−0.0031	0.0856
0.0114	86.96	0.5784	0.0694

data (see, for example, Chistyakov 1986). Hence the 210-year cycle may be a new important regularity of solar activity. If so, it may give valuable guidance in forecasting the level of the next cycle. It seems possible that the maximum sunspot number of the forthcoming no. 22 cycle will be substantially more than 100.

REFERENCES

Chistyakov, V. F. 1986 *Sol. Data* **6**, 88–95.
Kanda, S. 1932 *Annls Obs. Tokyo* **5**, 1–18.
Wittmann, A. D. & Xu Zhentao 1987 *Astron. Astrophys. Suppl. Ser.* **70**, 83–94.
Xu Zhentao 1979 *Acta astr. sin.* **20**, 416–418.
Xu Zhentao 1987 *History of oriental astronomy* (ed. G. Swarup, A. K. Bag & K. S. Shukla), pp. 51–56. Cambridge University Press.
Xu Zhentao 1990 *Chinese Sci.* (In the press.)
Xu Zhentao, Chen Bing & Jiang Yiaotiao 1988 *Vistas Astr.* **31**, 119–122.
Xu Zhentao & Jiang Yiaotiao 1982 *Chin. Astron. Astrophys.* **2**, 84–90.
Xu Zhentao, Zhang Peiyu & Lu Yang 1985 *Proc. Kunming Workshop Solar Phys. Interplan. Travel Phenomena*, pp. 480–487.

Phil. Trans. R. Soc. Lond. A **330**, 517–527 (1990)

Printed in Great Britain

Relevance of medieval, Egyptian and American dates to the study of climatic and radiocarbon variability

By R. Berger

*Institute of Geophysics and Planetary Physics, Departments of Geography and Anthropology,
Interdisciplinary Archaeology Program, University of California, Los Angeles,
California 90024, U.S.A.*

Basic radiocarbon dating and dendrochronology have been combined to yield calibrated dates that are more accurate than conventional radiocarbon dates. This has been shown to be true for medieval and Egyptian dynastic dating. Because radiocarbon is a cosmogenically produced radioisotope, heliomagnetic and geomagnetic fields play a major role in its synthesis in the Earth's upper atmosphere. Inasmuch as a calibrated radiocarbon record exists for nearly 10000 years, we now seem to possess in the short-time variations of the production rate a history of solar activity expressed via heliomagnetic fields carried by the solar wind. In turn, solar activity has a controlling effect on climate on Earth within modifications provided by the complex interactions of the atmosphere–Earth–ocean system. Both radiocarbon measurements and other empirical research methods agree on variations of climate during historically more recent periods on Earth. This leads to the suggestion that the radiocarbon calibration curve may be also a significant indicator or tracer for climatic changes for the Holocene or the Neolithic–Mesolithic.

Introduction

Historical dates and the classic chronological frameworks derived from them have been based typically on the interpretation of written records of known languages. However, with the advent of tree-ring dating chronologies could be extended much further into the past, covering now most of the Holocene or Neolithic–Mesolithic in certain parts of the world where dendrochronologic master chronologies were applicable. Careful comparison of historical and dendro-dates confirm their near equivalence thus lending confidence to this biological method based on seasonality.

Dendrochronology not only yields dates synonymous with historical ones, but provides also known-age wood samples that can be radiocarbon-dated. These samples are equivalent to historically well-dated organic materials such as timbers from known-age buildings or straw used in the manufacture of Egyptian mud-bricks. Hence radiocarbon dates can be calibrated to match historical time by either dendrochronologically accurately dated specimen or those whose historic age is well documented.

Today a long and continuous tree-ring calibration for radiocarbon dates is available beginning with the pioneering work of Suess (1965). Since then this has been refined and expanded for most of the past 10000 years by many researchers whose work is collected in the Calibration Issue of *Radiocarbon* (Kra 1986; Becker & Kromer 1986; De Jong *et al.* 1986; Kromer *et al.* 1986; Linick *et al.* 1986; Pearson *et al.* 1986; Pearson & Stuiver 1986; Stuiver & Becker 1986; Stuiver *et al.* 1986; Stuiver & Pearson 1986; Stuiver *et al.* 1986; Stuiver &

Reimer 1986; Vogel *et al.* 1986). Aside from the chronologic aspects this sequence provides as well for a continuous record of secular variations of cosmogenic radiocarbon production.

It has been pointed out by several authors such as Stuiver (1961), Suess (1968) and Damon (1968) that there exists a complex correlation between (1) radiocarbon production trends in the atmosphere based on the interaction of cosmic rays with atmospheric nitrogen, (2) modulation of the cosmic ray flux by helio- and geomagnetic fields and (3) solar activity that controls both heliomagnetic activity and in the end terrestrial climates. In essence, higher solar energy output at the time of sunspot maxima as reported on recently by Willson & Hudson (1988) translates to stronger heliomagnetic fields that interfere more with access of cosmic rays into the Earth's atmosphere. This scenario would result in lower radiocarbon production with concomitant lower levels of incorporation into the biosphere such as tree-rings. At the same time increased insolation of the Earth's surface would result in a warming trend. Conversely, lower solar energy levels would tend to increase radiocarbon nucleosynthesis as weaker heliomagnetic fields deflect fewer cosmic rays while the Earth as a whole experiences cooling. These reaction steps do not take into account the ameliorating effects of the oceans and other reservoirs that significantly affect these relations, but whose joint action do not seem to completely mask them. Thus radiocarbon measurements not only yield dates but also seem to provide climatic information.

RADIOCARBON DATES FROM THE MIDDLE AGES

Before the extensive isotopic tree-ring studies summarized in the Calibration Issue of *Radiocarbon*, the veracity of Suess's (1965) secular variations was checked by dating known-age buildings of medieval vintage located principally in England and France as discussed by Berger (1970a, 1985) and Horn (1970). Indeed their correct historical ages could only be calculated from radiocarbon ages if Suess's calibration was used.

Experimentally, all these timber samples were first analysed for their position within the construction members relative to the outer-most tree-ring to determine the felling date that starts the radiocarbon clock. In the laboratory, 10–15 g of sample were decontaminated with

TABLE 1. LIST OF KNOWN-AGE MEASUREMENTS FROM THE MIDDLE AGES

UCLA no.	location	in-tree correction/years	$\delta^{13}C$, PDB (‰)	^{14}C age years	historical age (A.D.)	calibrated age[a] (A.D.)
1307	Arpajon	40	−23.83	515±30	1450–1470	1450
1316	Enstone	40	−24.61	630±50	1382	1390
1048	Gt Coxwell	40	−24.92	870±40	1250	1250
1049	Gt Coxwell	15	−25.20	750±30	1250	1245
1309	Maubuisson	10	−22.58	860±40	1236	1050–1220
572	Mereville	20	none	490±40	1456–1472	1440
1304	Mereville	30	−24.69	400±50	1456–1472	1460 or 1600
1312	Milly	none	−22.80	420±45	1479	1450
1313B	Parcay–Meslay	10	−24.52	655±30	1211–1227	1250
570	Parcay–Meslay	none	none	735±50	1211–1227	1225–1250
1310	Questembert	40	−26.02	330±30	1675	1500, 1625–1670
1303	Richelieu	30	−24.54	380±50	1631–1640	1480 or 1600
1306	Sully	100+?	−23.04	930±30	1363	1300
1308	Troussures	none	−24.42	400±45	1609	1450 or 1600
1214	Buddha	none	none	820±50	13th century	13th century

[a] For calibration the Suess relation (1965) was used.

dilute, cold hydrochloric acid to remove all inorganic carbonate dust or lichen and moss residue from leaking roofs. After washing and drying samples were burnt in a stream of pure oxygen to carbon dioxide. This gas was then purified by treatment with aqueous silver nitrate to precipitate halogens, hot copper oxide to oxidize any carbon monoxide to dioxide and chromic acid solutions for drying and elimination of impurities. Electronegative contaminants such as traces of oxygen were removed by repeated passage of the gas over elementary copper at 600 °C. The resultant pure carbon dioxide was then assayed after radon decay by repeated counting for at least 1000 min in a 7.5 l proportional counter whose accuracy in this time range is ± 25 years. In addition, stable carbon isotope ratios were determined and the standard correction applied to exclude fractionation effects.

The medieval dates obtained are listed in table 1 and show very good correlation between historical and experimental ages once a tree-ring calibration is applied. In some cases alternate radiocarbon ages or age ranges are due to the nature of the fluctuations in the radiocarbon production rate as represented by calibration curves. Thereafter the success of this dating approach was tried on many buildings of uncertain age in continental Europe, England and Ireland as discussed by Berger (1970a, 1990) and Horn (1970).

RADIOCARBON DATING OF EGYPTIAN DYNASTIES

Inasmuch as these medieval samples provided verification of the calibration proposed by Suess for the past millennium or so, radiocarbon dating of Egyptian dynastic specimen offered a check by the earliest and best understood historical chronology of subsequent extensions of

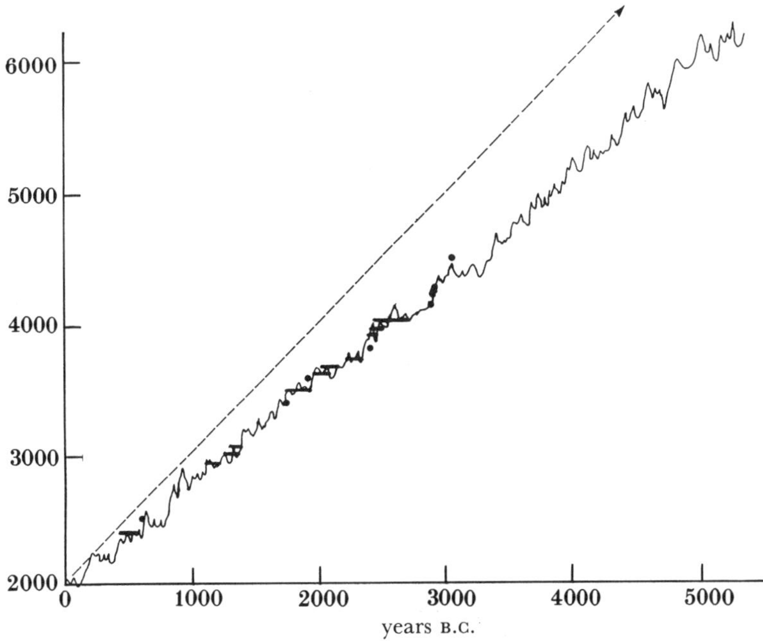

FIGURE 1. Calibration curve by Suess (1979) showing short-term deviations caused by heliomagnetic activity and long-term change due to geomagnetic effects. Calibrated radiocarbon dates from Egyptian dynastic sites are plotted as lines indicating dating ranges or circles that are specific dates. ----, Relation based on constant atmospheric ^{14}C levels. (After Berger (1985).)

TABLE 2. LIST OF KNOWN-AGE MEASUREMENTS FROM EGYPT

UCLA no.	location	dynasty	material	δ^{13}C, PDB (‰)	^{14}C age years	approximate historical age (years B.C.)	calibrated age[a] (years B.C.)
1200	Tomb 3357 Sakkara	I	matting	−22.65	4500±60	3100–3000	3030
1201	Tomb 3503 Sakkara	I	matting	−22.36	4290±60	3100–3000	2900
1202	Tomb 3035 Sakkara	I	matting	−21.95	4235±60	3000	2900
1203	Tomb 3505 Sakkara	I	matting	−22.71	4140±60	2890	2900 or 2600
739	Mastaba 2050 Tarkhan	I	linen	none	4265±80	3100–2890	2900
1204	Tomb 3046 Sakkara	II	matting	−23.05	4190±60	2890–2686	2900 or 2600
1205	Tomb 3030 Sakkara	III	matting	−23.36	4055±60	2670	2700–2400
1206	Tomb 3510 Sakkara	III	wood	−25.36	3965±60	2686–2613	2700–2400
1207	Tomb 3075–5 Sakkara	III	matting	−23.12	4050±60	2686–2613	2700–2400
1208	Tomb 3508/10	IV?	linen	−25.62	4015±60	2613–2494	2700–2400
1389	Cheops boat Gizah	IV	grass rope	−11.78	4210±60	2613–2494	2600 or 2900
928	Gizah 2220 Pit B, Gizah	V	linen	none	4120±60	2494–2345	2600 or 2900
1403	Mastaba of Haishetef, Zoser encl.	V- Late I. Inter P.	matting	−22.79	3935±60	2494–2173	2550–2400
1387	Tomb of Mereruka, Teti-Sakkara	VI	matting	−22.47	3860±60	2360	2450–2350
1388	Teti Sakkara	VI	wood	−22.13	3950±60	2360	2550–2400
1413	Gebelin	XI	wood	−25.59	3770±60	2133–1991	2400–2250
1211	Tomb 386 Thebes	XI	wood	−25.36	3500±60	2133–1991	1900–1750
1398	Tomb 386-T	XI–XII	linen	−25.49	3330±60	1991–1786	1700
1399	Tomb 386-I	XI–XII	charcoal	−26.52	3615±60	1991–1786	2100–1950
1400	Tomb 386-T	XI–XII	wood	−25.25	3700±60	1991–1786	2200–2000
1212	Sesostris II Pyramid, El-Lahun	XII	matting	−10.00	3500±60	1991–1786	1900–1750
900	Sesostris III Funerary boat	XII	wood	−25.10	3640±60	1991–1786	1800
1390	Ramesses II Funerary temple	XIX	matting	−11.42	3075±60	1320–1200	1400–1300
1393	Tomb 158 Thebes	XIX–XX	matting	−11.27	3060±60	1320–1085	1400–1300
1394	Tomb 158 Thebes	XIX–XX	wood	−24.02	3030±60	1320–1085	1400–1300
1395	Tomb 283 Thebes	XIX–XX	wood	−25.19	2880±60	1320–1085	1200–1100
1401	Tomb 386 Thebes	XXVI	wood	−24.96	2515±50	664–525	600
1391	Tomb 34 Thebes	XXV–XXVI	matting	−11.15	2530±60	751–525	600
1397	Gt Temple of Amun, Karnak	XXX	matting	−11.75	2335±60	380–343	600–400

[a] For tree-ring calibration Suess's (1979) correlations was used.

calibrated dating. The best historically dated and geochemically most-suited samples were collected with the cooperation of I. E. S. Edwards, British Museum, W. B. Emery, University College, London, W. F. Libby, UCLA, and G. T. Martin, Corpus Christi College, Cambridge. Most of the sample material consisted of plants growing only a single or few seasons such as straw, reed or flax.

All samples were decontaminated in this laboratory after treatment with dilute hydrochloric acid followed by sodium hydroxide leaching to remove humic acids from the archaeological environment below ground level. Conversion to carbon dioxide was identical to the method used before. Moreover, the same dating and correction procedures were used. The results of this measurement programme are listed in table 2 below. They were discussed earlier by Berger (1970a, 1985) in the light of the slowly emerging calibration curves.

As evident from this list, calibration of Egyptian radiocarbon dates by Suess's extended correlation and other subsequent calibrations results in dates closely resembling the historical estimates except that there occur occasionally ranges of time or alternate dating possibilities that need to be resolved by other methods.

In general Suess's calibration over the past several thousand years is very similar to those reported on in the Calibration Issue. Yet a unified standard calibration curve needs yet to be devised and agreed upon by the world's radiocarbon research community.

SOLAR ACTIVITY, CLIMATE AND RADIOCARBON VARIATIONS

Recent observations by Willson & Hudson (1988) of solar activity changes during solar cycle 21 with the Active Cavity Radiometer Irradiance Monitor (ACRIM I) on board the *Solar Maximum Mission (SMM)* satellite have provided a virtually continuous record of total solar irradiance. Long-term variations during a downward trend and subsequent upturn of the sunspot cycle suggest a direct correlation between luminosity and sunspot populations. At its peak the Sun was 0.1 % brighter than at its minimum. A model of solar luminosity modulation by magnetic activity between 1954 and 1984 as discussed by Lean & Foukal (1988) is successful in matching slow variations of solar irradiance measured simultaneously from the *Nimbus*-7 and *SMM* satellites. In fact this model extending back over three 11-year solar cycles predicts that the Sun is always brighter at activity maximum than at minimum.

The distribution of solar energy radiated is illustrated in figure 2, which shows that ultraviolet, visible and infrared radiation together account for virtually 100 % of solar luminosity. The energy of the corpuscular radiation emanating from the Sun as solar wind amounts to only about 10^{-6} of total solar output. Major differences in temperature of the terrestrial atmosphere during sunspot maximum and minimum are illustrated in figure 3 demonstrating a large effect at altitudes in excess of 100 km and reaching a full 30 % increase above 300 km. Near the surface of the Earth the differences are very small numerically yet large enough to influence climate.

The most comprehensive collection of empirical climatic observations has been assembled by Lamb (1977). Recently a more specialized and very extensive summary of the 'Little Ice Age' has been published by Grove (1988). In 1983 Leona Libby summarized our understanding of what was then known about the interrelation between climate, stable isotopes and historical economic trends (figure 4). La Marche (1974) related the departure of tree-ring widths of bristlecone pine (*Pinus longaeva*) in the White Mountains of California to precipitation and also

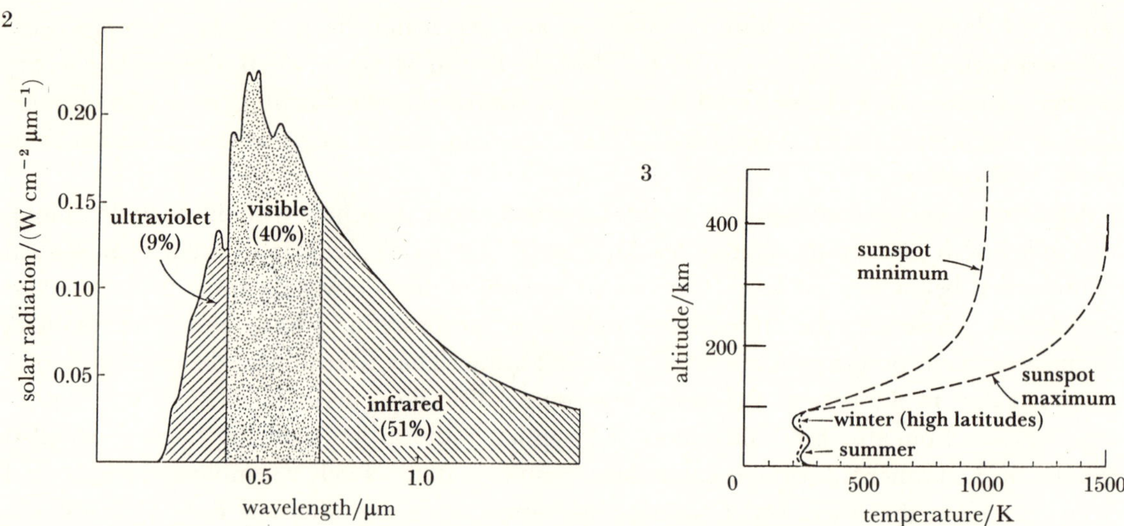

FIGURE 2. Distribution of solar energy between ultraviolet, visible and infrared radiation. (After Glasstone (1965).)
FIGURE 3. Response of the atmosphere to solar activity at sunspot minimum and maximum. (After Glasstone (1965).)

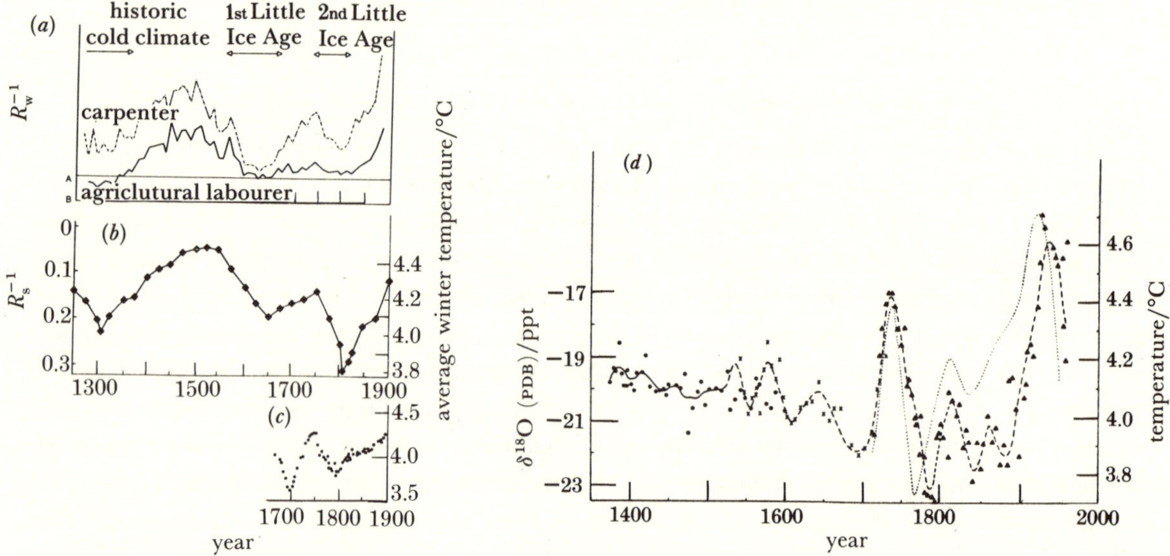

FIGURE 4. Correlation between economic data, climate and oxygen isotopic data for England and Central Europe (after Libby (1983)). (a) The amount of wheat purchaseable in England with the daily wages of a carpenter and a farmhand. (b) The price of wheat in central England (black diamonds). (c) A short curve to A.D. 1650 illustrating the average winter temperature in central England. (d) Oxygen isotopic data for oaks from central Germany. Spessart oak 1: ▲, 7 sample (30 year) running average; – – –, smoothed by eye. Spessart oak 2: ×, 5 sample (30 year) running average. Marburg oak: ——, 5 sample (30 year) running average; ●, 4–7 year sample; · · · · · ·, average January, February and March temperatures in England. All curves show a minimum at A.D. 1800 and further substantial agreement in the latter part of the sixteenth century, periods having been called Little Ice Ages.

temperature anomalies represented by local glacial advances. A broader comparison by Suess (1968) included sunspot numbers, variations of the specific activity of radiocarbon in wood, mean-temperatures in England and winter severity indices for Europe over the past 1000 years (figure 5). A much longer correlation covering several millennia is possible today because a longer tree-ring calibrated radiocarbon record is available coupled with a much more extensive history of glacial advances and retreats. In a world-wide survey Röthlisberger (1986) published a remarkable comparison of glacial responses on nearly every continent. Apparently, glaciers behave very similarly world-wide in their reaction to major climatic changes leading to the belief that they really represent global climatic effects.

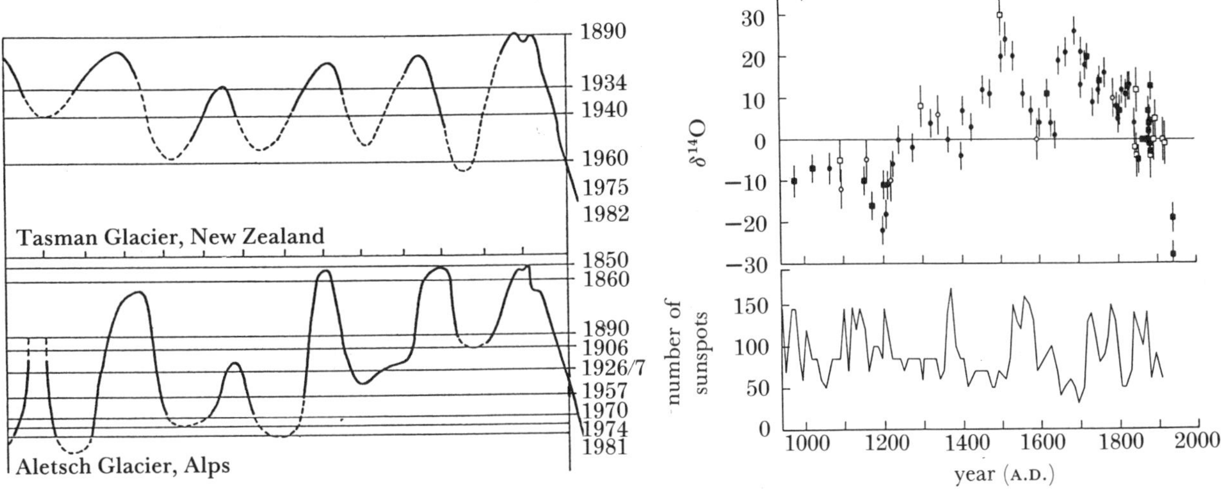

FIGURE 5. Comparison of solar activity expressed by sunspot number (Schove 1955), variations in the radiocarbon levels of wood (Suess 1965) and fluctuations of the Tasman glacier, New Zealand and the Grosser Aletsch glacier, Alps as combined in Grove (1988). Low sunspot number conditions and high radiocarbon production go hand in hand with glacier retreats.

A comparison of these glacial fluctuations with secular changes in cosmogenic radiocarbon production adapted from Neftel *et al.* (1981) is presented below in figure 6. Inspection suggests a sense relation between glacial advances and higher than average radiocarbon production. On the other hand, glacial retreats occur at times when radiocarbon synthesis is inhibited by reduced cosmic ray flux caused by increased solar activity. Thus the operation of the natural processes involved can be characterized as follows: when the Sun is more active it increases insolation on Earth, yet shields it more effectively from cosmic rays by stronger heliomagnetic fields. This leads to a reduction in radiocarbon production and subsequent lower specific activity levels in the terrestrial biosphere. Conversely, during periods of minimum solar activity, a cooler Earth–atmosphere system is exposed to higher concentrations of radiocarbon. As a consequence radiocarbon production trends seem to be able to serve as an indicator or tracer of climatic variability on the Earth. Because the tree-ring calibration record is continuously being extended toward the Holocene–Pleistocene boundary we have now already a climatic indicator chart covering most of the Neolithic and Mesolithic. In fact it should be possible to test its validity if climatic excursions are known to exist in certain prehistoric time intervals.

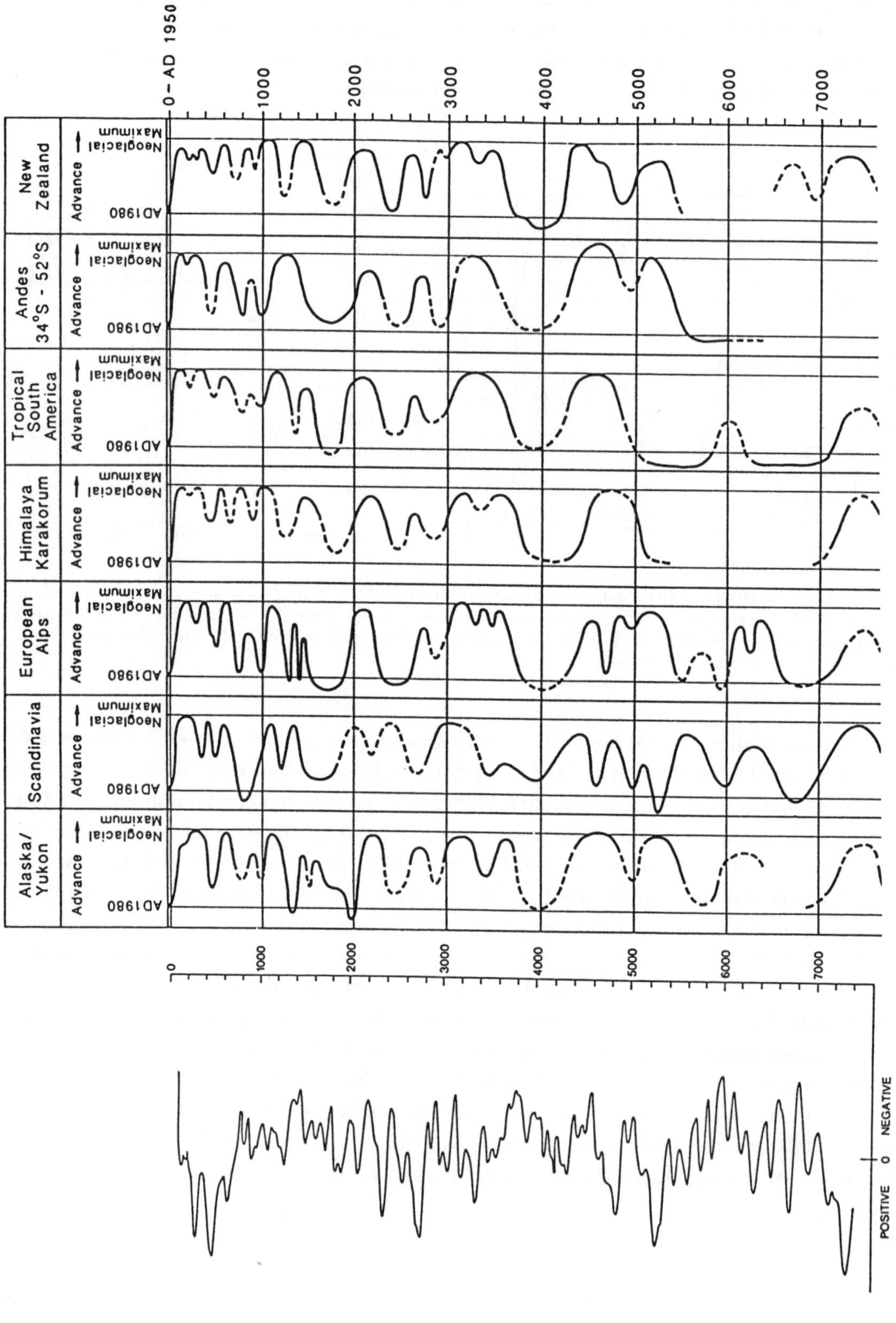

Figure 6. Comparison of world-wide glacial responses (Röthlisberger 1986) with radiocarbon production variations (Neftel *et al.* 1981). Positive radiocarbon production occurs at times of glacial advance signalling a cooling climate. Negative radiocarbon production trends indicates greater solar activity and glacial retreat, i.e. warming climate.

RADIOCARBON DATES FROM THE AMERICAN SOUTHWEST

This region is one of the most intensively investigated archaeologically in the United States. There are thousands of radiocarbon dates available, typically published in *Radiocarbon*, that concern human and climatic activity. The first study to test the use of the radiocarbon calibration as a climatic tracer involves timberlines. They respond to changes in climate by moving up or down mountains to accommodate the best environment in which particular tree species will grow. Typically, a warming trend will cause trees to move higher on mountain ranges in terms of new tree generations they produce. Cooling will induce the reverse effect, the upper tree line will gradually recede downward until a new environmental–biological equilibrium has been established.

The first study of a timberline advance to higher elevations in the White Mountains of California and the nearby Snake Range, Nevada, in itself signals a warming trend that has been observed in the southwest and called the Altithermal. A discussion of post-glacial climate and archaeology is found in a publication by Baumhoff & Heizer (1965), which addresses specifically the American desert West. Samples for this timberline study were collected by La Marche & Mooney (1967) and most radiocarbon dated at UCLA. All specimens were treated for the removal of contamination as before.

The data (table 3) show a subalpine forest advance of bristlecone pines upward of the order of 120–150 m. When the probable dates for the establishment for these trees are inspected, they often fall into date ranges because the basic radiocarbon dates need calibration, which then results in time ranges. However, when these ranges are compared to the isotopic curve in figure 5, and fitted into the glacier reaction curves, the probable date of first establishment of the trees can perhaps be further refined. In general the data agree well with the concept of the Altithermal, Climatic Optimum or Hysithermal first suggested by Antevs (1948).

TABLE 3. AGE AND ALTITUDE OF BRISTLECONE PINES

UCLA no.	location	altitude sealevel/m	above tree-line/m	in-tree correct/years	^{14}C age years	calibration age[a]/(years BP)	tree start (years BP)
1070F	White Mtns	3480	120	2260	840±80	850	3110
1070B	White Mtns	2490	130	660	2540±80	2550–2850	3210–3510
1070G	White Mtns	3510	150	735	3565±80	3750–4050	4485–4785
1070A	White Mtns	3360	120	507	4000±80	4400–4850	4907–5357
1070C	Snake Range	3540	120	138	2350±80	2400–2750	2538–2888
1070D	Snake Range	3540	120	809	2240±80	2150–2350	2559–3159
LJ-1336	Snake Range	3540	120	87	3930±80	4200–4450	4287–4587

[a] For calibration the Suess relation (1965) was used because this is also the base for the correlation in figure 4. LJ-1336 was determined in the La Jolla laboratory of Suess.

A later and more extensive study recognizes upward tree-line motion by as much as +200 m coupled with a temperature increase of about 2 °C above today's July average as noted by La Marche (1973). The same study also notes a dramatic decline of the timberline at the beginning of the Little Ice Age around A.D. 1500.

Another study covers the entire Mohave Desert, which is principally located in California. This work is based on the radiocarbon dating of packrat (Neotoma) middens that are made from collections of plants in the limited foraging area of these animals. Botanical analysis permits determination of the plant species used and therefore archaeoclimatic evaluation as discussed by Wells & Berger (1967). A large geographic area was analysed confirming the existence of woodland where today remain only disjunct stands of trees in the higher elevations separated below by desert. More than 20 UCLA radiocarbon dates ranging from about 8000 to more than 40000 years indicate how xerophilous juniper-dominated woodland descended some 600 m below the present lower limit of woodland during the waning millennia of the Pleistocene. The overall data show a slow spreading of the desert from what is now Sonora northward into California and adjacent states. Because, unfortunately, the radiocarbon calibration curves do not yet reach significantly into this time range, no correlations are possible. It goes without saying that the transition period from the Pleistocene to the Holocene is one of the most interesting in terms of the isotopic response to a major change in insolation and attendant environmental effects.

I very much appreciate the efforts of S. K. Runcorn, F.R.S., in organizing this meeting. Moreover, I have benefitted greatly from discussions with D. O. Gough, G. M. Grove, the late Leona Libby, J. A. Matthews, C. T. Russell, D. D. Sentman, C. P. Sonett and H. E. Suess. Also I would like to thank the staff of the Royal Society for their support of this conference.

This is publication No. 3240 of the Institute of Geophysics and Planetary Physics, University of California, Los Angeles.

REFERENCES

Antevs, E. 1948 *Univ. Utah Bull.* **38**, 168.
Baumhoff, M. A. & Heizer, R. F. 1965 In *The quaternary of the United States* (ed. H. E. Wright Jr & D. G. Frey), pp. 697–707. Princeton University Press.
Becker, B. & Kromer, B. 1986 *Radiocarbon* **28**, 961.
Berger, R. 1970a (ed.) *Scientific methods in medieval archaeology.* Berkeley, Los Angeles and London: University of California Press.
Berger, R. 1970b *Phil. Trans. R. Soc. Lond.* A **269**, 23.
Berger, R. 1985 *Meteoritics* **20**, 395.
Berger, R. 1990 PACT. *Early Medieval Irish buildings: radiocarbon dating by Mortor.* (In the press.)
Damon, P. E. 1968 *Meteorological Monographs* **8**, 151.
De Jong, A. F. M., Becker, B. & Mook, W. G. 1986 *Radiocarbon* **28**, 939.
Glasstone, S. 1965 *Sourcebook on the space sciences.* Princeton: Van Nostrand.
Grove, J. M. 1988 *The Little Ice Age.* New York: Methuen.
Horn, W. 1970 In *Scientific methods in medieval archaeology* (ed. R. Berger), pp. 23–87. Berkeley, Los Angeles, London: University of California Press.
Kra, R. S. (ed.) 1986 *Radiocarbon* **28** (Calibration Issue).
Kromer, B., Rhein, M., Bruns, M., Schoch-Fischer, H., Münnich, K. O., Stuiver, M. & Becker, B. 1986 *Radiocarbon* **28**, 954.
Lamb, H. H. 1977 *Climate: present, past and future.* New York: Methuen.
La Marche, V. C. Jr 1973 *Quaternary Res.* **3**, 632.
La Marche, V. C. Jr 1974 *Science, Wash.* **183**, 1043.
La Marche, V. C. Jr & Mooney, H. A. 1967 *Nature, Lond.* **213**, 980.
Lean, J. & Foukal, P. 1988 *Science, Wash.* **240**, 906.
Libby, L. M. 1983 *Past climates.* Austin: University of Texas Press.
Linick, T. W., Long, A., Damon, P. & Ferguson, C. W. 1986 *Radiocarbon* **28**, 943.
Neftel, A., Oeschger, H. & Suess, H. E. 1981 *Earth planet. Sci. Lett.* **56**, 127.
Pearson, G. W., Pilcher, J. R., Baillie, M. G. L., Corbett, D. M. & Qua, F. 1986 *Radiocarbon* **28**, 911.
Pearson, G. W. & Stuiver, M. 1986 *Radiocarbon* **28**, 839.
Röthlisberger, F. 1986 *10000 Jahre Gletschergeschichte der Erde.* Aarau: Verlag Sauerlande.
Schove, D. J. 1955 *J. geophys. Res.* **60**, 127.
Stuiver, M. 1961 *J. geophys. Res.* **66**, 273.
Stuiver, M. & Becker, B. 1986 *Radiocarbon* **28**, 863.
Stuiver, M., Kromer, B., Becker, B. & Ferguson, C. W. 1986 *Radiocarbon* **28**, 969.

Stuiver, M. & Pearson, G. W. 1986 *Radiocarbon* **28**, 805.
Stuiver, M., Pearson, G. W. & Braziunas, T. F. 1986 *Radiocarbon* **28**, 980.
Stuiver, M. & Reimer, P. J. 1986 *Radiocarbon* **28**, 1022.
Suess, H. E. 1965 *J. geophys. Res.* **70**, 5938.
Suess, H. E. 1968 *Meteorological Monographs* **8**, 146.
Suess, H. E. 1979 In *Radiocarbon dating* (ed. R. Berger & H. E. Suess), pp. 777–784. Berkeley, Los Angeles and London: University of California Press.
Vogel, J. C., Fuls, A. & Visser, E. 1986 *Radiocarbon* **28**, 935.
Wells, P. V. & Berger, R. 1967 *Science, Wash.* **155**, 1640.
Willson, R. C. & Hudson, H. S. 1988 *Nature, Lond.* **332**, 810.

Discussion

J.-C. Pecker (*Collège de France, Paris, France*). I would like to draw attention to a study by F. Link, a few decades ago, of medieval and antiquity records. This study was very thorough, and has concerned astronomical observations of auroras, novae, planets, etc. It strongly suggests a period of 300–400 years in the atmospheric transparency, this being an indication, according to Link, of a 300–400-year period in the solar activity. I think that one should not evoke that kind of problem without paying due tribute to the pioneering work of the late F. Link.

R. Berger. This is a very interesting comment in the light of recent work by Sonett & Suess (1984) and Sonett (1984), who discuss long-term periods in bristlecone pine ring-widths and atmospheric radiocarbon variations as well as very long solar periods and the radiocarbon record respectively.

Additional references

Sonett, C. P. 1984 *Rev. Geophys. Space Phys.* **22**, 239.
Sonett, C. P. & Suess, H. E. 1984 *Nature, Lond.* **307** 141.

J. M. Grove (*Girton College, Cambridge, U.K.*). The diagram of climatic change for Scandinavia by Ahlmann is now completely out of date. It shows no cool phases during the Holocene (Röthlisberger 1986). There is now extensive evidence that the Little Ice Age was a global phenomenon, this pointing to a fall in global temperature, though with some complication in timing and regional impact (Grove 1988).

R. Berger. The diagram by Ahlmann was shown during the oral presentation as a historical and scientific illustration of interest in the Little Ice Age. In this publication reference is made to Röthlisberger's work and your own massive compilation of data that are much more up to date.

Phil. Trans. R. Soc. Lond. A **330**, 529–541 (1990)
Printed in Great Britain

529

Evolutive spectral analysis of sunspot data over the past 300 years

By A. Berger, J. L. Mélice and I. van der Mersch

*Université Catholique de Louvain, Institut d'Astronomie et de Géophysique Georges Lemaître,
chemin du Cyclotron 2, B-1348 Louvain-la-Neuve, Belgium*

We analyse the series of the Wolf sunspot number in the frequency domain to determine the dimension of the solar cycle system by using the properties of its strange attractor and to study the stability in time of this dimensionality and of the main quasi-periodicities.

The two classical methods of time series analysis, Fourier harmonic and Blackman–Tukey spectral analysis, have been applied first to the series of the annual Wolf sunspot numbers to determine its overall character. To detect stationarity, periodic regression based upon the three most statistically significant quasi-periods and especially a moving form of the maximum entropy spectrum analysis (mesa) have been used. Both analyses show a splitting of the 11-year cycle before 1800, when a ± 55-year cycle is dominant, and a single 11-year and ± 100-year peak after 1800. Moreover, these quasi-periods are very sensitive to the time interval over which the analysis is carried out. The reason is that the sunspot numbers constitute a widely non-stationary process, which therefore implies that Fourier techniques are not useful to predict solar activity and must be used as fitting procedure only.

The minimum cross-entropy method serves to improve the maximum entropy spectrum. With a good *a priori* estimate and data containing a low noise level, this method allows the detection of very close peaks and the refinement of the main frequencies; it does not split nor introduce artificial peaks. The Thomson model was also applied for its superior bias control, its excellent leakage resistance and a better statistical information.

The same methods were then used to study the 22-year magnetic cycle, which is formed by taking into account the change in polarity of the succeeding 11-year cycle. The moving form of mesa confirms the 22-year cycle to be highly stable in contrast to the instability in the period of the 11-year sunspot series. This suggests the importance of working with the more invariant 22-year magnetic series to explain the complex, non-stationary behaviour of the sunspot series and of the solar–terrestrial interactions.

Finally, we tried to see if the system generated by the sunspot data was allowing the existence of an attractor and tried to determine the minimum number of variables necessary to describe this system. It is shown that the dimension of the attractor is highly unstable varying from 2.21 to 4.95 in a quasi-cyclic way.

Introduction

Attempts to analyse the time series of Wolf sunspot numbers have a long history dating back to the work of Yule (1927). With the advent of powerful computers and the construction of modern efficient techniques of data analysis in the frequency domain, many authors have attempted to understand better its spectral structure (Cohen & Lintz 1974; Sonett 1984; Sneyers & Cugnon 1986).

When reviewing all these results, one is puzzled by the lack of predictability of the future sunspot numbers using just statistical treatment of the sunspot numbers alone without reference

[131]

to any external predictor. Moreover, the periods found in these statistical investigations are different. The question that naturally arises, besides recognizing that physical models of solar activity are most probably the key to the problem, are therefore:

(i) are these results significantly different from each other?

(ii) are these differences due to the nature of the techniques used and/or to the instability of the time series itself, all the time intervals investigated being different?

The goal of this study is precisely to try to answer these questions by analysing the spectral characteristics of the sunspot numbers time series over a prescribed interval of time by using six different spectrum models, namely, periodic regression (Bloomfield 1976), harmonic analysis, Blackman–Tukey spectrum analysis (Jenkins & Watts 1968), Thomson (multitaper) spectrum analysis (Thomson 1982), maximum entropy spectrum analysis (Burg 1972) and minimum cross-entropy spectrum analysis (Shore 1981). Our results are not expected to be fundamentally different from others but our approach, comparing the results obtained from different techniques and different intervals of time, aims to clarify the problem.

Spectra of the 1700–1986 sunspot time series

To study the general characteristics of the annual Wolf sunspot numbers, the classical fitting procedure of harmonic analysis was first applied to the observations. Figure 1 shows this harmonic analysis of the whole sunspot series (1700–1986); a total of 27 harmonics (all above the 80% confidence level) are needed to explain 90% of the total variance. The most important peaks appear around 11, 96 and 57 years.

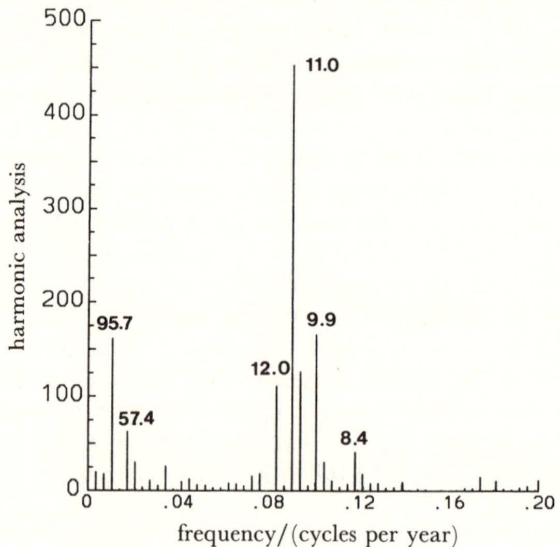

Figure 1. Harmonic analysis of the annual Wolf's sunspot number from 1700 to 1986.

It is interesting to note that these harmonics reach their maxima close to 1957, which contributes to the high sunspot numbers of 1957–58. A periodic regression using the 27 most significant components has also shown a good agreement with the results obtained by others provided the same interval of time is used, which demonstrates that the frequencies found in both models are independent of the techniques.

To improve the statistical behaviour and stability of the above detected spectral estimates, the same data-set was then analysed by using 'Blackman–Tukey' (BT) spectrum analysis (figure 2) for a maximum lag of 96 years, which corresponds to one-third of the total length of the series. As the first three computed autocorrelation coefficients of the series show that the analysed record does not behave as a Markov process, a 'white noise' continuum was chosen as a null hypothesis. The 97.5 % upper confidence limit at the 0.05 significance level drawn on figure 2 shows that only the 11-year period is significant. In addition, a weak non-significant 100-year peak is also found.

Thomson's 'multitaper' spectrum analysis was next performed to the same series for its high-frequency resolution and its amplitude estimate properties. As seen in figure 3, the significant 11-year period of the BT analysis splits into two peaks with a corresponding period fluctuating between 11 and 10 years. These results compare well with those obtained from the harmonic analysis. They show a clear non-stationarity of the frequency of the 11-year oscillation. In the low-frequency part of the spectrum, the broad 100-year period of the BT analysis splits also in two peaks with a period of respectively 100 and 56 years, again closely related to the harmonics 3 and 5 of the harmonic analysis.

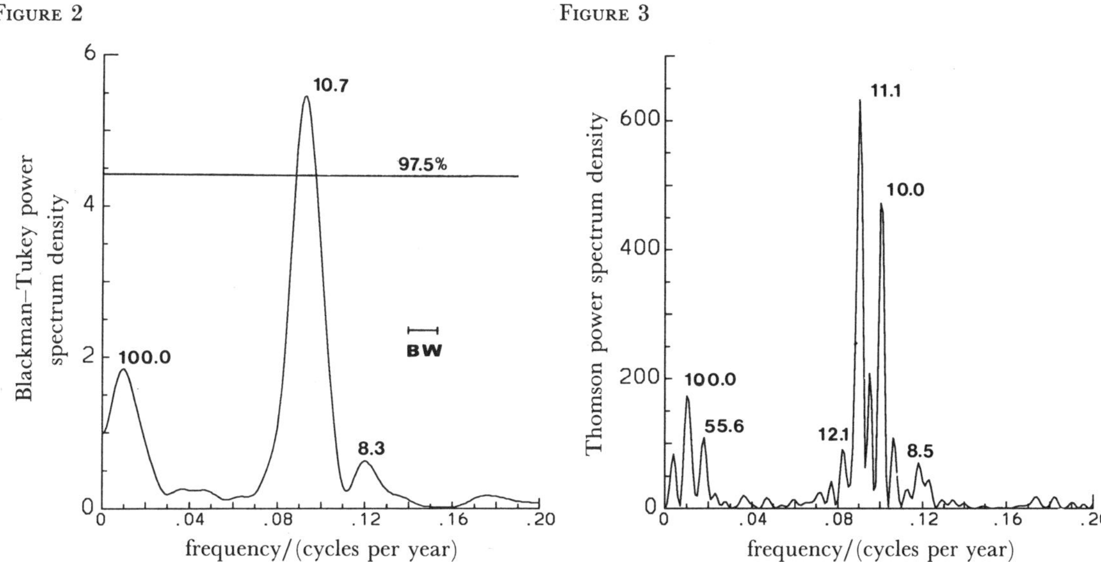

FIGURE 2. Blackman–Tukey spectrum analysis of the annual Wolf's sunspot number from 1700 to 1986. The bandwidth and the 97.5 % upper confidence limit for a 'white noise' continuum are also provided.

FIGURE 3. Thomson 'multitaper' spectrum analysis of the annual Wolf's sunspot number from 1700 to 1986.

To improve the frequency resolution for trying to solve this problem of splitted peaks, the maximum entropy spectrum analysis (MESA) technique was applied. It allows also to investigate the low-frequency part of the spectrum in much greater detail. Remarkably, as in the Thomson analysis, the 11-year oscillation becomes a broad bimodal peak with two main periods of 11.1 and 10 years (figure 4). This definitely indicates that the 11-year quasi-periodicity has slightly changed with time. The 100 and around 52–56-year peaks are also clearly present in the low-frequency part of the spectrum.

Finally, a minimum cross-entropy spectrum analysis performed to the same data-set is presented in figure 5. The advantage of this method compared with MESA is its ability to detect very close spectral peaks with a higher accuracy. The results obtained, compared with those given by Thomson and MESA analyses, confirm definitely the non-stationarity of the 11-year quasi-periodicity.

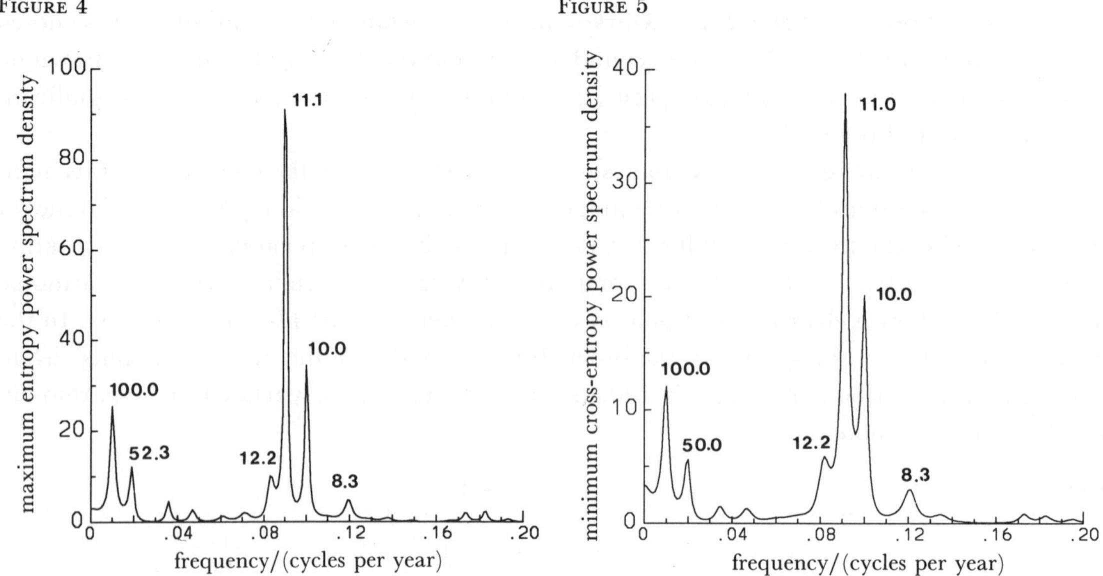

FIGURE 4. Maximum entropy spectrum analysis of the annual Wolf's sunspot number from 1700 to 1986.

FIGURE 5. Minimum cross-entropy spectrum analysis of the annual Wolf's sunspot number from 1700 to 1986.

NON-STATIONARITY OF THE WOLF SUNSPOT NUMBERS

An instability in frequency was thus broadly diagnosed with the above performed spectral analyses. From a simple plot of the Wolf number, it is also obvious that the cycles of activity vary in amplitude, the cycles of higher activity having a maximum Wolf number that can be four times as large as the cycle of lowest activity. Thus the question arises, as to how to resolve this instability in time of both the amplitudes and frequencies. Periodic regression and evolutive MESA will be used together with an evolutive Thomson analysis and complex demodulation to answer this question.

A moving form of MESA (Radoski *et al.* 1976) was used to answer this question of stability. The spectrum of the series made of the first 70 years is computed. Then this 'data-window' is moved forward 10 years and the next spectrum is computed for 1710–1780 and so on. This moving spectral analysis gives an overview of the shift or the persistence of the spectral peaks with time. The three-dimensional plot of these figures both in perspective (figure 6) and from the front (figure 7) gives the spectral density as a function of frequency for the 22 different 70-year subperiods analysed. In addition to the variation of the amplitude of some stationary peaks, note the large shift in the frequency of the 11-year quasi-period that indeed varies between 8 and 14 years.

As the maximum entropy amplitude estimates have to be regarded with caution, an evolutive Thomson analysis was performed and presented in figure 8 in a similar way, to give a more precise idea of the amplitude variation of all the quasi-periods. For 1700–1810, we observed a single peak around 58 years and a split in two peaks of the 11-year cycle around 14 and 8 years. From 1810 to 1950, the 58-year peak has disappeared and the 11-year periodicity becomes more stable in frequency. A very weak five-year peak appears also

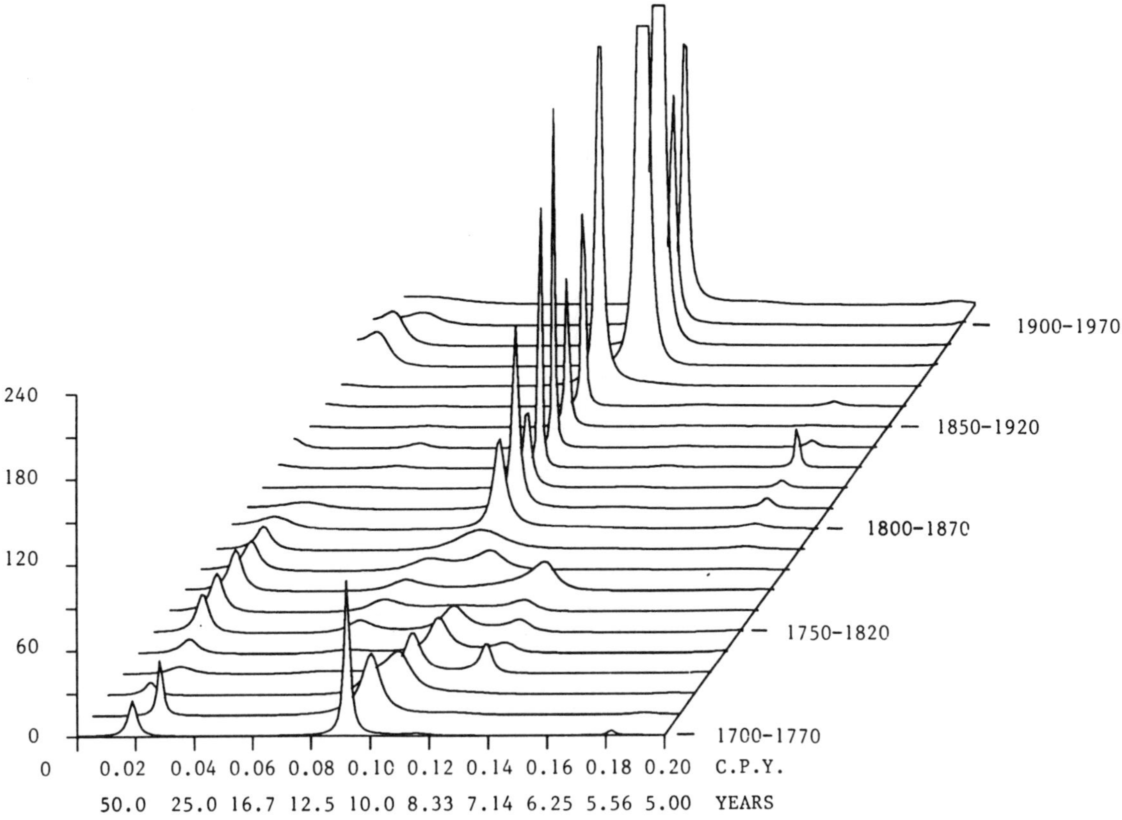

FIGURE 6. Evolutive maximum entropy spectrum analysis of the annual Wolf's sunspot number from 1700 to 1986. Note that the quasi 11-year peaks of the spectra corresponding to the series 1870–1940, 1880–1950, 1890–1960 and 1910–80 are truncated.

sometimes during the same interval. The instability of the 11-year cycle thus tends to occur when the 58-year peak is present and its amplitude is low during this interval of time (1700–1810).

THE 22-YEAR MAGNETIC SERIES

Transformation of the sunspot series can be carried out by considering the reversal of the sign between the alternating half magnetic cycles. This transformed series has a much stronger physical basis as it accounts not only for the reversal of the magnetism of the leading spots in alternate 11-year cycles, but also for some variation in the length and amplitude of this 11-year cycle. Indeed, there is a tendency for long and short cycles to alternate but mainly there is a clear antisymmetry between the period of increasing sunspot numbers and the period of decreasing values. The duration of the magnetic cycle, which combines successive 11-year cycles, should thus be more stable than that of the 11-year cycle.

Harmonic analysis, Blackman–Tukey, Thomson, maximum entropy and minimum cross-entropy spectral analysis were performed to this 22-year magnetic series from 1700 to 1986 by using the same parameters as for the preceding spectral analyses. The MESA is given in figure 9 (but all the other technics give the same result). As it can be seen, the main statistical advantage of this magnetic version of the Wolf sunspot number time series is the high stability

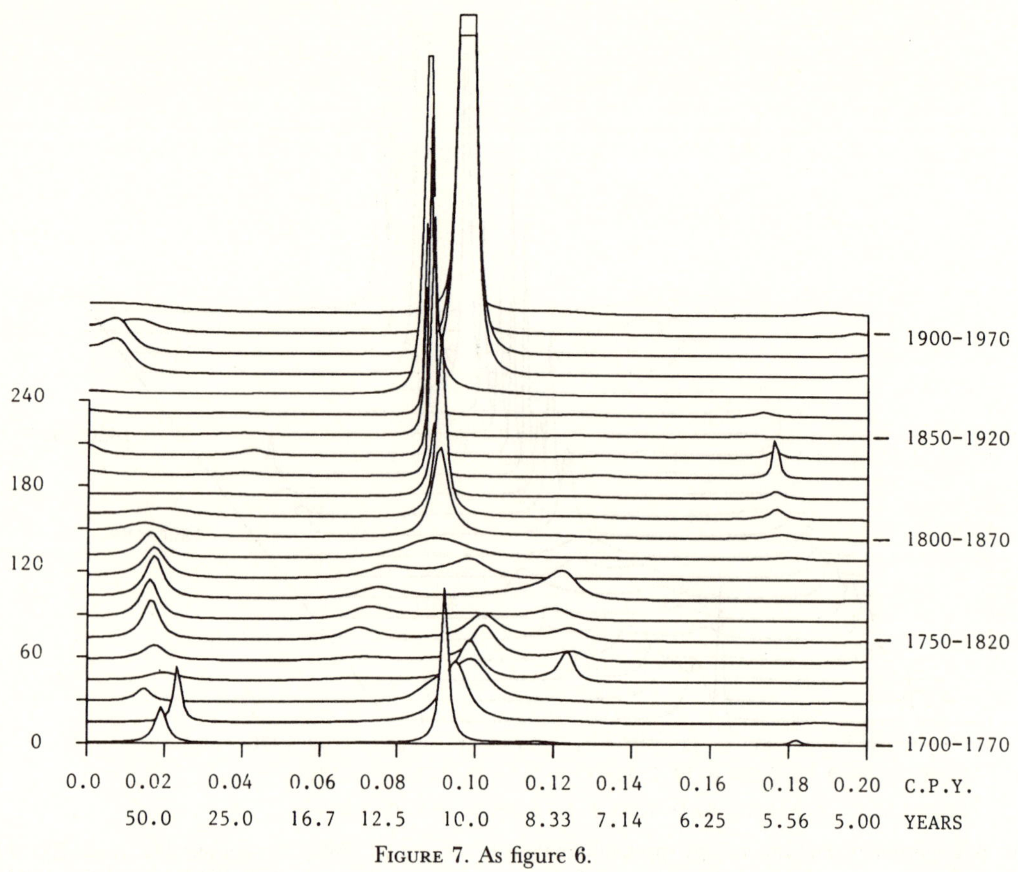

FIGURE 7. As figure 6.

of the 22-year cycle which explains 62 % of the total variance over the period 1700–1986 in contrast to the frequency instability of 11-year sunspot amplitude time series.

To demonstrate this higher stability in frequency of the 22-year cycle, an evolutive MESA (figures 10 and 11) and evolutive Thomson analysis (figure 12) were calculated based on sub-series having a length of 110 years and a time shift of 10 years. The results confirm this stability of the 22-year cycle.

The evolutionary MESA does not show any difference between early and late data: the 1700–1810 data window and the subsequent nineteenth- and twentieth-century windows are quite similar; i.e. there is no evidence that the early data are actually different from the later data, on a magnetic basis, which is also shown by the evolutive Thomson analysis

This result has interesting implications for eliciting physical mechanisms. It shows the existence of an invariant structure in what is otherwise complex and unstable data and suggests the importance of working with the 22-year magnetic series to provide reliable extrapolative patterns. Moreover, the change in the 11-year amplitude cycle after 1800 could suggest a complex internal 'oscillation' within the basically quite stable 22-year magnetic cycle instead of the inadequacy of early data.

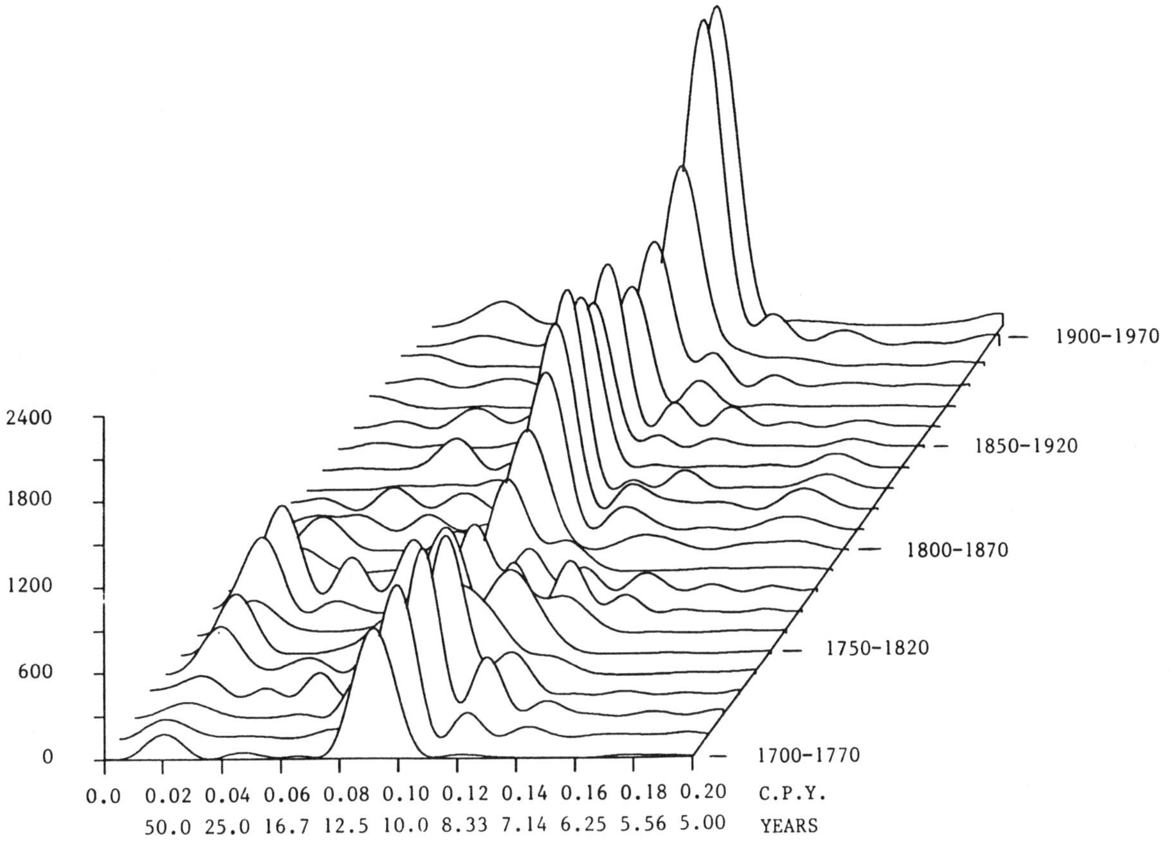

FIGURE 8. Evolutive Thomson 'multitaper' spectrum analysis of the
annual Wolf's sunspot number from 1700 to 1986.

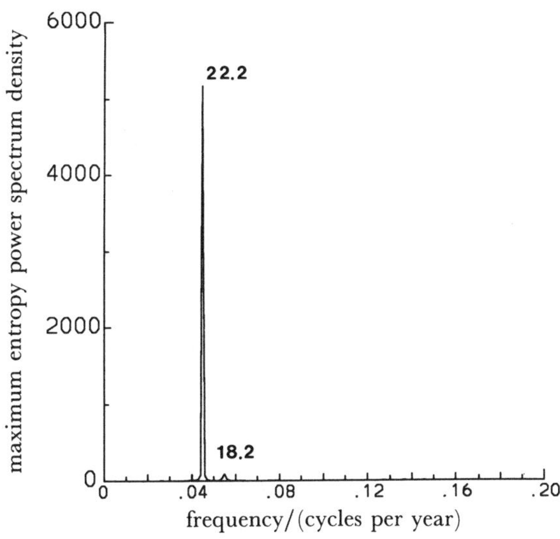

FIGURE 9. Maximum entropy spectrum analysis of the annual 22-year magnetic sunspot series from 1700 to 1986.

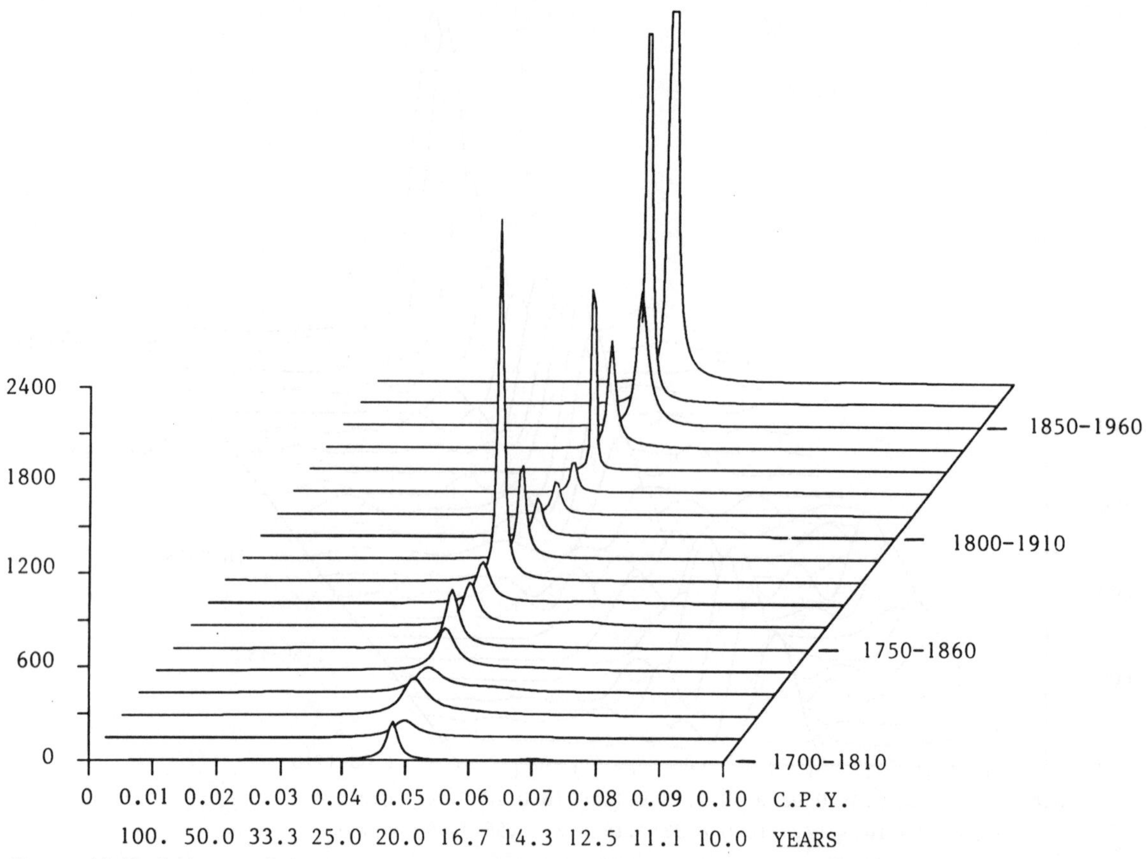

FIGURE 10. Evolutive maximum entropy spectrum analysis of the annual 22-year magnetic sunspot series from 1700 to 1986. Note that the quasi 22-year peaks of the spectra corresponding to the series 1860–1970 and 1870–1980 are truncated.

COMPLEX DEMODULATION

Complex demodulation (Bloomfield 1976) is a useful technique to describe the non-stability in frequency and amplitude of quasi-periodic data such as the Wolf sunspot number series. The amplitude curve ((a) in figure 13) shows that there are indeed substantial variations in amplitude of the 11-year quasi-period with a range of about 3:1.

The phase curve ((a) in figure 14) also displays substantial variations. These variations of the instantaneous phase may be interpreted in terms of a corresponding shift of the frequency around the 11-year quasi-oscillation. As an example, the increase in instantaneous phase between 1763 and 1790 compared with the preceding time interval 1724–62 corresponds to a decrease in period of the 11-year quasi-period, whereas the following instantaneous phase drop between 1790 and 1829 corresponds to an increase in the period of this oscillation.

Complex demodulation was then applied to the magnetic 22-year sunspot series. The corresponding results are given in figures 13 and 14 (curves (b)). As expected, the instantaneous phase shows a higher stability in frequency of the 22-year oscillation when compared with the highly unstable 11-year quasi-periodicity but the instantaneous amplitude variation of the 22-year magnetic curve is comparable with the 11-year amplitude variation.

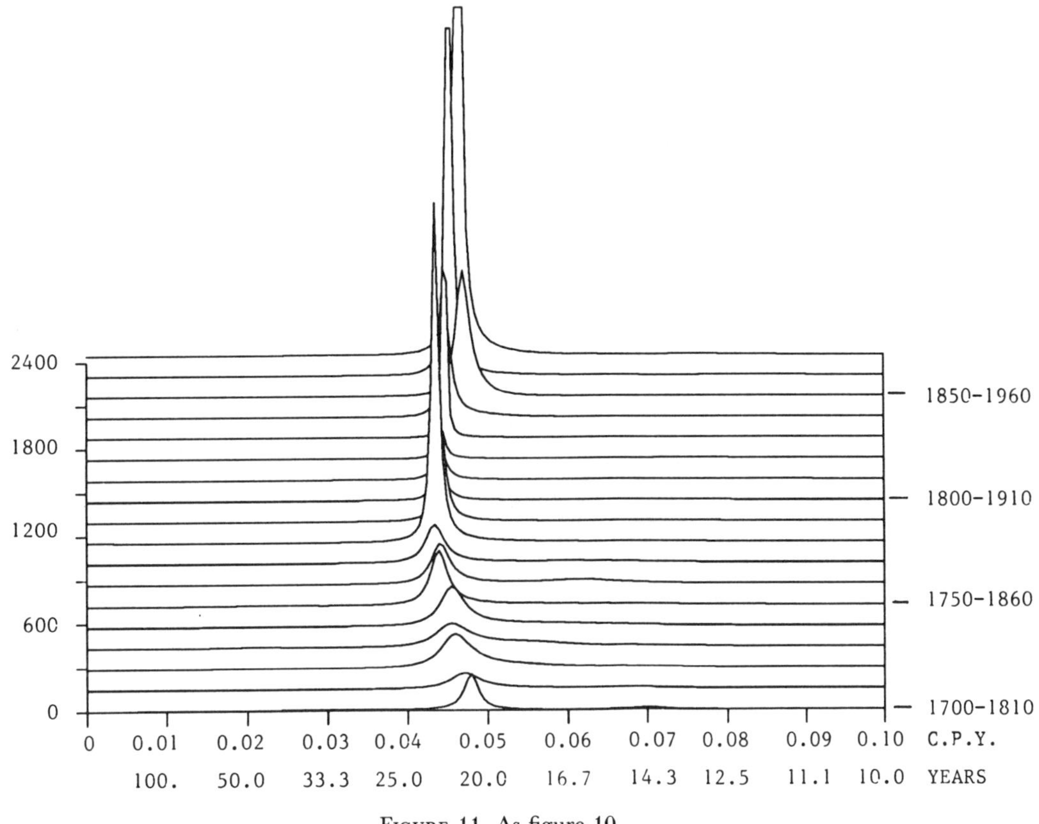

FIGURE 11. As figure 10.

DIMENSION OF THE STRANGE ATTRACTOR OF THE SOLAR CYCLE

Finally, we try to see if the system generated by the sunspot data allows the existence of an attractor and to determine the minimum number of variables necessary to describe this system. Our data will be the monthly sunspot numbers from 1750 to 1986. In figure 15, $C(r)$ is the logarithm of the correlation function of the attractor and r is the radius of the ball around any point of the phase space defined from the lagged variables of the time series. The saturation value of the slope $C(r)$ against r is here 3.15 which means that four variables should be necessary to describe the behaviour of the sunspot system. However, an evolutive version of this technique used for successive 40-year-long intervals, shows that this dimension of the attractor is highly unstable varying from 2.21 to 4.95 in a quasi-cyclic way.

CONCLUSIONS

This study applying six different spectral techniques to the Wolf sunspot time series confirms that the physical process producing sunspots is not stationary as already pointed out by Eddy (1977). The spectra indicate that the whole time series (1700–1986) is characterized by the existence of three main peaks which periods are ± 100, ± 55 and ± 11 years. The 11-year quasi-period is, however, the only one to be statistically significant following the Blackman–Tukey criteria. Moreover, it has two main components which are 11.1 and 10 years.

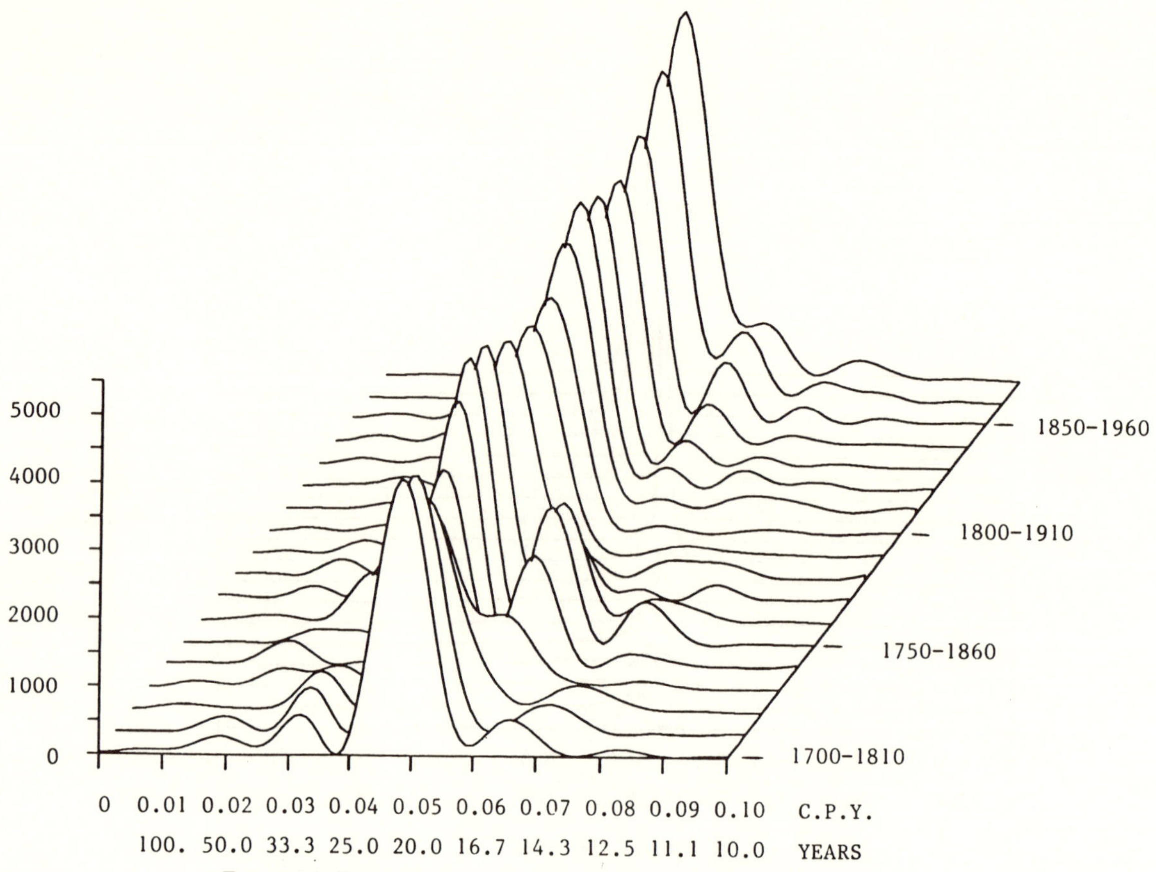

FIGURE 12. Evolutive Thomson 'multitaper' spectrum analysis of the
annual 22-year magnetic sunspot series from 1700 to 1986.

FIGURE 13   FIGURE 14

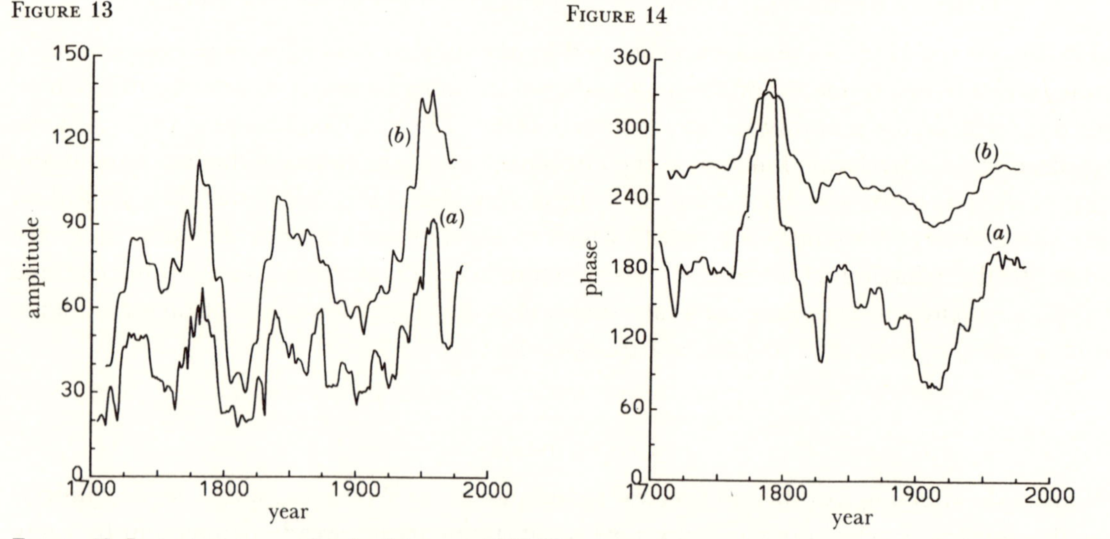

FIGURE 13. Instantaneous amplitude of the annual Wolf's sunspot number (a) and of the annual 22-year magnetic
sunspot series (b) from 1700 to 1986 (smoothed by taking a simple moving average of respectively 11 and 22
years).

FIGURE 14. Instantaneous phase of the annual Wolf's sunspot number (a) and of the annual 22-year magnetic
sunspot series (b) from 1700 to 1986 (smoothed by taking a simple moving average of respectively 11 and
22 years).

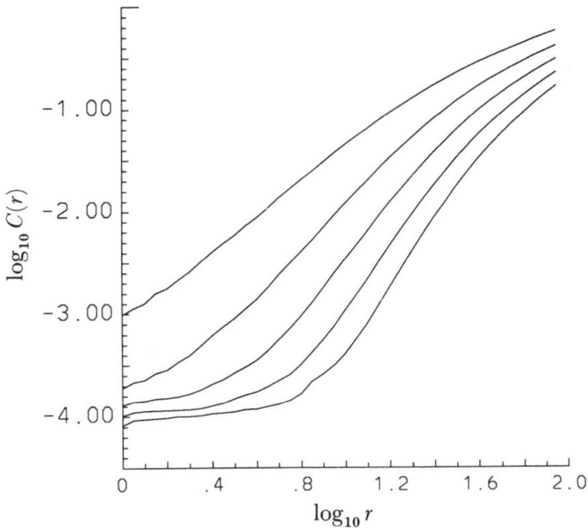

FIGURE 15. $\log_{10} C(r)$ against $\log_{10} r$ for the monthly sunspot data from 1750 to 1986. The curves are computed for $n = 2$ (first left curve) until $n = 6$ (last right curve).

This bimodal characteristic of the 11-year cycle might also be interesting in the frame of the existence of a long period ondulation (*ca.* 176–180 years), which is usually suggested by the planetary theory of sunspots and which is a beat phenomenon not caused by a primary long-term excitation function. The best correlation occurs at frequencies that correspond to periods of 11.2 and 9.9 years, but numerical values found in the literature for different time intervals may generate a long-period ranging from 167 to 220 years. Finally, we think that the sensitivity of this long-term period to the accuracy of the double 11-year peak and its potential meaning in terms of the physical origin of the sunspots and of the impact of the solar activity on climate justifies this kind of a better determination of the statistical properties of this Wolf sunspot time series.

We are grateful to Professor R. D. Rosen, AER, Cambridge (Massachusetts, U.S.A.) for his critical remarks on an early version of this paper, and to Dr A. Koeckelenbergh, Belgian Royal Observatory, Brussels, and Sunspot Index Data Center of the Federation of Astronomical and Geophysical Data Analysis Services (funded by the International Council of Scientific Unions under the auspices of the International Union of Geodesy and Geophysics, International Astronomical Union and the Union Radioscientifique Internationale). We also thank Dr C. Nicolis (Institute of Space Aeronomy, Brussels, Belgium) for providing us with her program on the strange attractor.

REFERENCES

Bloomfield, P. 1976 *Fourier analysis of time series: an introduction.* New York: Wiley.

Burg, J. P. 1972 The relationship between maximum entropy spectra and maximum like-hood spectra. *Geophys.* **37**, 375–376.

Cohen, T. J. & Lintz, P. R. 1974 Long-term periodicities in the sunspot cycle. *Nature, Lond.* **250**, 398–399.

Eddy, J. A. 1977 Climate and the changing sun. *Climatic Change* 1(2), 173–190.

Jenkins, G. & Watts, D. 1968 *Spectral analysis and its applications.* San Francisco: Holden-Day.

Radoski, H. R., Zawalich, E. J. & Fougere, P. F. 1976 The superiority of maximum entropy power spectrum techniques applied to geomagnetic micropulsations. *Phys. Earth planet. Inter.* **12**, 208–216.

Shore, J. L. 1981 Minimum cross-entropy analysis. *IEEE Trans. Acoust. Speech signal processing*, **ASSP-29**, 230–237.

Sneyers, R. & Cugnon, P. 1986 On the predictability of the Wolf sunspot number. *Annls. Geophysicae* **4**, 81–86.
Sonett, C. P. 1984 Very long solar periods and the radiocarbon record. *Rev. Geophys. Space Phys.* **22**, 239–254.
Thomson, D. J. 1982 Spectrum estimation and harmonic analysis. *IEEE Proc.* **70**, 1055–1096.
Yule, U. G. 1927 On a method of investigating periodicities in disturbed series, with a special reference to Wolfer's sunspot numbers. *Phil. Trans. R. Soc. Lond.* A **226**, 267–298.

Discussion

M. BERAN (*Institute of Hydrology, Wallingford, U. K.*). I note the somewhat arbitrary and subjective definition of the Wolf sunspot number. Is Professor Berger concerned that he may be applying his numerical analysis to what looks like the results of a Rorschach ink blot test?

A. BERGER. An evolutive maximum entropy spectral analysis (MESA) of the yearly sunspot data series from 1700 to 1986 had indeed indicated that the 11-year cycle is particularly unstable during the early part of the record (before 1810). At that time, a ±55-year peak is more significant. This difference between the MESA results for the more recent observations and the early ones could suggest that the early sunspot data must be treated with care.

However, the splitting in the pre-1800 period of the 11-year peak could be linked to the combination of the three most important harmonics instead of just suggesting early data inadequacy. Moreover, the high-frequency stability of the 22-year cycle through the whole time span (1750–1986) seems to support the reliability of these data, at least for the magnetic series.

These results show thus clearly that adequate statistical analyses of *a priori* non-reliable data is a worth while exercise in the sense that these analyses may question the data and the validity of the non-reliability assumption in an objective manner.

A. PROVENZALE (*Instituto di Cosmogeofisica, Torino, Italy*). Are the fluctuations (in amplitude and frequency) of the 11-year spectral peak in sunspot dates statistically signifiant? I caution the attractor dimension calculations because (*a*) the number of points in the sunspot number can be insufficient for the dimension evaluation (see Smith 1988); (*b*) I don't see a clean scaling range in the correlation curves Professor Berger has shown; (*c*) the calculation of a finite spectral dimension does not imply the existence of low-dimensional chaos, because there are simple stochastic processes with finite fractal dimension (see Osborne & Provenzale 1989).

A. BERGER. Harmonic analysis of the yearly sunspot data from 1700 to 1986 shows that for the 11 harmonics that are detected in the interval between 11.96 and 8.44 years, five are significant at the 0.01 significance level (with periods of respectively 11.96, 11.04, 10.63, 9.90 and 8.44 years) and one is significant at the 0.02 significance level (with a period of 9.57 years). This shows clearly that the splitting of the 11-year peak is not only a result of statistical fluctuations.

J. L. STANFORD (*Department of Atmospheric, Oceanic and Planetary Physics, University of Oxford, U.K.*). Dr Ribes (this Symposium) discusses solar diameter oscillations with periods near that of the quasi-biennial oscillation (QBO), and Dr Labitzke (this Symposium) discusses apparent correlations between sunspots and atmospheric phenomena at similar periods. Has Professor Berger performed spectral analyses with sunspot time series spaced at less than yearly intervals, and if so, does he find a statistically significant signal near QBO periods (slightly more than two years)?

A. BERGER. We have indeed performed harmonic analysis and Blackman–Tukey spectrum analysis of the monthly sunspot series from 1750 to 1986. When looking at the period interval from 24 to 36 months, the results are the following:

(1) no statistically significant signal at the 0.05 significance level in the interval 24 to 36 months;

(2) one harmonic with a period of 33.5 months detected as significant at the 0.1 significance level with the harmonic analysis;

(3) two harmonics with periods of respectively 25.86 and 25.62 months detected as significant at the 0.2 significance level with the harmonic analysis and which could be related to the QBO signal.

Z. REUT (*Bath, U.K.*). The dynamics of the solar system may have to be taken into account in explaining the variability of the Sun. The Sun revolves round the centre of mass of the Solar System that does not coincide with the Sun's own centre of mass due to the gravitational effect of planets (mainly Jupiter). The Sun's revolution relates to the 11-year solar cycle and to the 12-year period of the Jupiter's revolution round the Sun.

A. BERGER. Indeed the Sun revolves round the centre of mass of the Solar System. This is taken into account in the computation of the planetary motion and in the related Milankovitch theory. This could play a role for simulating the physics of the Sun but according to Meeus (1975) the planetary tides on the Sun are negligible and the Jupiter effect does not exist.

Additional references

Meeus, J. 1975 *Icarus* **26**, 257–267.
Osborne, A. R. & Provenzale, A. 1989 Finite correlation dimension for stochastic systems with power-law spectra. *Physica* D35, 357–381.
Smith, A. L. 1988 *Phys. Lett.* A **133**, 283–288.

Phil. Trans. R. Soc. Lond. A **330**, 543–545 (1990)

Printed in Great Britain

543

Some thoughts on Sun–weather relations

By J. A. Eddy

University Corporation for Atmospheric Research, Boulder, Colorado 80307, U.S.A.

Recent measurements of variations in the total solar irradiance now offer a quantitative mechanism through which year-to-year changes in solar activity may influence surface temperature. It follows that at least a part of the global warming of the last century could be ascribed to changes in solar output, and that effects of solar radiative forcing may need to be taken into account in predictions of greenhouse warming. A number of questions still remain, however, before this thesis rests on a firm foundation.

I am one who believes in a real and physical connection between long-term trends in what we know of climate through the past several hundred years and similar long-term changes in the level of solar activity through the same period. I also carry a prejudice that the solar effects on climate will show a marked frequency dependence: more pronounced and of greater practical importance for changes of longest term. This kind of purported long-term connection between the Sun and the climate has many fathers, advocated or at least pointed to by almost any in the past who believe in a Little Ice Age (*ca.* 1400–1800) and who accept the realities of the Spörer and Maunder Minima of solar behaviour (A.D. 1450–1540, 1645–1715). Once that leap of faith is taken, it is an easy step to accept as further evidence the coincidence of the subsequent Dalton Minimum in solar and auroral activity (*ca.* 1790–1830) and the cold decades that characterized climate the turn of the nineteenth century, and the striking correspondence between what has happened on the Sun through the present century (a rise of a factor of three since 1890 in the long-term envelope of recorded sunspot numbers) and at least a part of the coincident rise of about 0.6 °C in land surface temperature in the Northern Hemisphere.

These connections, however tempting, have lacked until recently a convincing demonstration of a mechanism, related to solar surface activity, that invokes sufficient energy to drive decadal changes in the climate system.

What is new, and worth the attention of this Symposium, are the emerging bulwarks of observational and analytical research that now make the connection more credible. What has moved the subject into a new realm of importance is the possibility that solar-forced changes in surface temperature in the past (and hence potential, solar-forced changes in the future) are of the same order of magnitude (though smaller, by a factor of two to four) as the changes now predicted through greenhouse warming in the first half of the next century.

Dr Foukal's interpretation (this Symposium) of our new knowledge of activity-related changes in the most energetic output of the Sun provides a quantitative mechanism that links decadal changes in sunspot numbers to decadal trends in the so-called solar constant: by far the most energetic of known changes in the output of the Sun. Professor Wigley's work (T. M. L. Wigley & P. M. Kelly, this Symposium) tests this relation through the device of quantitative climate models, against the most carefully compiled record that climatology can offer: the collected and corrected surface temperature record of the modern era. His results strengthen what he and independently, George Reid, had more tentatively claimed at a similar meeting in Durham several years ago.

As with most apparent, simple connections in science, there are parts of the thesis that are weak or missing altogether, some important unknowns, and for those who would seek it, room for doubt. Given the potential of solar forcing in decadal climate forcing, and current concerns regarding effects of future climate change, we need to give priority to clarifying the case.

Among outstanding questions are the following.

1. The measurements from spacecraft tie observed trends in the solar constant to changes in sunspot number, made since 1980, possibly in phase with the 11-year cycle. How secure is the conclusion that the effect is one of 11 years, as opposed to 22 years, or a longer period? How does the observed effect extrapolate quantitatively to solar changes of longer term, such as the twentieth century rise in activity, or the prolonged depression of the Maunder Minimum? Is it possible that secular or long-term trends in surface activity are related to proportionately greater modulations in total solar flux than are the year-to-year changes now associated with the 11-year solar cycle?

2. The value to climatology of a long-term connection between solar activity and solar constant depends upon an ability to forecast long-term trends in solar behaviour over more than one solar cycle. On what basis can this be done? Is the so-called 'Gleissberg cycle' of 80–90 years a predictable, periodic feature of the Sun? How firm is the evidence for an overriding cycle of 200–210 years? Is long-term solar behaviour periodic or chaotic?

3. Our longest, quantitative records of solar activity come from proxy records (principally tree-ring ^{14}C and ice-borne ^{10}Be) that are linked to solar surface activity (sunspot number, now tied to variations in the solar constant) through a chain of connection that is less than perfectly known. The mechanism through which galactic cosmic rays are modulated by conditions in the solar wind is imperfectly known; nor is there a simple relation between solar wind flux or velocity and sunspot number. How reliably may we substitute these indicators for the solar surface parameters that provide the apparent new linkage with changes in photon flux?

4. The purported connections linking solar activity to solar constant changes to variations in surface temperature are of the coarsest, spatial nature. The effect of changes in the solar constant on terrestrial climate can be expected to be a function of duration or persistence, and to be globally heterogeneous, greater for persistent or long-term changes and more pronounced in the centres of large land masses, where the thermal intertia of the oceans is less felt. To what extent do available climate data bear this out? Can spatial differences be used as a tool to separate the effects of solar forcing from that due to past changes in globally well-mixed greenhouse gases? What was the spatial pattern of the climatology of the Little Ice Age?

5. What fraction of the gradual global warming since the fading of the Little Ice Age can be attributed to the solar forcing that must have accompanied the gradual rise in the level of solar activity, and how much to the coincident increase in CO_2 that followed, through the same period, the dawn of the industrial age?

The case for a significant decade-to-century connection between solar activity and terrestrial surface temperature may seem to many of us convincing. But until we can answer these and other fundamental questions it rests on a less than solid foundation.

Discussion

Jenny Allsop (*Deep Geology Research Group, British Geological Survey, Nottingham, U.K.*). The importance of the historical background in relating changes in the ^{14}C calibration curve to variations in the Earth's climate has already been illustrated for the medieval period: fourteenth-century documents describe a dramatic change from arable to sheep farming on a large scale that was purely due to social and economic events and not to climatic change. Therefore, the use of earlier archaeological data in assessing climatic changes may be misleading in the absence of documentary evidence. A few excavated sites with dateable material may show evidence of climatic variation (e.g. crop changes), but such evidence cannot necessarily be extrapolated over large geographical areas. Furthermore, the human influence on these changes may also be more difficult to assess. The deforestation of large areas of Europe, before the end of the Bronze Age, may have had as profound an effect on climate as the clearance of large areas in South America today. Further difficulties might occur if a variation in climate due to human activity were to coincide with any rhythm or cycle recognized in the ^{14}C calibration curves or similar stable isotope profiles before the discovery of new archaeological evidence.

J.-C. Pecker (*Collège de France, Paris, France*). I want to add two additional words of warning about the use of R (Wolf's number) as a unique index of solar activity.

1. The solar data are far from simple. Spots have different heliographic latitudes. Their average latitude changes during the cycle, as shown by the 'butterfly' diagram, in an equatorward trend. Effects on the Earth are affected by this average heliographic latitude, not only by the Wolf's number.

2. If one wants a single, unique, index for solar activity, is R the best one? One has used also the 'spotted area' and both are very highly correlated. However, in both cases, any given spot is counted at every of its successive passages, in successive rotations. Perhaps only newly born spots are really interesting; or at least one might conceive that any spot may have different effects at different stages of its development.

Sir William McCrea, F.R.S. (*Astronomy Centre, University of Sussex, U.K.*). With regard to the historical evidence from records of auroras, Professor Wilfried Schröder of Bremen-Roennebeck has published recent critical discussions (Schröder 1984, 1988*a*, *b*). He considers that 'the absence of auroral observations [in some of the years] during 1645–1715 is not conclusive evidence for a general minimum of solar activity during this period' (Schröder 1988*a*).

References

Schröder, W. 1984 *Das Phänomendes Polarlichts* Darmstadt: Wissenschaftliche Buchgesellschaft.
Schröder, W. 1988*a* Estimating the Maunder Minimum from auroral data. *Nature, Lond.* **335**, 676.
Schröder, W. 1988*b* Aurorae during the Maunder Minimum. *Meteorology atmos. Phys.* **38**, 246–251.

Phil. Trans. R. Soc. Lond. A **330**, 547–560 (1990)

Printed in Great Britain

547

Holocene climatic change, ^{14}C wiggles and variations in solar irradiance

By T. M. L. WIGLEY AND P. M. KELLY

Climatic Research Unit, School of Environmental Sciences, University of East Anglia, Norwich NR4 7TJ, U.K.

Evidence from the advances and retreats of alpine glaciers during the Holocene suggests that there were at least 14 century-timescale cool periods similar to the recent Little Ice Age. Here, we examine the hypothesis that these cool periods were caused by reductions in solar irradiance. A statistically significant correlation is found between the global glacial advance and retreat chronology of Röthlisberger and variations in atmospheric ^{14}C concentration. A simple energy-balance climate model is used to show that the mean reduction of solar irradiance during times of maximum ^{14}C anomaly like the Maunder Minimum would have to have been between 0.22 and 0.55% to have caused these cool periods. If a similar solar irradiance perturbation began early in the 21st century, the associated climate effects would be noticeable, but still considerably less than those expected to result from future greenhouse gas concentration increases.

INTRODUCTION

Although they have been far less spectacular than the longer, ice-age timescale changes, significant changes in climate have occurred during the Holocene. The evidence for these changes comes from a variety of forms of indirect or 'proxy' evidence: fossil pollen records (COHMAP members 1988), glacial advance and retreat chronologies (Röthlisberger 1986; Grove 1987), lake level fluctuations (Kutzbach & Street-Perrott 1985), tree line changes (LaMarche 1973; Karlén 1976), ice core data (Dansgaard *et al.* 1982; Jouzel *et al.* 1987), and so on. These data, together with information from modelling studies (Kutzbach & Guetter 1986), suggest that there are at least two main causal factors operating.

On 1000-year and longer timescales, the primary control appears to be the Milankovitch effect, changes in the seasonal and spatial character of incoming solar radiation due to variations in the Earth's orbit around the Sun. Nine thousand years ago, for example, the Northern Hemisphere received up to 17% more solar radiation in the summer and up to 12% less solar radiation in the winter compared with today. Averaged over the year, high latitudes received up to 2% more solar radiation, whereas low latitudes received less (see Mitchell *et al.* 1988, fig. 4). The zero change points were around 43° N and S. Globally, the net change was close to zero. Over subsequent millennia, these radiation differences slowly relaxed towards present-day conditions.

Because the largest changes are on the seasonal timescale, the effects are manifest mainly in variables which respond to seasonal climatic influences, with vegetation and lake levels being two of the most widely studied (COHMAP members 1988). Fossil pollen data indicate that, during the early Holocene, summer temperatures in many parts of the globe were noticeably warmer than today, by up to 2 °C (where 'today' refers to the early twentieth century), and lake level data suggest that the early Holocene was considerably wetter than today in low latitudes. These conditions have been successfully simulated by using general circulation models of the climate system forced with appropriate insolation values (see, for example, Kutzbach & Guetter 1986; Mitchell *et al.* 1988).

Because most available proxy data tend to reflect summer climate conditions (Williams & Wigley 1983), it is not known with certainty whether any noticeable changes in annual-mean temperature occurred on the 1000-year timescale. The Milankovitch effect clearly produces annual mean changes on the ice age (10 000-year) timescale, but these occur through a variety of feedback effects involving ice sheets, CO_2 and CH_4 concentration changes, cloudiness changes, tropospheric aerosol changes, and so on. Most of these feedbacks would have had substantially smaller influences during the Holocene. Because winter insolation over most of the globe was much less than today during the early Holocene, one might argue that this may have compensated for the greater summer insolation. Model results, however, together with the zonal, annual-mean insolation data which show a net increase north or south of 43° N or S, suggest that annual-mean temperatures were greater in the early Holocene, at least in high latitudes (Mitchell *et al.* 1988).

On shorter timescales, it is clear that substantial global-scale annual-mean temperature changes did occur during the Holocene. The main evidence for this comes from glacial advance and retreat chronologies. Small mountain glaciers respond to both precipitation and temperature changes (Porter 1981), and their movements are controlled by the integrated effects of these parameters over the seasonal cycle. Both empirical evidence (Meier 1984) and modelling studies (Oerlemans 1988) indicate that advances and retreats reflect annual-mean temperature changes. Thus the chronologies produced by Röthlisberger (1986), which are supported by a number of other studies (see, for example, the comprehensive review by Grove (1988)), may be cautiously interpreted as indicators of annual-mean temperature changes. These data show that, during the Holocene, there were a number of globally near-synchronous cold periods lasting for centuries and interspersed by longer warmer intervals. The most recent of these cold periods was the Little Ice Age.

The resulting picture of the Holocene is therefore as follows. On the 1000-year timescale, slow changes occurred associated with Milankovitch orbital effects. Although these are manifest mainly in seasonal data, noticeable annual-mean temperature changes on this timescale are likely to have occurred in mid to high latitudes. Superimposed on this there have been a number of century-timescale Little Ice Age events, apparently occurring at random. Mainly through the work of Kutzbach and colleagues, we are reasonably certain of the cause of the slower changes. But what is the cause of the series of Little Ice Ages? Grove (1988) has reviewed the possibilities. Here we concentrate on just one of these, solar variability.

There is no direct evidence of changes in solar irradiance during the Holocene. However, we do know that some characteristics of the Sun's output did change during the Holocene on the century timescale. For example, historical evidence shows that the incidence of sunspots has varied markedly, with prolonged periods of sunspot minima occurring on a number of occasions during the past thousand years (Eddy 1976). The Maunder Minimum of the seventeenth century is the most recent and well known of these. These sunspot minima are also associated with periods of enhanced ^{14}C productivity in the upper atmosphere (Stuiver & Quay 1980; Stuiver & Braziunas 1987).

Changes in ^{14}C production rate cause changes in atmospheric ^{14}C concentration that can be measured by comparing radiocarbon and dendrochronological dates of tree-ring samples. The difference between these dates is called the ^{14}C anomaly. Over the past 10 millennia, the ^{14}C anomaly record shows both slow, 1000-year timescale fluctuations and more rapid changes. Of the latter, on the century timescale there are a number of 'wiggles' in the record. We now know that these wiggles are the result of changes in the solar wind (Stuiver & Quay 1980; supported by ^{10}Be evidence, Siegenthaler & Beer 1987) and that they are largely contemporaneous with periods of sunspot minima. What we do not know is whether these ^{14}C anomalies were associated with changes in solar irradiance.

In the following, we examine the relation between atmospheric ^{14}C changes and changes in climate during the Holocene. On the assumption that they are related, we use a simple climate model to estimate the solar irradiance changes that would be required to produce the observed climate changes (namely, the Little Ice Age events described above). Finally, we consider the implications that a future Little Ice Age might have, given that it is likely to occur against a backdrop of anthropogenic climatic change due to the greenhouse effect.

COMPARING THE ^{14}C ANOMALY AND CLIMATE RECORDS

In Wigley (1987a), the time histories of atmospheric ^{14}C anomaly changes and climate changes were compared both visually and statistically, and it was concluded that there was a statistically significant similarity between the two. Here we repeat this analysis, but with the following differences. First, the analysis is extended back beyond 4000 B.C. to include the earliest available data. Secondly, the dates of the major ^{14}C anomalies are determined by using an objective filtering method, rather than by visual inspection of the curve given by Stuiver et al. (1986a). Thirdly, the calendar dates for the Holocene cold intervals and temperature minima are re-estimated by using all of the most recently available calibration curves (namely Stuiver & Pearson 1986; Pearson & Stuiver 1986; Stuiver & Becker 1986; Pearson et al. 1986; de Jong et al. 1986; Linick et al. 1986; Kromer et al. 1986; Stuiver et al. 1986b). Fourthly, more statistical tests will be carried out.

For the climate data, the record we use is as in Wigley (1987a); namely, a composite of the regional records of glacial advance and retreat produced by Röthlisberger (1986). Röthlisberger's glacial data come from 12 regions in both the Southern and Northern Hemispheres. He has combined these into six larger-region time series for the following locations: southeastern Alaska, tropical western South America (around 10° S), southern South America (35–55° S), the European Alps, the Himalayas, and southern New Zealand. In addition, he has included data for nothern Scandinavia obtained from Wibjorn Karlén. These have been combined with rough area weights by Wigley (1987a) to produce hemispheric- and global-mean time series. Because all time series are highly correlated (an interesting result in itself), the mean series do not depend much on the actual values used for the weights. The series are shown in figure 1.

To identify cold intervals, an arbitrary threshold was used (see figure 1). Minima were read directly from the curve. Some minima occur as isolated values rather than within a cold period because they are at times when prevailing conditions were apparently warmer than the chosen threshold value. These occur particularly in the early Holocene, suggesting that Milankovitch effects may have caused low-frequency, annual-mean changes upon which the shorter cooling events were superimposed. The radiocarbon dates for the minima and the interval boundaries were converted to calendar dates by using the above-cited references. In some cases more than

[151]

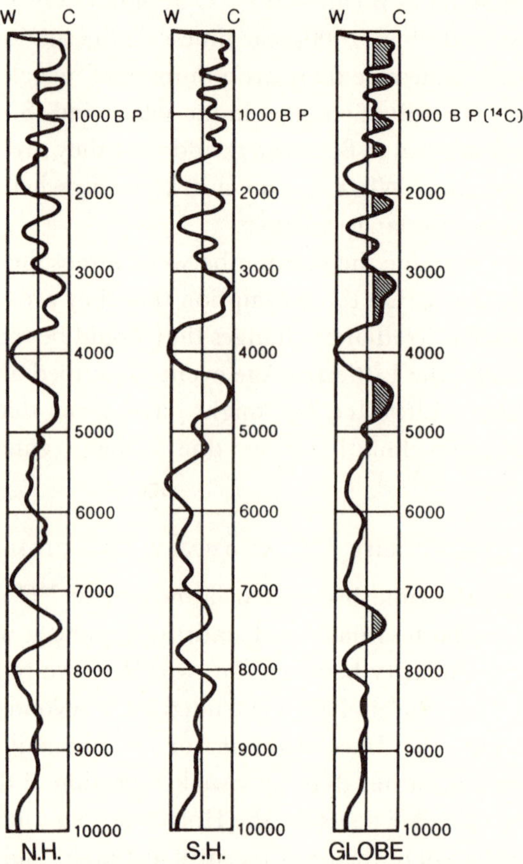

FIGURE 1. Alpine glacier advance and retreat chronologies for the Northern Hemisphere (N.H.), Southern Hemisphere (S.H.) and globe based on Röthlisberger (1986). W and C refer to warm (glaciers less advanced) and cold (glaciers more advanced).

one calibration curve could be used giving slightly different calendar dates. Single 'best' calendar dates were then chosen, mostly as the mean of a range of possibilities, but with greater emphasis on results derived by using the calibration curves with smaller error bars. In a few cases it was impossible to distinguish between two alternative dates. The results are shown in table 1.

The calendar dates given here are subject to considerable uncertainty for a number of reasons, not least being uncertainties in the original data and dating, the method used for averaging the data from different regions, the extraction of radiocarbon dates from the composite curve, and the conversion from radiocarbon to calendar dates. These uncertainties produce an inherent 'noise' in the climate record which would probably tend to obscure any ^{14}C–climate link, if one exists.

For the ^{14}C anomaly record we used data kindly supplied by Minze Stuiver (personal communication). These are essentially a digitized version of the information contained in figure 1 of Stuiver et al. (1986a). Major ^{14}C anomalies were identified in the following way. First, two gaussian filters were used to band-pass filter the data, retaining frequencies within the periods L_1 to L_2. These results were filtered again to produce a roughly constant base level by applying the L_2 filter to the negative values of the record (i.e. if Z is the band-pass filtered data, then Z was replaced by zero if $Z > 0$, and the results filtered by using L_2 to give Z^*). The curve finally

TABLE 1. CALENDAR DATES OF COLD INTERVALS, TEMPERATURE MINIMA AND ^{14}C MAXIMA

| | climate[a] | | ^{14}C maxima | |
cold intervals	minima	this work[b]	Stuiver[c]
—	—	—	7490 B.C.
—	7110 B.C.	7010 B.C.	7030 B.C.
6430–6080 B.C.	6190 B.C.	6370 B.C.	6370 B.C.
—	—	5970 B.C.	5950 B.C.
—	5190 B.C.	5190 B.C.	5190 B.C.
—	—	4710 B.C.	—
—	—	4250 B.C.	4330 B.C.
4030–3940 B.C.	3980 B.C.	3930 B.C.	3930 B.C.
—	—	3590 B.C.	3590 B.C.
—	—	—	3490 B.C.
3550–2920 B.C.	3320 B.C.	3290 B.C.	3290 B.C.
—	—	2830 B.C.	2850 B.C.
—	—	2410 B.C.	—
2040–1250 B.C.	1430 B.C.	1350 B.C.	—
900–800 B.C.	820 B.C.	730 B.C.	740 B.C.
350 B.C.–0	150 B.C.	330 B.C.	340 B.C.
A.D. 570–660	A.D. 640	A.D. 730	—
A.D. 850–1040	A.D. 930	A.D. 1050	A.D. 1040
—	A.D. 1190	—	—
A.D. 1280–1400	A.D. 1330	—	A.D. 1340
A.D. 1450–1890	A.D. 1460	A.D. 1510	A.D. 1460–1540
—	A.D. 1680	A.D. 1690	A.D. 1720

[a] From figure 1.
[b] From figure 2b.
[c] From Stuiver & Braziunas (1987, figures 1–4).

considered was $Z-Z^*$. Various combinations of L_1 and L_2 were used. Examples are shown in figure 2. Dates of major ^{14}C anomaly maxima were then extracted (from figure 2b). These are listed in table 1.

Also shown in table 1 are dates of ^{14}C maxima identified by Stuiver & Braziunas (1987). The differences, which are generally small, arise mainly from the methods used to filter the raw data. We have identified four maxima (at 730 A.D., 1350 B.C., 2410 B.C. and 4710 B.C.) that are not identified by Stuiver & Braziunas, while these authors have identified two maxima which are not evident in figure 2b (at 1340 A.D. and 3490 B.C.). These two maxima are evident in figure 2a. Note, however, that figure 2a contains an additional maximum at 5450 B.C. not identified by Stuiver & Braziunas. Apart from these differences, the only other discrepancy occurs around 4250 B.C. This peak occurs at 4330 B.C. in Stuiver & Braziunas. In figure 2a it appears as a split peak with maxima at 4210 B.C. and 4310 B.C.

Is there any relation between the climate data (figure 1) and the ^{14}C data (figure 2)? We can examine this possibility visually by superimposing the two time series. This has been done in figure 2b where the climate timescale has been converted to calendar years. It is very difficult to judge from this figure whether or not a link exists. If one accounts for an inherent timing uncertainty of, say, ± 100 years in the climate data, then most of the cold periods (taking isolated minima as cold periods of zero length) contain ^{14}C peaks (11 out of 13), and most of the ^{14}C peaks occur within cold periods (14 out of 18). In the absence of this ± 100-year flexibility, however, the direct correspondences are less convincing, and some cold periods correspond better with ^{14}C minima than with maxima.

To examine the possible relation further, we consider only the climate minima. Climate

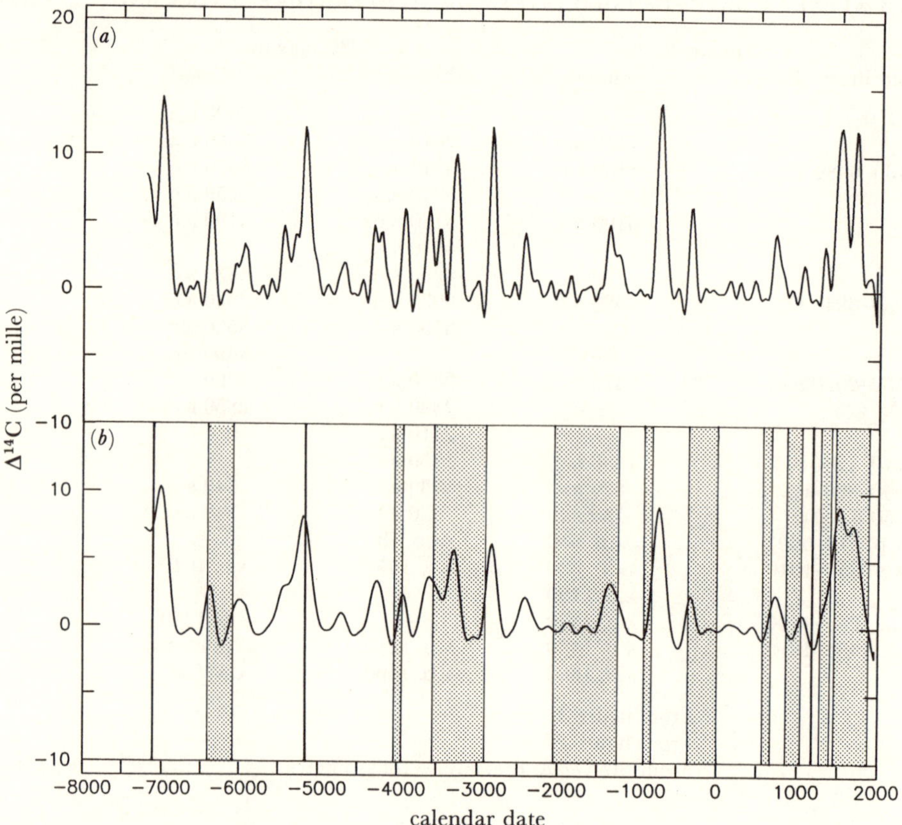

FIGURE 2. Atmospheric ^{14}C anomalies and climate. The two curves show band-pass filtered values of the atmospheric ^{14}C anomaly record using (a) 1000- and 100-year filters and (b) 1000- and 200-year filters. In (b), cold intervals (from figure 1, converted to calendar years) are marked by screening or as vertical lines (for the three shortest-duration events).

minima and ^{14}C maxima are highlighted and compared in figure 3. The visual agreement is better than in figure 2b, partly because the $\pm$100-year uncertainty is represented. There are, nevertheless, some striking correspondences. Six out the the seven strongest ^{14}C maxima correspond closely to climate minima.

The statistical significance of the correspondence can be assessed in the following way. The ^{14}C maxima can be considered as 'bullets' that are fired at a set of 200-year wide 'targets', namely the climate minima. If the maxima were fired randomly at the whole 9160-year record, how many 'hits' would we expect to occur? As a reasonable approximation, this can be considered as a binomial problem. The total target width is 2670 years (i.e. 14 minima by 200 years, less a small amount of overlap because of closely spaced minima), so the probability of a single hit occurring by chance is $2670/9160 = 0.291$. The observed number of hits is 9 out of 18, and the probability of this occurring by chance is 0.033.

Alternatively, the roles of ^{14}C maxima and climate minima can be reversed. In this case the target area is 3580 years so the probability of a hit in any one trial is 0.391. The observed number of hits is 9 out of 14, and the probability of this occurring by chance is 0.036. Both of these results are significant at the 5% level, implying a significant ^{14}C–climate relation. This is in accord with the slightly different analysis carried out in Wigley (1987a).

[154]

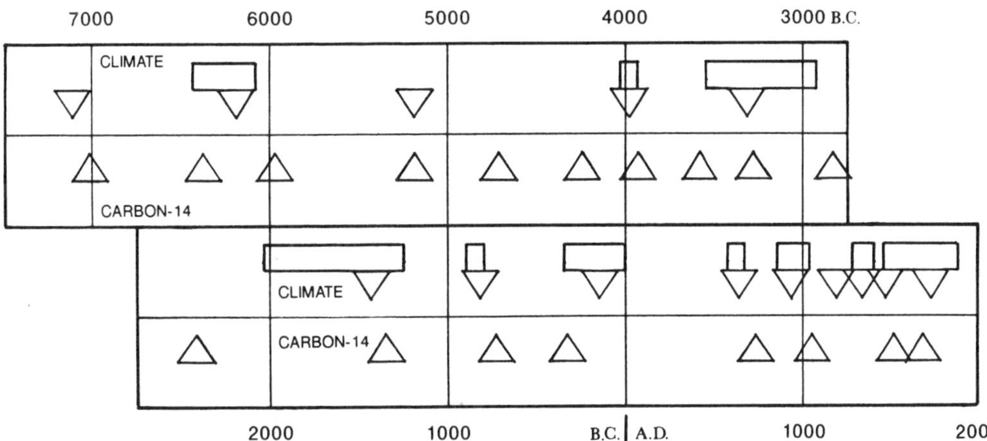

FIGURE 3. Simplified atmospheric ^{14}C and climate records. The upper halves show temperature minima (triangles) and cold periods (rectangles) based on figure 1. The lower halves show ^{14}C maxima based on figure 2b.

To further test this hypothesis, we considered the correspondence between ^{14}C *minima* and climate minima. By using the same filtering method as used for ^{14}C maxima, 18 ^{14}C minima were identified and located. For ^{14}C minima as 'bullets', the results were as follows: probability of a hit in a single trial, 0.291 as before; probability of the observed number of hits (7 out of 18), 0.129. For climate minima as bullets, the results were the following: probability a hit in a single trial, 0.384; probability of the observed number of hits (8 out of 14), 0.078. These results are not statistically significant, but they do indicate some sort of correspondence between ^{14}C minima and climate minima. This arises because the major ^{14}C minima identified by our filtering method tend to lie close to the identified maxima. The example exposes some of the problems in proving a relation, and should act as a warning in the interpretation of the ^{14}C maxima results.

Overall, we consider these results to be highly suggestive of a ^{14}C–climate link, but not ultimately convincing. For some ^{14}C maxima, Röthlisberger's data show no evidence of a climate minimum, although this does not preclude the existence of an unidentified cold period. Alternatively, potential cold periods may on occasion be masked by other processes, including internally generated natural climatic variability. There is also a lack of accord between the duration of ^{14}C maxima and the duration of cold periods. The former generally last from 100–200 years (see Stuiver & Braziunas 1987, figures 5–7), while the latter are generally substantially longer, by amounts much larger than can be explained by the lag effect of oceanic thermal inertia and/or the response times of alpine glaciers. A 'believer' may explain this duration difference as resulting from uncertainties in the dating of the climate events, but to the unbiased observer there must be nagging doubts. Clearly, there are complexities that remain to be resolved.

Although we have described this analysis as a test of the hypothesis that there are major solar irradiance perturbations that occur in parallel with the solar events that lead to major ^{14}C anomalies, there may be alternative explanations. For example, the ^{14}C anomalies could, themselves, be the result of the climate perturbations. We consider this to be unlikely, because the ^{14}C anomaly link with the Sun has been convincingly demonstrated (Stuiver & Quay 1980; Siegenthaler & Beer 1987). However, if major cold periods were associated with a large change in the rate of oceanic bottom water formation (as has been hypothesized for the late glacial), then parallel perturbations in the atmospheric ^{14}C content would certainly occur. This is a topic that deserves further attention.

Irradiance changes during ^{14}C maxima

We now consider the implications of a ^{14}C–climate link and estimate the change in solar irradiance that is likely to be associated with ^{14}C–anomaly-producing solar events like the Maunder Minimum. To do this, it is first necessary to estimate the global-mean temperature deviation associated with the major glacial advances documented in figure 1. This can be done (albeit with some uncertainty) by reference to the most recent period of the record. Over the past 100–150 years, the globe has warmed by approximately 0.5 °C (Jones *et al.* 1986). This interval corresponds to that part of the most recent warming trend in figure 1 that lies above the 'zero' reference line. From this it can be deduced that the glacial advance maxima correspond roughly to an annual, global-mean cooling of 0.4 °C below the reference line, or about 0.6 °C below the average level of the warm intervals. We therefore assume that the cooling associated with ^{14}C maxima is 0.4–0.6 °C.

The solar forcing required to produce such a cooling depends on a number of factors. The most important of these is the equilibrium climate sensitivity, i.e. the temperature change what would eventually occur for a forcing of 1 W m^{-2} at the top of the troposphere. The observed response to any given forcing, however, is only a fraction of the equilibrium response because of the lag or damping effect of oceanic thermal inertia. Just how much of the equilibrium response is realized depends on the timescale of the forcing, the mixed layer depth, the rate of ocean mixing below the mixed layer, and the climate sensitivity. For example, for sinusoidal forcing with a 10-year period, only 13–23 % of the potential response is observed, while for forcing with a 200-year period, 57–75 % of the potential response is observed (see Wigley 1987*a*, table 1). Finally, the observed cooling clearly depends on the precise time history of the forcing.

Stuiver & Braziunas (1987) have given details of the average shape of ^{14}C maxima events. They distinguish two types of event, similar to either the Maunder Minimum or the Spörer Minimum. For each, the ^{14}C fluctuation spans a period of 100–200 years and is roughly parabolic in shape. We have therefore assumed that the solar forcing is quadratic in time and spans 200 years. To be specific we have assumed that the mean forcing over the 200-year period is 2 W m^{-2} (i.e. with a peak value of 3 W m^{-2}). The response is directly proportional to the forcing, so the results may be scaled to any similar forcing value. Temperature changes were calculated by using the energy-balance climate model of Wigley & Raper (1987), which takes account of oceanic thermal inertia by modelling vertical ocean mixing as an upwelling-diffusion process. The results are shown in table 2.

For a mean cooling of 0.4 °C, the implied mean forcing depends critically on the climate sensitivity. Based on table 2, for a sensitivity of 0.33 °C (W m^{-2})$^{-1}$ the forcing would be 1.32 W m^{-2}, for 0.67 °C (W m^{-2})$^{-1}$ it would be 0.72 W m^{-2}, whereas for a sensitivity of 1 °C (W m^{-2})$^{-1}$ the forcing would be 0.53 W m^{-2}. These values are slightly smaller than those estimated previously, based on slightly different assumptions, by Wigley (1987*a*); namely, 1.1 W m^{-2} and 0.7 W m^{-2} for climate sensitivities of 0.5 °C (W m^{-2})$^{-1}$ and 1.0 °C (W m^{-2})$^{-1}$.

When expressed as percentages of the net incoming solar radiation at the top of the troposphere (namely 240 W m^{-2}), the implied forcing is 0.55, 0.30 or 0.22 % for sensitivities

TABLE 2. GLOBAL-MEAN TEMPERATURE CHANGES

(Changes result from a parabolic solar irradiance fluctuation lasting 200 years with mean radiative forcing of -2 W m^{-2} ($\Delta Q = 3(0.01t-1)^2 - 1$) W m^{-2}). The results were obtained using the model of Wigley & Raper (1987) with a mixed layer depth of 100 m and a diffusivity of 1 cm^2 s^{-1}.)

climate sensitivity °C (W m^{-2})$^{-1}$	equilibrium response to mean forcing	mean change over forcing interval	maximum equilibrium response	maximum change	residual changes at		
					$t = 200$	$t = 300$	$t = 400$
0.333	-0.667	-0.604	-1.000	-0.915	-0.180	-0.024	-0.011
0.667	-1.333	-1.106	-2.000	-1.664	-0.559	-0.107	-0.049
1.000	-2.000	-1.515	-3.000	-2.276	-0.996	-0.243	-0.117

of 0.33, 0.67 or 1.0 °C (W m^{-2})$^{-1}$. Thus to produce a cooling similar to that of the Holocene Little Ice Age events, solar irradiance would have to be reduced by an average of 0.22–0.55 % over a period of order 200 years. For the parabolic time dependence assumed here, the maximum depression of solar irradiance is 1.5 times these values. These changes are up to an order of magnitude greater than the changes in irradiance that have been observed over the past decade by using satellite-based instruments (Kyle *et al.* 1985; Willson *et al.* 1986; Lean & Foukal 1988).

The only comparable estimate of possible century-timescale changes in solar irradiance is a reduction by 0.14 % during the Maunder Minimum, attributed by Kerr (1987) to Lean & Foukal and based on their model relating solar irradiance changes to changes in sunspots, faculae and network radiation (Lean & Foukal 1988). There is a difference of a factor of two to four between this estimate and ours. Apart from the quantitative differences in these estimates, however, there are important qualitative differences. The Maunder Minimum (i.e. the period of near-zero sunspot activity) lasted only from around 1645–1715 (Eddy 1976), whereas the corresponding ^{14}C anomaly spanned the period 1600–1800 (Stuiver & Braziunas 1987). The required comparison, therefore, is between a 70-year-long reduction in irradiance of 0.14 %, and a 200-year period with an average irradiance reduction of 0.22–0.55 %.

On the basis of this comparison, it would appear that the Lean–Foukal model does not give a large enough, nor long enough, drop in irradiance to explain the observed cooling during the Little Ice Age. Thus if the Little Ice Age (and earlier, similar cold periods) were the result of solar irradiance reductions, either some currently unsuspected mechanism must operate to cause these reductions, or the form of the relation between sunspot number, sunspot area, facular area and network radiation deduced from recent data by Lean & Foukal (1988) must break down during and around periods of anomalously low sunspot activity.

CONCLUSIONS AND IMPLICATIONS FOR THE FUTURE

The atmospheric ^{14}C evidence of solar variability on the century timescale during the Holocene is indisputable. Because major ^{14}C anomalies occur throughout the Holocene (17–21 in 9500 years), spanning the whole of the available data, this type of solar variability must be considered to be a permanent feature of the Sun's behaviour. Global-scale climate fluctuations of the Little Ice Age type appear to be associated with these ^{14}C anomalies. This implies that significant irradiance variations occur in parallel with the solar fluctuations responsible for the ^{14}C anomalies.

This link has been examined previously by a number of authors (e.g. de Vries 1958; Damon 1968; Suess 1968; Denton & Karlén 1973; Eddy 1977; Williams *et al.* 1980), but until recently the data were of insufficient quality to offer convincing evidence either for or against. Here, in expanding on a previous analysis (Wigley 1987a), we have shown that temperature minima and ^{14}C maxima are significantly correlated at the 5% level. Although some measure of doubt must remain because of uncertainties in the climate record, and because there are still many inconsistencies when the two records are compared in detail, we consider these new results to be highly suggestive. Indeed, the dating uncertainties in the climate record, together with the undoubted natural variability of the climate system that must have been superimposed on and would tend to obscure any solar forcing effects, make it somewhat surprising that a significant result can be obtained at all.

On the assumption that the Little Ice Age events of the Holocene are the result of solar irradiance changes, we have used an appropriate, time-dependent climate model to estimate the magnitude of these changes. The results depend on the cooling assumed to occur during the Little Ice Ages, which we have taken to be 0.4–0.6 °C, and on the assumed climate sensitivity. For sensitivities in the range 0.33–1.0 °C (W m^{-2})$^{-1}$ (which span the known range of uncertainty in this parameter), the implied solar irradiance changes averaged over the 200-year interval range between 0.55 and 0.22%.

These results allow Man's potential influences on future climate to be put into perspective. In the future the dominant climate forcing mechanism is likely to be the effect of increasing concentrations of greenhouse gases, primarily carbon dioxide, methane, nitrous oxide, ozone and the chloroflurocarbons (CFCs). Over the past few centuries, the forcing due to this effect has amounted to about 2.2 W m^{-2} (Wigley 1987b), equivalent to increasing solar irradiance by over 0.9%. Already, therefore, Man's activities have perturbed the Earth's radiation balance by appreciably more than the perturbations that lead to the Holocene Little Ice Ages. This result is independent of the feedback mechanisms whose uncertain magnitude leads to uncertainties in the climate sensitivity.

Future greenhouse-related forcing changes are certain to exceed those that have occurred in the past. Between now and 2030, the best estimate for future forcing is about 2.4 W m^{-2} (Wigley 1987b; modified to account for reduced CFC concentrations due to the Montreal Protocol, (see Wigley (1988)). If a new Little Ice Age began soon, as has been suggested by some authors (e.g. Landscheit 1983; Fairbridge & Shirley 1987), how would this influence future climate in the face of the onslaught of the greenhouse effect?

Let us suppose a Little Ice Age begins in the year 2000 A.D., and that the solar irradiance decline lasts for 200 years. As a backdrop, we use a projection for greenhouse-gas forcing that follows Wigley (1987b, 1988), to 2030, increases at half the 2000–30 rate of increase until 2100, and then remains unchanged. Under this scenario, the forcing between now (1988) and 2100 is 4.6 W m^{-2}, a not unreasonable possibility. This amounts roughly to a further doubling of the effective CO_2 level (i.e. accounting for all other anthropogenic greenhouse gases) between now and 2100. These greenhouse projections, shown in figure 4, are given only for illustrative purposes. They should not be taken too seriously beyond 2050, because, on the 50-year timescale, Man has the ability to modify future greenhouse-gas concentration changes and significantly reduce future forcing changes.

The solar perturbation during the Little Ice Age depends on the assumed climate sensitivity. For low sensitivity the perturbation is larger, although the influence in terms of global-mean

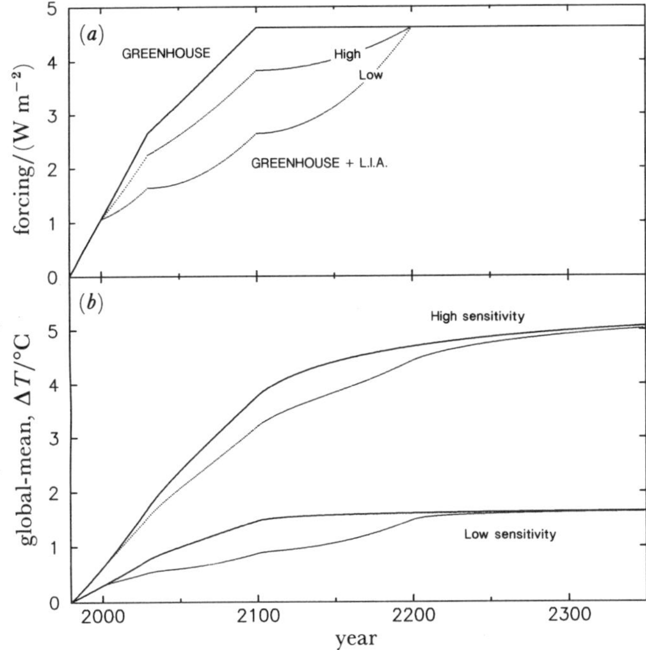

FIGURE 4. Future climatic change if greenhouse-gas-induced warming were partly offset by a Little Ice Age (L.I.A.) event. Part (a) shows the forcing: top curve due to greenhouse gases alone and the lower two curves for greenhouse plus L.I.A. forcing. 'High' and 'low' give the range of uncertainty for the L.I.A. perturbation, and correspond to high and low climate sensitivities, as explained in the text. Part (b) gives the global-mean temperature response to the assumed forcing. The upper curves of the two pairs are for greenhouse forcing alone assuming either high or low climate sensitivity (namely $1.0\ °C\ (W\ m^{-2})^{-1}$ or $0.33\ °C\ (W\ m^{-2})^{-1}$). The lower curves of the pairs show the relative cooling that would be induced by a Little Ice Age event. The model used was that of Wigley & Raper (1987) with a diffusivity of $1\ cm^2\ s^{-1}$. The simulation was begun with the system in equilibrium in 1765 and forced by observed greenhouse gas concentration changes to 1988 and projected changes subsequently. Future greenhouse forcing is subject to considerable uncertainty; the scenario assumed here is close to the current best estimate.

temperature changes remains the same. Figure 4 shows projected temperature changes for climate sensitivites of 0.33 and $1.0\ °C\ (W\ m^{-2})^{-1}$. In the low sensitivity case the solar perturbation is clearly appreciable relative to future greenhouse forcing, but still far from sufficient to offset it completely. In this case the rate of future warming is roughly halved. For high climate sensitivity, however, the effect is, in relative terms, much smaller.

Given the difficulty of detecting the greenhouse effect to date, detecting the onset of a future Little Ice Age in the climate record would be no easy task. The solar irradiance changes would be more easily detectable. At the low end of the estimated irradiance change, where the mean change is 0.22%, the rate of change during the early stages (based on the parabolic forcing assumed above) is 0.044% per decade. Over the last sunspot cycle, solar irradiance varied at about 0.07% per decade, so it would probably require the better part of a full cycle to be sure that some new mechanism had begun to operate.

On uniformitarian grounds, another Little Ice Age must be expected. Past anomalous ^{14}C events were separated by an average of 250–350 years, with the last such event ending around 1800. In one way, therefore, the anthropogenic greenhouse effect might be considered a benefit, because it will certainly offset the next and possibly future Little Ice Ages. This would be an exceedingly optimistic viewpoint, however. Given the disruption caused to society by the cold conditions of the last Little Ice Age, and given that this was a mere 'blip' compared with expected future climatic change, the results of this paper should be seen as further reinforcement for the concerns already felt with regard to the future.

We thank Minze Stuiver for providing the raw ^{14}C anomaly data and Mike Salmon for help in preparing diagrams.

REFERENCES

COHMAP members 1988 *Science, Wash.* **241**, 1043–1052.

Damon, P. E. 1968 *Meteorological Monographs* **8**, 106–111.

Dansgaard, W., Clausen, H. B., Gundestrup, N., Hammer, C. U., Johnsen, S. F., Kristinsdottir, P. M. & Reeh, N. 1982 *Science, Wash.* **218**, 1273–1277.

de Jong, A. F. M., Becker, B. & Mook, W. G. 1986 *Radiocarbon* **28**, 939–942.

de Vries, Hl. 1958 *Koninkijk Nederlandse Akademie Von Wetenschappen, Amsterdam, Proc., Series B* **61(2)**, 94–102.

Denton, G. H. & Karlén, W. 1973 *Quaternary Res.* **3**, 155–205.

Eddy, J. A. 1976 *Science, Wash.* **192**, 1189–1202.

Eddy, J. A. 1977 *Climatic Change* **1**, 173–190.

Fairbridge, R. W. & Shirley, J. H. 1987 *Sol. Phys.* **109**, 191–210.

Grove, J. M. 1988 *The Little Ice Age.* London: Methuen.

Jones, P. D., Wigley, T. M. L. & Wright, P. B. 1986 *Nature, Lond.* **322**, 430–434.

Jouzel, J., Lorius, C., Petit, J. R., Genthon, C., Barkov, N. I., Kotlyakov, V. M. & Petrov, V. M. 1987 *Nature, Lond.* **329**, 403–408.

Karlén, W. 1976 *Geografiska Annaler* A **58**, 1–34.

Kerr, R. A. 1987 *Science, Wash.* **236**, 1624–1625.

Kromer, B., Rhein, M., Bruns, M., Schoch-Fischer, H., Münnich, K. O., Stuiver, M. & Becker, B. 1986 *Radiocarbon* **28**, 954–960.

Kutzbach, J. E. & Street-Perrott, F. A. 1985 *Nature, Lond.* **317**, 130–134.

Kutzbach, J. E. & Guetter, P. J. 1986 *J. atmos. Sci.* **43**, 1726–1759.

Kyle, H. L., Ardanuy, P. E. & Hurley, E. J. 1985 *Bull. Am. meteorolog. Soc.* **66**, 1378–1388.

LaMarche, V. C. Jr 1973 *Quaternary Res.* **3**, 632–660.

Landscheidt, T. 1983 In *Weather and climate responses to solar variations* (ed. B. M. McCormac), pp. 293–308. Boulder: Colorado Associated University Press.

Lean, J. & Foukal, P. 1988 *Science, Wash.* **240**, 906–908.

Linick, T. W., Long, A., Damon, P. E. & Ferguson, C. W. 1986 *Radiocarbon* **28**, 943–953.

Meier, M. F. 1984 *Science, Wash.* **226**, 1418–1421.

Mitchell, J. F. B., Grahame, N. S. & Needham, K. J. 1988 *J. geophys. Res.* **93**, 8283–8303.

Oerlemans, J. 1988 *J. Glaciology* **34**, 333–341.

Pearson, G. W. & Stuiver, M. 1986 *Radiocarbon* **28**, 839–862.

Pearson, G. W., Pilcher, J. R., Baillie, M. G. L., Corbett, D. M. & Qua, F. 1986 *Radiocarbon* **28**, 911–934.

Porter, S. C. 1981 In *Climate and history* (ed. T. M. L. Wigley, M. J. Ingram & G. Farmer), pp. 82–110. Cambridge University Press.

Röthlisberger, F. 1986 *10000 Jahre Gletschergeschichte der Erde.* Aarau: Verlag Sauerländer.

Siegenthaler, U. & Beer, J. 1987 In *Secular solar and geomagnetic variations in the last 10 000 years* (ed. F. R. Stephenson & A. W. Wolfendale), pp. 315–328. Dordrecht: Kluwer.

Stuiver, M. & Quay, P. D. 1980 *Science, Wash.* **207**, 11–19.

Stuiver, M. & Becker, B. 1986 *Radiocarbon* **28**, 863–910.

Stuiver, M. & Pearson, G. W. 1986 *Radiocarbon* **28**, 805–838.

Stuiver, M., Pearson, G. W, & Braziunas, T. F. 1986*a* *Radiocarbon* **28**, 980–1021.

Stuiver, M., Kromer, B., Becker, B. & Ferguson, C. W. 1986*b* *Radiocarbon* **28**, 969–979.

Stuiver, M. & Braziunas, T. F. 1987 In *Secular solar and geomagnetic variations in the last 10 000 years* (ed. F. R. Stephenson & A. W. Wolfendale), pp. 245–266. Dordrecht: Kluwer.

Suess, H. E. 1968 *Meteorological Monographs* **8**, 146–150.

Wigley, T. M. L. 1987*a* In *Secular solar and geomagnetic variations in the last* 10000 *years* (ed. F. R. Stephenson & A. W. Wolfendale), pp. 209–224. Dordrecht: Kluwer.

Wigley, T. M. L. 1987*b* *Geophys. Res. Lett.* **14**, 1135–1138.

Wigley, T. M. L. 1988 *Nature, Lond.* **335**, 333–335.

Wigley, T. M. L. & Raper, S. C. B. 1987 *Nature, Lond.* **330**, 127–131.

Williams, L. D., Wigley, T. M. L. & Kelly, P. M. 1980 In *Sun and climate*, pp. 11–20. Toulouse: Centre National d'Études Spatiales.

Williams, L. D. & Wigley, T. M. L. 1983 *Quaternary Res.* **20**, 286–307.

Willson, R. C., Hudson, H. S., Fröhlich, C. & Brusa, R. W. 1986 *Science, Wash.* **234**, 1114–1117.

Discussion

J. A. Eddy (*UCAR/OIES, Boulder, Colorado, U.S.A.*). Does Dr Wigley agree that the weaker link in the connection he proposes between solar irradiation and terrestrial surface temperature response is that of climate history? Is it not true that for the Holocene period we now know the solar total irradiance forcing better than we know the corresponding climate history?

T. M. L. Wigley. It is true that the climate history is a weak link in inferring that significant solar irradiance changes occurred in parallel with century-time scale atmospheric ^{14}C fluctuations. However, the irradiance history is far less certain: we simply have no direct information at all. The atmospheric ^{14}C record is now well established, but one cannot assume that this is a proxy for irradiance changes; and, even if one could, the magnitude of these changes would still be unknown. The only way we can estimate irradiance changes during the Holocene is by using climate data, as we have done.

A. Berger (*Université Catholique de Louvain, Belgium*). One fundamental hypothesis of Dr Wigley's model is to identify the sensitivity of the climate system to a doubled CO_2 forcing with its sensitivity to solar forcing. Can he comment to what extent this is allowed given the difficult physical mechanisms within the climate system that would be involved in these two experiments.

T. M. L. Wigley. In my verbal presentation, I expressed climate sensitivity in terms of the equilibrium global-mean temperature change for a CO_2-doubling merely to use a concept more familiar to non-specialists. In the written text, climate sensitivity is expressed in the conventional way. There is a close relatinship between the two. If f is the climate sensitivity (in degrees Centigrade per (watt per square metre)), ΔT_d is the equilibrium warming for $2 \times CO_2$, then

$$f = \Delta T_d / \Delta Q_d.$$

ΔQ_d is approximately 4.4 W m^{-2} (uncertainty, $\pm 10\%$ at least), while ΔT_d is usually thought to lie in the range 1.5–4.5 °C with about 90% confidence. Hence, f lies roughly in the range 0.33–1.0 °C (W m^{-2})$^{-1}$. Model studies suggest that f is similar for solar and greenhouse forcing.

A. C. Renfrew, F.B.A. (*Department of Archaeology, University of Cambridge, U.K.*). The paper deals with short-timescale (up to century-timescale) events. But it would be helpful if Dr Wigley would clarify the assumptions that he is making about the major, long-term factors that were responsible for the ice ages themselves and for the succeeding onset of the Holocene. Is it clear that these first-order factors (on the 1000-year timescale and beyond) are irrelevant to Holocene changes? To put the question another way, are these first-order variations unaccompanied by shorter-term, second-order changes? If the major effects are seen as governed by orbital variations, usually expressed by Milankovitch curves, is it appropriate to omit all considerations of these variations when surveying Holocene changes?

T. M. L. Wigley. The Milankovitch context of Little Ice Age type fluctuations during the Holocene is discussed in our written paper. The crucial differences are in the timescale and in the seasonal character of the radiative forcing perturbations. Milankovitch effects are changes in *seasonal* insolation that occur on the 1000-year timescale, while I hypothesize that the Little Ice Age climate fluctuations are caused by perturbations in *annual-mean* incoming radiation occurring on the century timescale. Even during the Holocene, there is striking evidence of Milankovitch effects, and I see these as a slowly moving backdrop against which Little Ice Age events are superimposed. Although this could be related to the difficulty of discerning evidence of global advances so far back in the past, the glacier fluctuation record (figure 1) indicates that Little Ice Age events may have been less pronounced in the early Holocene and that the prevailing background conditions were warmer then. Indeed, in the mid- to high-latitude zones that most of the glacial data come from, the annual-mean incoming solar radiation in the early Holocene was up to 2% more than today, and model results (Mitchell *et al.* 1988) suggest that the climate was warmer in these zones throughout the year.

Phil. Trans. R. Soc. Lond. A **330**, 561–574 (1990)

Printed in Great Britain

561

Modelling the climatic response to solar variability

By J.-C. Gérard

Institut d'Astrophysique, Université de Liège, B-4200 Ougrée-Liège, Belgium

The Sun, the primary energy source driving the climate system, is known to vary in time both in total irradiance and in spectral composition in the ultraviolet. According to solar interior evolution models, the solar luminosity has increased steadily by 25–30 % over the past 4×10^9 years. Periodic variations are also suspected with characteristic timescales of 11 or 22 years, 80–90 years and possibly longer periods. The ultraviolet radiation below 300 nm also exhibits significant changes over the 27-day solar rotation period as well as the 11-year solar cycle.

Variations in the solar constant are expected to produce both direct and indirect (feedback) perturbations in the global surface temperature. A hierarchy of zero- to three-dimensional models have been used to study the complex couplings involved by such effects. The response of a zonally averaged model to possible total irradiance changes associated with the Gleissberg cycle is investigated and compared with measurements of the sea-surface temperature made since 1860.

Changes in the solar ultraviolet irradiance modulate the amount and distribution of atmospheric ozone, which is predicted to change by several percent in the stratosphere. These perturbations directly affect the middle atmospheric thermal structure, but may also generate indirect effects that could possibly account for some short-term geophysical signatures of solar activity. The cycle-modulated energetic particle interaction with the middle atmosphere is also a possible source of global climatic perturbations.

Introduction

The growing concern about climatic changes induced by anthropogenic perturbations of the energy balance of the planet has prompted a re-examination of the natural causes of climatic variability. The most extensively studied of the external varying forcings is the change in solar insolation due to the periodic changes in the orbital characteristics of the Earth's motion (Milankovitch theory). In addition to these effects, intrinsic variations of the solar output have also been observed and have become increasingly well documented through the use of space-borne instruments. Variations in the total solar irradiance and in the short-wavelength portion of the solar spectrum on timescales from seconds to years have been observed and modelled during the past 10 years. Longer-term variations are also suspected, but proxy data and/or solar physics modelling are the only means of investigating long-period modulations.

In this paper, I summarize the available evidence for intrinsic solar variability and describe the climatic consequences of such variations as predicted by current models. I present model calculations of the possible temperature signature of the Gleissberg solar cycle and climatic effects of ultraviolet irradiance variations and intense solar charged particle interaction with the Earth's atmosphere.

Variability of the total solar irradiance

One of the key questions in climate prediction and interpretation of palaeoclimates concerns the importance of the intrinsic solar luminosity changes in time and the response of the climate system to these variations of energy input.

Stellar evolution theories, based on firm theoretical physical grounds, predict a steady increase of the solar luminosity on a timescale of thousands of millions of years. This change in total solar irradiance is an unescapable consequence of the conversion of hydrogen into helium during the Sun's evolution on the main sequence. Sun interior models predict that the solar luminosity increased almost linearly by about 25–30 % in 4.5×10^9 years. Climate models classically indicate that if the solar constant is decreased by a few percent, the ice-albedo feedback resulting from the temperature drop is strong enough to destabilize the system and produce a completely ice-covered Earth. This configuration with ice-caps extending to the Equator is fairly stable and solar luminosities well in excess of the present value would be required to unfreeze the planet. Various scenarios have been proposed to explain the long-term apparent stability of the past climate. One is based on the enhanced greenhouse effect due to larger atmospheric CO_2 concentrations at the early phases of the planet's history (cf. Gérard 1989). No numerical model of the CO_2 global cycle extending far enough in distant past has been developed so far. However, Marshall *et al.* (1988) showed that the reduced extent of the total continental area and the subsequent smaller silicate weathering rate could have resulted in a larger content of the atmospheric CO_2 reservoir than presently. The resulting positive heat increment would have maintained the globally averaged temperature at a sufficient value to prevent complete freezing.

The short-term variability of the solar output is increasingly better observed by sensitive space-borne instruments measuring total solar irradiances over periods of years with excellent stability (P. Foukal, this Symposium). These measurements indicate that, in addition to the blocking effect of sunspots and its compensation by facular energy release, a positive correlation is observed between the solar output and 11-year solar cycle (Willson & Hudson 1988). For solar cycle 22, a total variation of about 1 W m^{-2} was observed with the ACRIM instrument, corresponding to a peak to peak variation of about 0.09 %. Because of the limited timespan of accurate solar radiometric data, a direct assessment of the presence and amplitude of longer periodicities is presently impossible.

Proxy data and indirect evidence need to be used for longer periods and, consequently, lead to speculative and more qualitative results. The best-documented and probably most widely used indicator is the sunspot index time series, which dates back to 1700. This time series was extensively studied by using various analysis methods (cf. Berry 1987; A. Berger, J. L. Mélice and I. van der Mersch, this Symposium). Besides the ubiquitous 11-year periodicity, an amplitude modulation of this cycle by an 80–90-year component (the 'Gleissberg cycle') has been suggested by various studies of the sunspot index series. A 180-year period was also indicated by some analysis but the length of the record and the intrinsic noise do not allow an unambiguous identification of these long periods. Support for the existence of a Gleissberg cycle signature in proxy data mostly originates from indirect climate records, auroral activity and isotropic composition measurements of ice cores. Observations of 80–90 periodicities in the diameter of the Sun have been identified (Gilliland 1981). Ribes *et al.* (1987) indicate that the solar radius was larger and the rotation slower during the Little Ice Age associated with the

Maunder Minimum. The analysis of the thermoluminescent profile of a recent sedimentary core spanning a period of over 18 centuries shows the presence of four main periodicities. The sum of the two low-frequency signals may be considered (Cini Castagnoli *et al.* 1988) as a 82.6-year period component modulated by a second wave with a period of 206 years. The location of the last two maxima corresponds fairly well with the dates of past solar diameter maxima, suggesting a possible link between the two sets of observations.

In summary, converging evidence suggests the existence of a variability of the solar total irradiance with periods of *ca.* 11 years, 80–90 years and possibily longer characteristic times. The reality of claimed statistical Sun–climate correlations and the search for physical mechanisms linking the solar radiative and corpuscular 11- or 22-year cycle to climatic effects has been a subject of considerable debate and speculation (Pittock 1978, 1983). We now concentrate on the description of the global temperature variations induced by changes in the total solar energy input to the climate system.

GLOBAL CLIMATE SENSITIVITY

The sensitivity of climate models to radiative perturbations (including changes in solar output) is frequently expressed as the sum of a direct radiative response and various indirect contributions (feedbacks). Major feedbacks include the effect of the surface-albedo response to changes in surface temperature (extent of the polar ice and snow cover and vegetation changes), water vapour content of the atmosphere, vertical temperature lapse rate, cloud temperature and optical depth. In the study of climate model sensitivity to solar constant variations, it is of common use to specify the sensitivity parameter β defined as

$$\beta = S\,\mathrm{d}\overline{T}_{s}/\mathrm{d}S,$$

where S denotes the total solar irradiance and $\overline{T}_{s}$ the globally averaged mean surface temperature. Another global feedback parameter λ (Dickinson 1985) was also introduced to characterize the relative importance of direct and indirect forcing factors on the global response. This parameter links the perturbations in the global surface temperature ΔT_{s} to an externally prescribed change in the net radiative flux across the tropopause ΔQ:

$$C\partial(\Delta T_{s})/\partial t + \lambda\Delta T_{s} = \Delta Q,$$

where C denotes the system heat capacity. The total sensitivity factor λ may be considered as the summation of a blackbody sensitivity $\lambda_{B}(=3.75\ \mathrm{W\ m^{-2}\ ^{\circ}C^{-1}})$ and other contributing feedback factors:

$$\lambda = \lambda_{B} + \sum_{i}\lambda_{i}.$$

The main components λ_{i} are the ice–albedo and the water–vapour feedback. Another potentially important contribution to λ is the cloud-albedo feedback. However, conflicting results have been obtained about the cancellation of the albedo and long-wave absorption effects resulting from perturbations of the physical characteristics of the clouds. A classical and useful test of a model sensitivity is to change the solar constant by 1 % and calculate the resulting steady-state surface temperature perturbation $\Delta\overline{T}_{s}$.

A hierarchy of simple to very sophisticated models have been developed, mostly as a response to the growing concern about climate perturbations induced by the anthropogenic production of radiatively active trace gases. The simplest (zero-dimensional) model expresses radiative equilibrium between the global surface emitting as a blackbody and absorption of solar radiation,

$$4\pi R_{\mathrm{E}}^2 \sigma T_{\mathrm{e}}^4 = S(1-a)\pi R_{\mathrm{E}}^2, \tag{1}$$

where R_{E} is the Earth radius, T_{e} the effective radiating temperature of the planet, a the global planetary albedo and σ the Stefan–Boltzman constant. In this approximation, the sensitivity parameter β is simply equal to $\frac{1}{4}T_{\mathrm{e}} \approx 64$ °C, implying a temperature change of 0.6 °C for a 1 % variation of the solar irradiance. This value is generally used as a reference to quantify the importance of feedbacks.

One-dimensional models are classified as zonally averaged (za) energy-balance models (ebms) if the spatial coordinate is latitude and radiative–convective (rc) models if variables depend explicitly on the vertical dimension (altitude or pressure). Two-dimensional models consider both altitude and latitude explicitly, whereas three-dimensional models (gcms) solve the full coupled energy, momentum and mass conservation equations. The heat storage and spatial redistribution by the ocean is parametrized by using various degrees of complexity ranging from box models, which may include eddy mixing (bdoms), advection and diffusive mixing (badoms) to three-dimensional representations.

TABLE 1. CLIMATE MODEL SENSITIVITY TO THE TOTAL SOLAR IRRADIANCE VARIATIONS

model type	authors	β parameter/°C
zero-dimensional		
black body		64
EMB[a]	North *et al.* (1981)	112
one dimensional		
EBM[b]	North *et al.* (1981)	160–400
RC	Manabe & Wetherald (1967)	128[a]
RC	Cess (1974)	125[a]–231[b]
RC	Ramanathan (1976)	121[a]–197[b]
RC	Wang & Stone (1980)	110[a]–188[b]
RC	Gérard & François (1988)	93[a]–233[b]
ZA	Thompson & Schneider (1979)	145[a]–230[b]
ZA	Harvey (1988)	168[b]
two dimensional		
	Peng *et al.* (1982)	155[a]
three dimensional		
GCM	Wetherald & Manabe (1975)	180[b]
GCM	Wetherald & Manabe (1980)	205[b]
GCM	Hansen *et al.* (1984)	205[b]

[a] No surface-albedo feedback.
[b] With surface-albedo feedback.

Table 1 lists the value of β obtained with various classical or recent models of each category. The predicted temperature increases for a 1 % change of the solar constant and no ice-albedo–temperature feedback is of the order of 1.1 ± 0.2 °C, a value nearly twice as large as the pure blackbody case. The difference stems mostly from the water–vapour feedback: increased contents of atmospheric H_2O are associated with higher tropospheric temperatures as expressed by the Clausius–Clapeyron law. Surface-albedo–temperature feedback results from the decrease in the extent of the polar cap associated with a temperature increase (positive ice albedo feedback). Another possible component is the vegetation feedback associated with the changes of radiative, sensible and latent heat fluxes stemming from various land-surface processes.

Secular variation of the surface temperature

Climatic models indicate that a global temperature increase of a few tenths of a degree could have occurred in this century as a result of the CO_2 increase due to extensive use of fossil fuel and the resulting enhanced greenhouse. Attempts to detect this climatic signature are based on the reconstruction of the recent records of near-surface continental air temperature (CAT), marine air temperature (MAT) and sea-surface temperature (SST). Reconstruction of these time series are made difficult by changes in the instrumentation and observing methods, presence of spurious measurements (sampling errors) and the intrinsic variability of the climate system. CAT and MAT compilations since 1850 were discussed by Jones *et al.* (1986). Over 46 million SST data from the British Meteorological Office Main Marine Data Bank were used by Folland *et al.* (1984) to estimate the magnitude of the temperature fluctuations for the period 1856–1981. The measurements concentrate at low and middle latitudes and various corrections were applied to account for changes in the sampling technique and experimental method. They found a worldwide temperature fluctuation of *ca.* 0.6 °C, with the coldest period centred around 1905–10 and the warmest values occurring in the 1940s. A three-dimensional plot of the 10-year mean of the temperature anomaly suggests the presence of a 80–90-year modulation reminiscent of the solar Gleissberg cycle (figure 1). The possibility of a climatic signature of the Gleissberg cycle was discussed by Gilliland (1982) and Reid & Gage (1987) using ocean box models. It is further discussed below on the basis of numerical simulations obtained with a zonally averaged seasonal model.

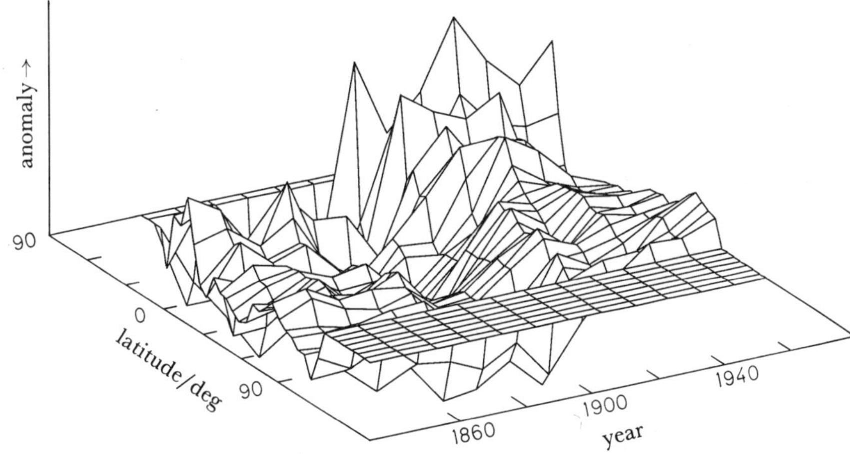

FIGURE 1. Three-dimensional plot of the 10-year average sea-surface anomaly from 1860 to 1980 determined by Folland *et al.* (1984). The empty high-latitude bins correspond to regions with insufficient data coverage.

The possible effect of the secular solar irradiance variability on the climate of the past 120 years is investigated using a two-layer zonally averaged model together with parametrized results from a radiative–convective model. The ocean model is composed of a variable-depth mixed layer overlying a deeper ocean layer with seasonal vertical heat exchange and latitudinal diffusion. This formalism was introduced by Thompson & Schneider (1979) and will only be briefly described here. The latitudinal grid extends from pole to pole in 5° latitude bins. The temperature of the top layer represents a composite between continental and oceanic surface temperatures. The bottom layer provides a thermal sink or source exchanging energy

566   J.-C. GÉRARD

with the top layer over a prescribed annual cycle. It accounts for heat redistribution by vertical
exchanges. The energy-balance equations of the two layers are written

$$\partial(RT)/\partial t = Q(1-a)-I-\operatorname{div}\phi+E \qquad (2)$$

for the top layer, and
$$\partial(R_0\,T_0)/\partial_t = -E \qquad (3)$$

for the bottom layer, where R and R_0 are the thermal inertias of the top and bottom layers
respectively, T and T_0 the surface and bottom layer temperatures, Q the extraterrestrial
insolation, a the planetary albedo, I the outgoing infrared radiation and ϕ the total meridional
heat flux.

The expression of the last term E of equations (2) and (3) depends on the temporal variation
of the effective thickness of the top layer:

$$E = T\partial R/\partial t \quad \text{if} \quad \partial R/\partial t < 0,$$

and
$$= T_0\partial R/\partial t \quad \text{if} \quad \partial R/\partial t > 0.$$

The total thermal inertia in each bin R_t is constant in time and R is written as a sum of two
time-dependent terms, which each correspond to one of the vertical layers:

$$R_t(\theta) = R(\theta, t) + R_0(\theta, t).$$

The values of the coefficients R_t and the R/R_t ratio are adjusted to best reproduce the
observed seasonal variation of the zonally averaged surface temperatures. The amplitude of the
seasonal exchange cycle between the two layers is derived from observations. In the context of
the preceding section, the sensitivity parameter β of this model is equal to 145 °C.

I now discuss the parametrization of the secular variation of the total solar irradiance.
Gilliland (1982) postulated a sinusoidal variation of the solar constant with a 76-year period
shifted in phase by 19 years from the observed solar radius modulation. Reid & Gage (1987)
assumed a linear relation between the sunspot index averaged over 11 years and the solar
output. Schatten (1988) derived a linear expression between the value of S and the variation
of the upper envelope of the sunspot number R_e. Figure 2 illustrates the percentage variation

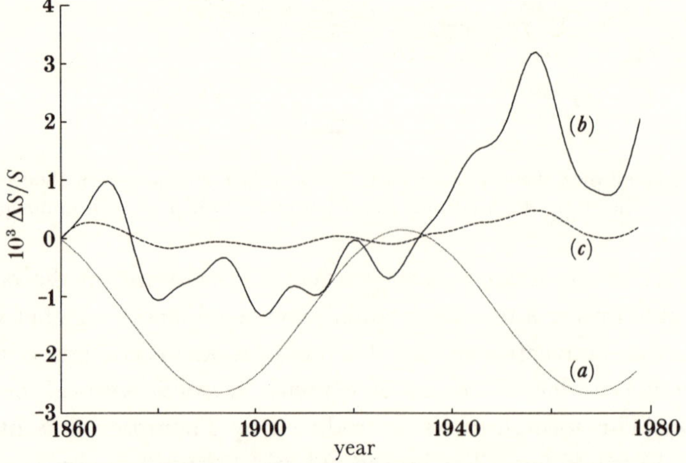

FIGURE 2. Example of assumed secular fractional variations of the solar constant $\Delta S/S$: (a) Gilliland (1982), (b)
Reid & Gage (1987), (c) Schatten (1988). The three curves are normalized to a common value of the solar
constant in 1860.

[168]

of the solar constant since 1860 proposed in these three studies. To test the plausibility of a solar modulation of the surface temperature, we adopt scenarios based on the last two assumptions:

$$S(t) = S_0 + \alpha S_0 (\bar{R} - R_0) \tag{4}$$

and

$$S(t) = S_0 + \gamma (R_e - R_0), \tag{5}$$

where $\bar{R}$ is the 11-year mean sunspot number, R_0 and S_0 reference values for R and the total irradiance respectively. The coefficient α is a proportionality constant to be determined, whereas we use the value $\gamma = 7.5 \times 10^{-3}$ determined by Schatten.

The increase of atmospheric CO_2 of anthropogenic origin generates a time-dependent enhanced greenhouse effect that also affects the long-term evolution of the surface temperature. We adopt the observed CO_2 variation as parametrized by Hoffert *et al.* (1980). The time-dependent CO_2 greenhouse warming is simulated by including a dependence on CO_2 in the coefficient B of the infrared radiation term $I = A + BT$ of equation (2). This dependence was calculated by running a full radiative–convective model (Gérard & François 1988) for various values of the tropospheric CO_2 mixing ratio. The possible contribution of high-altitude aerosols of volcanic origin is neglected in this simulation.

Figure 3*a* illustrates the global mean (area-weighted) decadal temperature anomaly (with respect to the 1945–75 average) derived from Folland *et al.*'s data and the calculated values by using equations (2)–(5). The general trend of the sst variations deduced from the 10-year average data is reasonably well reproduced by the model (figure 3*b*) with the total irradiance variation given by (4) for values of α of the order of 0.7–1.3×10^{-4}. The temperature minimum observed near 1910 is predicted by the model as a consequence of the minimum of solar activity at the beginning of the twentieth century. The peak-to-peak total irradiance variation during this period if the simple dependence of equation (4) is correct, would be on the order of 0.8%, on a timespan of about 50 years. This amplitude is significantly larger than the *ca.* 0.1% variation associated with the current 11-year solar cycle, but may not be unreasonable over the longer period discussed here. The fairly steady sst temperatures observed after 1960 are also

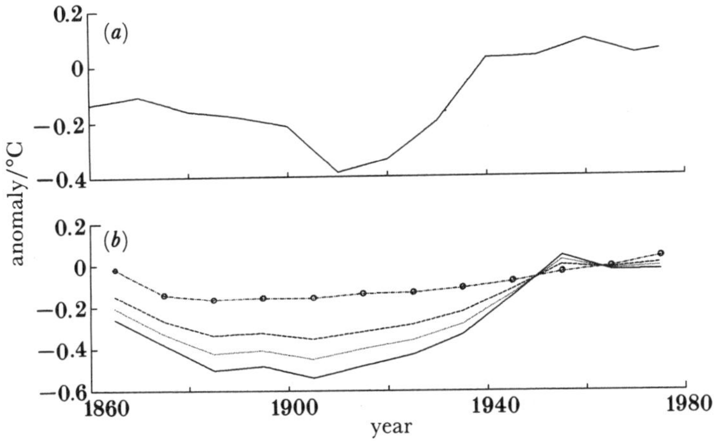

FIGURE 3. (*a*) Time-dependence of the globally averaged sea-surface temperature anomaly deduced from Folland *et al.*'s data. (*b*) Model calculation of the globally averaged anomaly evolution driven by the atmospheric CO_2 increase and assumed secular changes in the solar total irradiance. The four curves correspond to different solar forcing functions: ----, equations (4), $\alpha = 7.5 \times 10^{-5}$; ······, idem, $\alpha = 1.1 \times 10^{-4}$; ——, idem, $\alpha = 1.4 \times 10^{-4}$; ○—·—○, equation (5), $\gamma = 7.5 \times 10^{-3}$.

well reproduced by the model with the functional dependence given by (4) for all three values of the parameter α represented in figure 3b. By contrast, the anomaly calculated by using (5) fails to exhibit the correct anomaly amplitude and the fairly constant value observed during the second part of the twentieth century. The amplitude of the solar constant variation in this case is less than 0.1 %, which explains the weak response of the average surface temperature calculated with this expression.

We thus conclude that, if the secular variation of ssts observed since 1860 is of solar origin, a total solar irradiance of several tenths of a percent is needed to account for the *ca*. 0.6 °C observed temperature change. A similar conclusion was reached by Gilliland & Schneider (1984) using a seasonal 10-box energy-balance model with least-squares fitting of the amplitude and phase of various forcing functions. However, this observational data were based on hemispherically averaged air temperature over land instead of ssts and involved several degrees of freedom.

The agreement obtained by using an expression such as (4) does however not imply any causal relation between the number of sunspots and the solar constant. It would merely suggest that a common physical mechanism modulates the amplitude of the 11-year sunspot cycle and the value of the total irradiance with a 80–90-year periodicity.

Solar ultraviolet cycles

Solar observations

The two dominant periodicities observed so far in the ultraviolet (uv) solar irradiance below 300 nm correspond to the 27-day solar rotation period and the 11-year solar cycle. The 27-day variability and its spectral dependence are increasingly well known (Lean 1987; Simon 1990). An analysis of the 27-day variations observed with the *Solar–Mesosphere Explorer* (*SME*) satellite and their dependence on the solar cycle using Fourier spectral methods was recently obtained by Simon *et al.* (1987). It shows that the amplitude of the 27-day modulation strongly depends on the phase of the solar cycle.

The wavelength dependence of the modulation of the solar uv irradiance over a solar cycle is much more uncertain. Satellite observations require a very good intrumental stability over several years of observations and corrections for sensitivity changes due to drift in surface reflections, transmission of optical elements, detector sensitivity, etc. Rocket observations carried out with similar instruments at different phases of the solar cycle do not guarantee identical observing conditions and calibrations. Nevertheless, a general view of the amplitude of the 11-year solar cycle may be obtained by synthesizing available observations obtained over the past 15 years or so (figure 4).

Below 102.5 nm, the *Atmosphere Explorer* database has been and is still the most widely used source of information for aeronomical modelling of the neutral and ionized upper atmosphere. The curve labelled AE-E refers to the corrected reference solar maximum and minimum spectra (21 REFW and F79050N) (Torr & Torr 1985). The variability in the wavelength region of the Schumann–Runge continuum deduced from the *AE* data by Torr *et al.* (1980) is also indicated. The LASP rocket observations were made between 1972 and 1977 (minimum) by Rottman (1981) and in 1979–80 (maximum) by Mount & Rottman (1981) using similar instruments and calibration techniques. The SBUV curve was deduced from an empirical relation based on the temporal variation of the ratio between core and wing

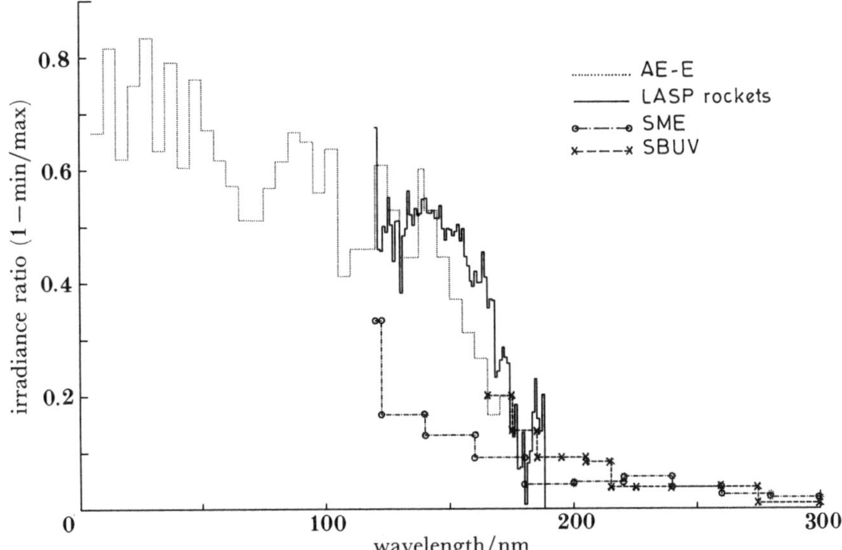

FIGURE 4. Wavelength dependence of the variability of the UV solar irradiance compiled from
various satellite and rocket experiments.

irradiances of the MgII doublet at 280 nm and is thus not a direct measurement. This method
was used because ageing problems of the instrument prevented the long-term observations to
be used directly for solar sycle studies. The amplitude of the variability deduced from these
observations is substantially larger than that derived from the *SME* mission. The UV solar
spectrometer on board the *SME* satellite obtained solar spectra from 1982 to 1988 that exhibit
a maximum:minimum ratio not exceeding 5% at 200 nm. The general behaviour of the UV
solar spectrum during a solar cycle is a decreasing amplitude of variability from the X-rays to
the visible. However, the exact quantitative wavelength dependence of this solar cycle
variation is still debated in spite of major instrumental and calibration improvements.
Virtually nothing is known concerning possible differences between individual solar cycles and
the secular variability of the UV solar output.

The atmospheric response

The climate response of the middle atmosphere to UV solar variations has been observed with
satellites and numerically simulated with various techniques. I briefly review its major
characteristics.

The response of the middle atmosphere climatology to the 11-year solar cycle was studied
in details by Brasseur & Simon (1981), Garcia *et al.* (1984) and more recently by Brasseur
et al. (1988) using two-dimensional chemical–dynamical models. They calculate a maximum
variation ranging between 2 and 3 °C near 50 km. The predicted variation depends on latitude
(and season), the highest temperature changes being associated with the high-latitude summer
regions.

At lower altitudes, the atmospheric temperature response decreases drastically, reaching less
than 0.5 °C at 20 km. The ozone column variation between 1986 and 1979 is about 2% in

Brasseur *et al.*'s simulation (including also changes in trace gas concentrations) and reaches about 3.5 % between solar maximum and minimum conditions in Garcia *et al.*'s model.

A part of the atmospheric response in Garcia *et al.*'s model is a result of the 11-year modulation of the ionization caused by the interaction of energetic particles with the high-latitude upper and middle atmosphere. The exact magnitude and the details of the changes in the contribution of the particle flux and energy spectrum of the precipitated particles are not well established. Of particular importance is the change over a solar cycle of the nitric oxide density in the lower thermosphere and in the mesosphere resulting from the N_2 impact dissociation by magnetospheric particles and the subsequent formation of NO molecules. These molecules flow downward with a higher efficiency in the polar night region and partly reach the stratosphere. Once in this region, they contribute to the catalytic recombination of ozone. The variability of the upper atmospheric NO source is predicted by two-dimensional studies of the distribution and downward flow of nitric oxide at solar minimum and maximum conditions (Gérard & Roble 1988). The results exhibit larger NO concentrations associated with high solar activity, in agreement with the *AE* and *SME* satellite observations. This thermospheric–stratospheric coupling may contribute to the seeked solar–climatic link.

As altitude decreases, the temperature structure becomes less and less sensitive to the direct effect of solar uv variations for two reasons. First, the increasing air mass density decreases the temperature variations associated with a given change of energy per unit volume. Secondly, only radiation above *ca.* 320 nm reaches the troposphere. This wavelength region exhibits unmeasurably small (less than 10^{-3}) fractional spectral variation. As described before, the total irradiance variations observed in the visible and near infrared (IR) affect the energy balance of the troposphere and the surface rather than the photochemistry. Therefore, only indirect mechanisms may influence the tropospheric climate. A wealth of such processes have been proposed to explain the claimed correlation between climatic or meteorological effects and the solar output. The reader is referred to the excellent reviews by Pittock (1978, 1983) on this subject. Taylor (1986) discussed possible links between solar wind and climatic short-term processes.

Reductions of the amount of atmospheric ozone produces competing effects on the thermal structure. More solar (visible and uv) radiation reaches the troposphere and the surface temperature tends to increase. However, this effect is counteracted by a cooling of the stratosphere and a subsequent drop of the downward flux of infrared radiation to the surface. This downward flux is further decreased by the reduction of the greenhouse efficiency in the 9.6 µm band. Numerical experiments have shown that the IR cooling effect dominates the surface solar warming by about an order of magnitude.

The climatological influence of ozone perturbations on the global surface temperature was analysed quantitatively by using radiative–convective models (Reck 1975; Wang *et al.* 1980; François 1988). A general conclusion of these studies is that a decrease of ozone in the troposphere and lower stratosphere cools the surface as a consequence of the reduced greenhouse efficiency. The predicted response of the higher-altitude perturbations is more complex and strongly depends on climatic (warm or cold) conditions themselves. However, the magnitude of the calculated induced temperature perturbation remains fairly small. The peak value is 0.08 °C/(percent O_3 perturbation) (or 1.82 °C/(O_3 mm atm† change)). These studies also demonstrate that perturbations in the O_3 vertical distribution may induce climatic

† 1 atm ≈ 10^5 Pa.

[172]

effects, even in the absence of any total column change. More important indirect effects may result from the changes in stratospheric temperatures and subsequent perturbations in dynamical processes.

THE ATMOSPHERIC RESPONSE TO PAST SOLAR EVENTS

The global response of the ancient oxygen-poor atmosphere to solar cycle effects was studied by Gérard & François (1987) with a one-dimensional photochemical–radiative–convective model. The aim of this study was to test quantitatively the atmospheric and surface temperature modulation as a response to the 11-year cycle in an atmosphere whose O_2 content was significantly lower than the present one. The model study indicated that the maximum temperature response is predicted for an O_2 mixing ratio of *ca.* 3×10^{-3} times the present level. The calculated temperature variation at the time of the Elatina Precambrian varves (640 Ma ago), where periodicities ascribed to the solar cycle have been observed. The recent re-examination of the Elatina deposits and analysis of other Precambrian varves (G. E. Williams, this Symposium) leads to the conclusion that the periodicities observed in the lake sediments are most likely of tidal origin. This conclusion would be in good agreement with the results of the model study indicating a weak direct solar-induced temperature effect.

The importance of the solar energetic particle interaction with the middle atmosphere was described in details by Thorne (1980). He showed that the galactic cosmic ray and the relativistic electron inputs to the atmosphere are modulated by the 11-year solar cycle. The interaction of these particles with atmospheric N_2 produces NO molecules that participate in the ozone photochemistry and NO_2 that absorbs a fraction of the solar radiation. Consequently, climatic effects may possibly accompany solar periodic activity cycles. Similarly, large solar flares responsible for major solar proton events tend to concentrate near solar maximum activity peaks.

The largest such event during the space era was observed in August 1972 and followed by a strong stratospheric ozone depletion with a lifetime of several weeks. It is possible that, during the Earth's history, much larger and more frequent proton events accumulated during high solar activity periods. The efficiency of such events would be enhanced during periods of geomagnetic field reversals when a larger fraction of the Earth's atmosphere would be in contact with solar wind particles (Reid *et al.* 1978). The potential photochemical and climate impact of such events was examined recently by Hauglustaine & Gérard (1990) using a coupled photochemical–radiative–convective model. An upper limit on the effects generated by the accumulation of strong solar proton events was simulated by calculating the steady-state perturbations induced by the August 1972 proton flux. A large increase of NO_2 is predicted, together with a substantial ozone depletion near the altitude of the peak of the proton energy deposition. The solar radiation spectrum at the surface exhibits a depletion in the visible region as a consequence of the enhanced NO_2 absorption. The perturbation to the thermal structure is important (figure 5), especially in the stratosphere where NO_2 becomes the major source of heating. The surface temperature drops by about 6 K and the top of the convective region moves from 20 to 16 km. The total planetary albedo increases by 6.5% as a result of the increase of the atmospheric component. As mentioned before, the calculated temperature perturbation is an upper limit that would be obtained rapidly in the stratosphere whereas the time-constant of the surface thermal inertia would add considerable delay until equilibrium is

[173]

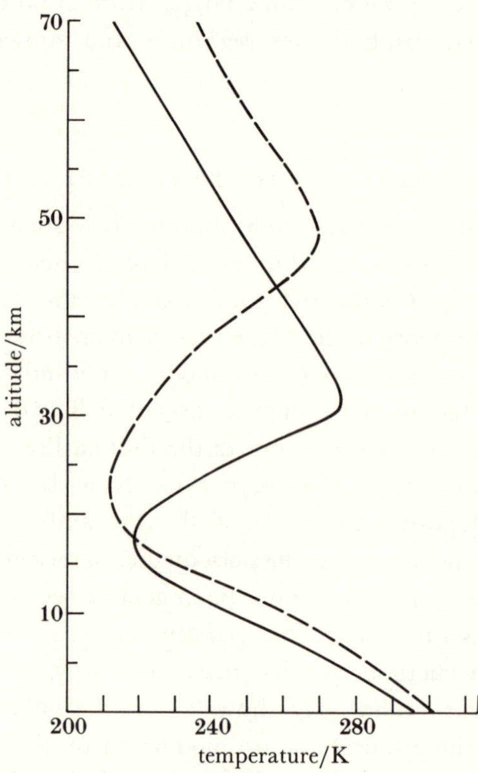

FIGURE 5. Vertical distribution of the atmospheric temperature (kelvins) calculated for solar–proton event conditions of August 1972 with a coupled chemical–radiative–convective model ran to steady state (solid line). The profile calculated for unperturbed conditions is also shown for comparison (dashed line).

obtained. This simulation, however, demonstrates the potential effect of the variability of the solar corpuscular output on the planet's climate.

CONCLUSIONS

Solar constant variations on timescales of minutes to years have been shown to exist with amplitudes less than 0.1 %. The associated predicted global temperature variations would be of the order of 0.1–0.3 °C, with the main uncertainty stemming from the difficulty to determine accurately the importance of the surface feedbacks and the role of heat exchanges inside the oceans. These temperature variations are small and probably insignificant to human activity. However, in this period of growing concern about the future of the climate, it is crucial to understand the details of the natural variability and its past and present response to external forcing. If surface warming due to the enhanced CO_2 and other trace gas concentrations is to be detected without ambiguity, it is necessary to properly account for even small natural causes of climate variability.

The spectral variations of the uv solar spectrum are increasingly better known. Their direct climate impact seems small but various indirect mechanisms may link for example the ozone solar cycle variation and global or regional climate. The interaction of energetic solar particles with the atmosphere may also have triggered climatic changes of fairly large magnitude.

I am grateful to P. Simon for providing me with unpublished material on solar variability, D. Delcourt for her help in the model development and D. Hauglustaine for his assistance with the manuscript. The author is supported by the Belgian Fund for Scientific Research. This research was partly financed by an FNRS grant. Decadal mean sst anomalies were provided by the U.K. Meteorological Office.

References

Berry, P. A. M. 1987 *Vistas Astr.* **30**, 97–108.

Brasseur, G. & Simon, P. C. 1981 *J. geophys. Res.* **86**, 7343–73–62.

Brasseur, G., Hitchman, M. H., Simon, P. C. & De Rudder, A. 1988 *Geophys. Res. Lett.* **15**, 1361–1364.

Cess, R. D. 1974 *J. quant. Spectrosc. Radiat. Transfer* **14**, 861.

Cini Castagnoli, G., Bonino, G. & Provenzale, A. 1988 *Solar Phys.* **117**, 187–197.

Dickinson, R. E. 1985 *Adv. Geophys.* A**28**, 99–129.

Folland, C. K., Parker, D. E. & Kates, F. E. 1984 *Nature, Lond.* **310**, 670–673.

François, L. M. 1988 *Atmos. Res.* **21**, 305–321.

Garcia, R. R., Solomon, S., Roble, R. G. & Rusch, D. W. 1984 *Planet. Space Sci.* **32**, 411–423.

Gérard, J. C. 1989 In *Contribution of geophysics to climate change studies* (ed. A. Berger, R. E. Dickinson & J. Kidson), Geophysical Monograph no. 52, pp. 139–148. Washington, D.C.: American Geophysical Union.

Gérard, J. C. & François, L. M. 1987 *Nature, Lond.* **326**, 577–580.

Gérard, J. C. & François, L. M. 1988 *Ann. Geophys.* **6**, 101–112.

Gérard, J. C. & Roble, R. G. 1988 *Planet. Space Sci.* **36**, 271–279.

Gilliland, R. L. 1981 *Astrophys. J.* **248**, 2144–2155.

Gilliland, R. L. 1982 *Clim. Change* **4**, 111–131.

Gilliland, R. L. & Schneider, S. H. 1984 *Nature, Lond.* **310**, 38–41.

Hansen, J., Lacis, A., Rind, D., Russell, G., Stone, P., Fung, I., Ruedy, R. & Lerner, J. 1984 *Climate processes and climate sensitivity*, pp. 130–163. Washington, D.C.: American Geophysical Union.

Harvey, L. D. D. 1988 *Clim. Change* **13**, 191–224.

Hauglustaine, D. & Gérard, J. C. 1990 *Annls. Geophys.* **8**. (In the press.)

Hoffert, M. I., Callegari, A. J. & Hsiek, C. T. 1980 *J. geophys. Res.* **85**, 6667–6679.

Jones, P. D., Wigley, T. M. L. & Wright, P. B. 1986 *Nature, Lond.* **322**, 430–434.

Lean, J. 1987 *J. geophys. Res.* **92**, 839–868.

Manabe, S. & Watherald, R. T. 1967 *J. atmos. Sci.* **24**, 241–259.

Marshall, H. G., Walker, J. C. G. & Kuhn, W. R. 1988 *J. geophys. Res.* **93**, 791–802.

Mount, G. H. & Rottman, G. J. 1981 *J. geophys. Res.* **86**, 9193–9198.

North, G. R., Cahalan, R. F. & Coakley, J. A. 1981 *Rev. Geophys.* **19**, 91–121.

Peng, L., Chou, M. D. & Arking, A. 1982 *J. atmos. Sci.* **39**, 2639–2656.

Pittock, A. B. 1978 *Rev. Geophys. Space Phys.* **16**, 400–420.

Pittock, A. B. 1983 *Jl R. met. Soc.* **109**, 23–55.

Ramanathan, V. 1976 *J. atmos. Sci.* **33**, 1330–1346.

Reck, R. 1975 *Atmos. Environ.* **10**, 611–617.

Reid, G. C. & Gage, K. S. 1987 *Nature, Lond.* **329**, 142–143.

Reid, G. C., McAffee, G. R. & Crutzen, P. J. 1978 *Nature, Lond.* **275**, 489–492.

Ribes, E., Ribes, J. C. & Barthalot, R. 1987 *Nature, Lond.* **326**, 52–55.

Rottman, G. J. 1981 *J. geophys. Res.* **86**, 6697–6705.

Schatten, K. H. 1988 *Geophys. Res. Lett.* **15**, 121–124.

Simon, P. C. 1990 In *Summary document* (ed. R. A. Vincent). MAP Handbook. (In the press.)

Simon, P. C., Rottman, G. J., White, O. R. & Knapp, B. G. 1987 In *Proc. Workshop on Solar Radiative Variation* (ed. P. Foukal). Cambridge, Massachusetts: CRI.

Taylor, H. A. 1986 *Rev. Geophys.* **24**, 329–348.

Thompson, S. L. & Schneider, S. H. 1979 *J. geophys. Res.* **84**, 2401–2414.

Thorne, R. S. M. 1980 *Pure appl. Geophys.* **118**, 128.

Torr, M. R. & Torr, D. G. 1985 *J. geophys. Res.* **90**, 6675–6678.

Torr, M. R., Torr, D. G. & Hinteregger, H. E. 1980 *J. geophys. Res.* **85**, 6063–6068.

Wang, W. C. & Stone, P. H. 1980 *J. atmos. Sci.* **37**, 545–552.

Wang, W. C., Pinto, J. P. & Yung, Y. L. 1980 *J. atmos. Sci.* **37**, 333–338.

Wetherald, R. T. & Manabe, S. 1975 *J. atmos. Sci.* **32**, 2044–2059.

Wetherald, R. T. & Manabe, S. 1980 *J. atmos. Sci.* **37**, 1485–1510.

Willson, R. C. & Hudson, H. S. 1988 *Nature, Lond.* **332**, 810–812.

Discussion

J.-C. PECKER (*Collège de France, Paris, France*). In the study of effects of solar UV irradiance, Dr Gérard has shown a diagram displaying the effects, at all altitudes and latitudes on Earth, of the cyclic variations of solar UV. I have noted in this diagram a clear N–S asymmetry. What is its origin?

J.-C. GÉRARD. The diagrams concern the solstice, not the equinox!

Phil. Trans. R. Soc. Lond. A **330**, 575 (1990)

Printed in Great Britain

575

Are there solar signals in the African monsoon and rainfall?

By H. Faure[1] and M. Leroux[2]

[1] *Laboratoire de Géologie du Quaternaire, Centre National de la Recherche Scientifique,
Faculté des Sciences de Luminy, Case 907, 13288 Marseille Cedex 9, France*
[2] *Université de Lyon, Lully, F-74890 Bons en Chablais, France*

A review of areas of Africa where a signal in meteorological, hydrological or sedimentological parameters can be suspected to be of solar origin shows a variety of deposition. Nowhere is a persistent signal clearly defined, but many areas are involved: Nile basin, Ethiopian high plateau, Sahel, North and South Africa, Central and East Africa. Each area shows a repartition of rainfall distributed by successive sequences of drought and inundation, each lasting several years. Nearly all areas show a strong quasi-biennial fluctuation; in the Sahel, for instance, the mean annual river run-off in Senegal is about 2.3 years. The 11.4-year signal is also seen in many records of extreme (positive or negative) river run-off.

To test if this can be considered as the consequence of a physical process we first examine the main rainfall mechanism in the Sahel. The mechanism of Sahelian individual rainfalls (during summer monsoon) is controlled mostly by the possibility and frequency of squall line formations. Among the necessary conditions for a reasonable frequency and intensity of rainfall, the African easterly jet wind must not be too strong, because these winds have a saw effect, stop the cumulo-nimbus development and inhibit rainfall. The simplified model shows that small changes in the general atmospheric circulation can enhance or diminish the frequency of rainfall during the summer monsoon. Rain could be related to the strengthening or slowing of the African and tropical easterly jet as part of the general atmospheric circulation.

At the timescale involved the possible solar action is not a result of a change in irradiance energy of the Sun, but it seems to be changing along with its corpuscular and electromagnetic activity.

The mechanism of this discontinuous solar activity, which has a possible effect on global climate, may be found in the interaction between the low ionosphere and the high troposphere through the gravity waves of the quasi-biennial oscillation. Because electric currents in the Earth's magnetic field have opposed latitudinal displacement with their zonal direction, they tend to compress or expand the polar vortex. And because the electrojets change their west to east direction along with solar activity, they are a possible link through the following simplified process.

When and where the ionospheric electrojet gravity waves come to add their momentum to high tropospheric jets, they speed the general circulation. This could give a 'cold global situation' with a strong polar vortex and a feeble monsoon, giving more frequent droughts in Sahel.

When the situation is reversed and the electrojet direction is opposite, it tends to slow down the high tropospheric wind: a 'warm global situation' is more frequent, the Northern Hemisphere summer monsoon can expand more and droughts are less frequent.

In this proposed mechanism, one momentum is of gravity origin in the general neutral atmospheric circulation (coriolis) and the other one takes its dynamics in the electromagnetic forces (ampère) of the ionized stratosphere and ionosphere.

Phil. Trans. R. Soc. Lond. A **330**, 577–589 (1990)
Printed in Great Britain

577

Associations between the 11-year solar cycle, the quasi-biennial oscillation and the atmosphere: a summary of recent work

By Karin Labitzke[1] and H. van Loon[2]

[1] *Institut für Meteorologie, Freie Universität Berlin, Dietrich-Schäfer-Weg 6–10, 1000 Berlin 41, F.R.G.*

[2] *National Center for Atmospheric Research, Boulder, Colorado 80307, U.S.A.*

Atmospheric elements at all levels from the surface to the top of the middle atmosphere show a probable association with the 11-year solar cycle that can be observed only if the data are divided according to the phase of the quasi-biennial oscillation. In either phase the range between solar extremes is as large as the interannual variability of the given element; and the correlations are statistically meaningful when tested both by conventional and Monte Carlo techniques. The sign of the correlations changes spatially on the scale of planetary waves or teleconnections. As the correlations tend to be of opposite sign in the two phases of the quasi-biennial oscillation, correlating a full time series of an atmospheric element with the solar cycle nearly always yields negligible correlation coefficients.

1. The discovery of the solar signal in the middle atmosphere

We must first introduce two features in the atmosphere that were essential to the discovery of the association between the variability of the Sun and of the atmosphere: (i) the major midwinter warmings in the polar stratosphere, and (ii) the quasi-biennial oscillation (QBO) in the equatorial stratosphere.

(i) The QBO is a tendency for the wind in the equatorial stratosphere to change from east to west about every second year. Its phase propagates downward, and the oscillation disappears near the tropical tropopause (*ca.* 16 km; Naujokat 1986). Studies of the QBO in the extratropical stratosphere by Holton & Tan (1980) and Labitzke (1982) found it useful to group the data into winters with easterly or westerly winds in the equatorial QBO at 50 mbar† (*ca.* 21 km), because this division brings out marked differences in the general circulation of the entire stratosphere in the northern winter.

(ii) The polar cyclonic vortex in the normally cold stratosphere of the Northern Hemisphere in winter is a vigorous westerly current around a low in the Arctic. In certain years the vortex breaks down in the middle of winter, the Arctic becomes warmer than middle latitudes, and the west winds are replaced by easterlies. Such a major midwinter warming usually lasts about a month (Labitzke 1981). The warmings are unique to the Northern Hemisphere, and they are associated with amplification of the planetary waves in the troposphere and stratosphere. The effect of the major midwinter warmings extends horizontally to the tropics and vertically into the upper mesosphere.

The idea to search for a solar–terrestrial relation through the QBO stems from an observation by Labitzke (1982) that in the westerly phase of the QBO, major midwinter warmings tend to occur only in maxima of the 11-year solar cycle. Labitzke (1987a) confirmed this observation

† 1 mbar = 10^2 Pa.

[179]

44-2

by plotting the 30 mbar (*ca.* 22 km) temperature at the North Pole in the west years of the QBO as a function of the sunspot number. We show an updated version of her diagram in figure 1, in which we substitute the 10.7 cm solar flux for the sunspots; the strength of this flux is closely related to the sunspot number, and it is a good, objective measure of the Sun's variability.

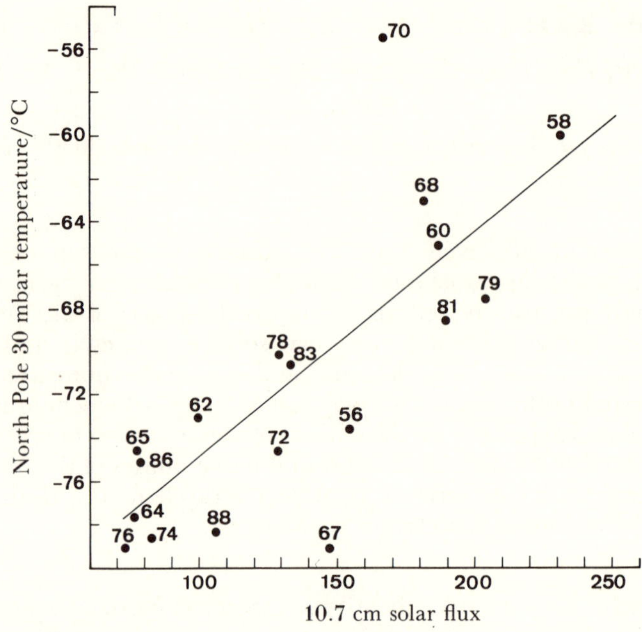

FIGURE 1. The 30 mbar (*ca.* 22 km) temperatures (degrees centigrade) at the North Pole in 18 winters (January–February) plotted against the 10.7 cm solar flux (units are 10^{-22} W m^{-2} Hz^{-1}) in the west years of the QBO. The line is the linear regression; the correlation is 0.77. In a bootstrap sample of 1000, 95% of the correlations lie between 0.54 and 0.91. The 99.9% confidence level for 16 degrees of freedom is 0.82.

The correlation between the two quantities in figure 1 is 0.77. Large correlation coefficients occur time and again in the following; we have tested their statistical significance both with conventional methods and with Monte Carlo methods, and they pass the tests without difficulty. The coefficient of 0.77, for instance, is significant between the 99% and 99.9% levels for 16 degrees of freedom. Even if one reduces the degrees of freedom to 10, considering that it takes four points to describe an 11-year solar cycle, the 0.77 remains well above the 95% confidence level. A Monte Carlo method called 'bootstrapping' tells us that the true correlation coefficient for our sample size is likely to lie between 0.54 and 0.91, and that it is almost certainly not zero. The statistical methods which we have applied are described in Labitzke & van Loon (1988) and van Loon & Labitzke (1988).

Our investigation has two obvious limitations: (1) we can use only slightly more than three solar cycles because the phase of the QBO cannot be defined before 1952; and (2) we cannot explain how the observed, comparatively small solar variability – $\frac{1}{10}$% of the total irradiance from solar maximum to solar minimum during the last solar cycle – can produce the often astonishingly large correlations with the atmospheric elements. It is therefore not certain that our results are universally valid. We must wait for confirmation, either through a physical explanation or through longer series of observations that can tell whether our statistics are stable.

The full series of 30 mbar temperature at the North Pole in January–February, 33 years, is

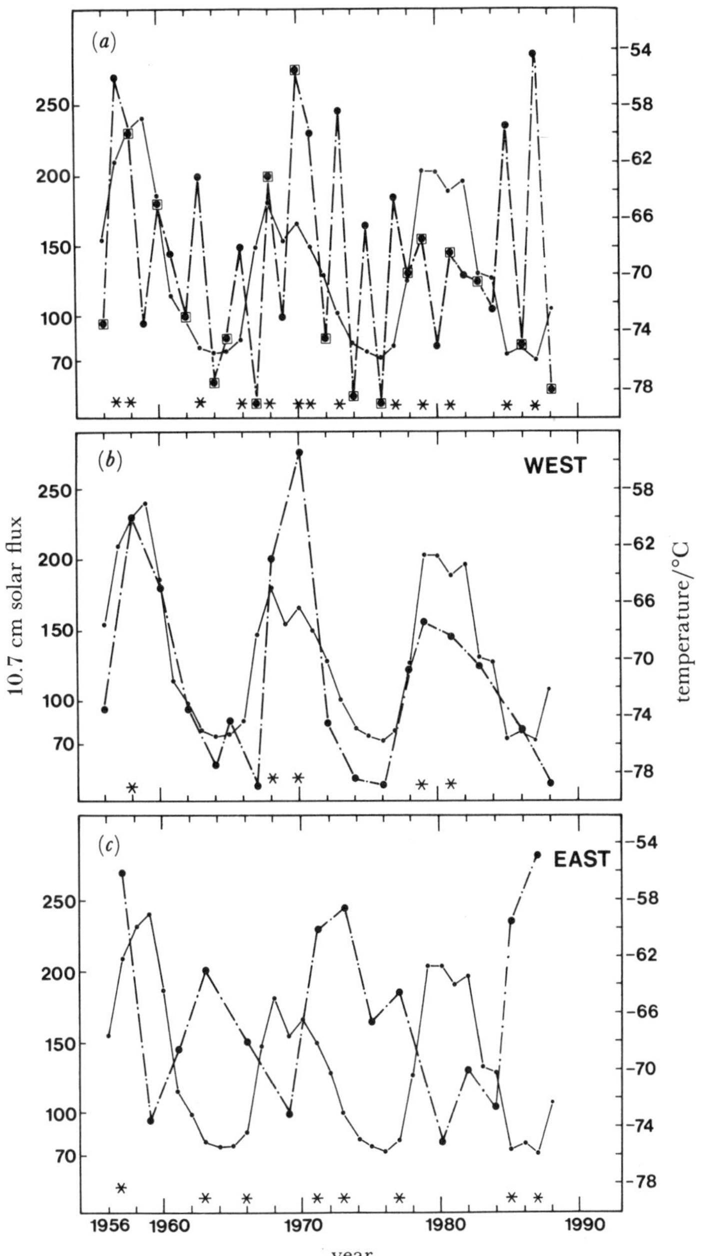

FIGURE 2. (a) Time series of the 10.7 cm solar flux and of the mean 30 mbar temperature (degrees centigrade) at the North Pole, January–February. Squares on the temperature curve denote winters in the west phase of the QBO. The asterisks are the years with major midwinter warmings: $n = 33$, $r = 0.13$ (1956–88). (b) The solar flux as in (a). The 30 mbar temperature curve is only for the winters in the west phase: $n = 18$, $r = 0.77$, 99 % c.l. (confidence level) = 0.65. (c) As (b), but for winters in the east phase: $n = 15$, $r = -0.45$, 90 % c.l. = -0.44. Update of figure 1 in Labitzke & van Loon (1988).

plotted in figure 2a together with the solar flux. The correlation between the two is only 0.13, but if the 33 years are divided into the 18 west and the 15 east years (figure 2b, c) the correlation is, as noted, 0.77 for the west and -0.45 for the east years. We examined how high is the probability for such a cross correlation with the solar flux to occur by accident if one divides a 33-year series into two with 18 years in one and 15 in the other. A Monte Carlo test

showed that this would happen by chance only four times out of 1000. The details of the test
are explained in Labitzke & van Loon (1988).

2. THE VERTICAL AND HORIZONTAL DISTRIBUTION IN THE MIDDLE ATMOSPHERE

The correlations in the east and west phase of the QBO at the North Pole in figure 2 have the
same sign as high as about 35 km all over the Arctic. Labitzke & Chanin (1988) extended the
correlations into the mesosphere by means of the rocketsonde observations from Heiss Island
(81° N, 58° E) In the upper stratosphere, figure 3a, the correlations change sign so that here
the west years are negative and the east years positive. Not until one reaches 60 km is the
correlation of the same sign in both phases of the QBO. Consequently, there is a positive
correlation in the mesosphere between the complete temperatures series and the Sun, whereas
there is little or no relation between the complete series and the solar flux in the stratosphere
where the east and west years offset each other.

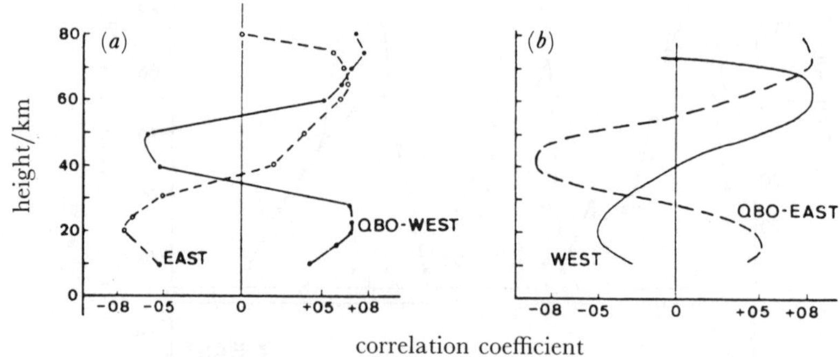

FIGURE 3. (a) Vertical distribution at Heiss Island (81° N, 58° E) of the correlation between the 10.7 cm solar flux
and temperature in January–February in the east and in the west years of the QBO. (b) As (a), but for Haute
Provence (44° N, 6° E). (Labitzke & Chanin 1988.)

The horizontal pattern of the correlations at the 30 mbar level is the same for the
temperature and geopotential height. The latter is shown in figure 4. In the west years, figure
4a, the positive correlations in the Arctic are surrounded by negative correlations in middle
and lower latitudes; in the east years (figure 4b) the distribution is the opposite. Translated into
geostrophic wind, it means that in the west years the polar vortex tends to be weak in solar
maximum and strong in solar minimum, and conversely in the east years.

The opposition in sign between the correlations in middle and high latitudes in figure 4
continues through the middle atmosphere. We illustrate this by means of the observations from
a lidar station at Haute Provence (44° N, 6° E). The series of observations is short (Labitzke
& Chanin 1988), but the fact that the vertical distribution is quite similar but opposite in phase
to the one at Heiss Island fits the pattern in figure 4 and is worth emphasizing (figure 3b).

The difference between winters in the east and west phase is further illustrated by figure 5.
We show (figure 5a) the mean difference of 30 mbar geopotential height between the two QBO
phases in solar minima (i.e. solar flux below 90 units; see ordinate in figure 2). The largest
difference in geopotential height is between the North Pole and the Aleutians and amounts to
1100 m. Where the gradient in this area is steepest it corresponds to a difference in geostrophic
wind of 25 m s^{-1}. In other words, in solar minima the polar night jet stream in the west years

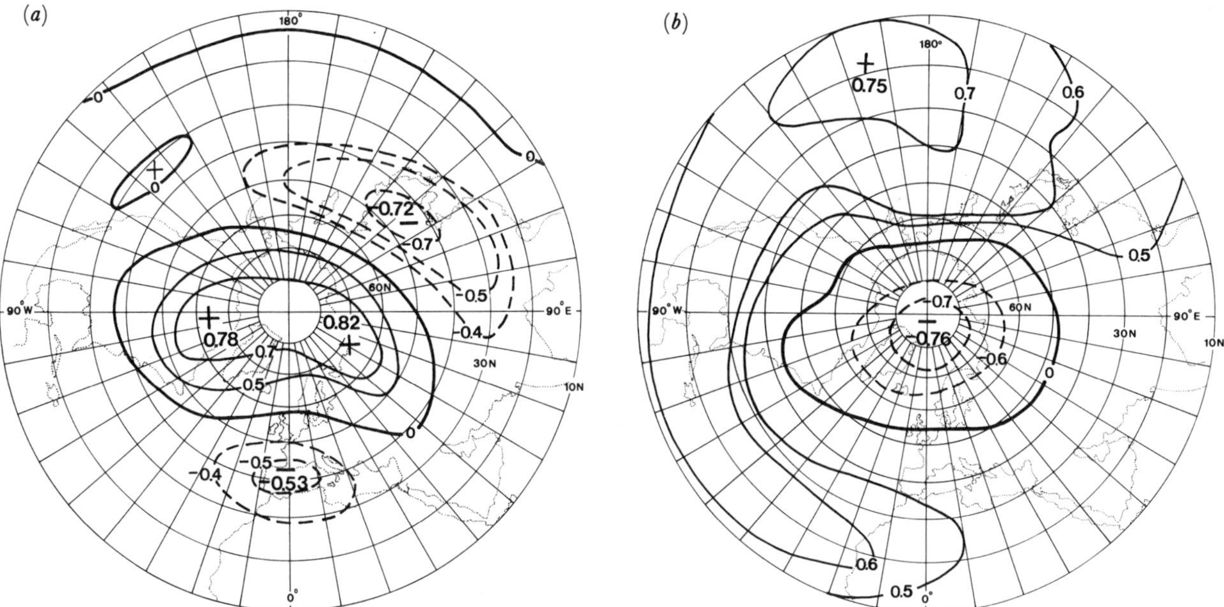

FIGURE 4. (a) Lines of equal correlation between 30 mbar geopotential height and 10.7 cm solar flux in January–February in the west phase of the QBO (n = 15). (b) As (a), but for the east years (n = 12). (Labitzke & van Loon 1988.)

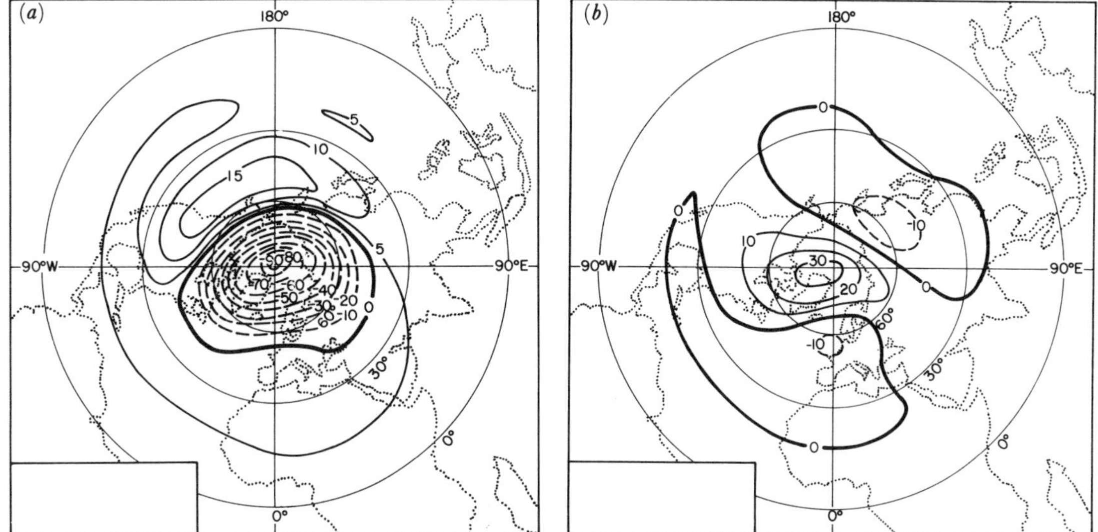

FIGURE 5. (a) The difference in 30 mbar heights (west minus east) in solar minima between January and February in west years (1964, 1965, 1974, 1976, 1986) and east years (1963, 1966, 1975, 1977, 1985); in geopotential dekameters. (b) As (a), but for solar maxima. West years: 1958, 1968, 1970, 1979, 1981. East years: 1959, 1969, 1971, 1980, 1982.

is here 25 m s^{-1} stronger than in the east years. The difference between the east and west years is not nearly so strong in the solar maxima, figure 5b, where it is only 40 % of that in the minima. One may interpret the difference between west and east years in solar maxima and solar minima as a difference in the influence of the QBO.

The response of the stratosphere on the Northern Hemisphere in winter to the 11-year solar cycle is then quite marked. As the following analysis indicates, the response to the cycle changes from winter to spring: figure 6 shows the correlation between the solar flux and the 30 mbar height in spring (March–April) in 16 west years of the QBO. The pattern is the opposite of that in winter (figure 4a). This change in pattern is associated with the well-known observation (Labitzke 1982, table 2) that winters during which a major midwinter warming has taken

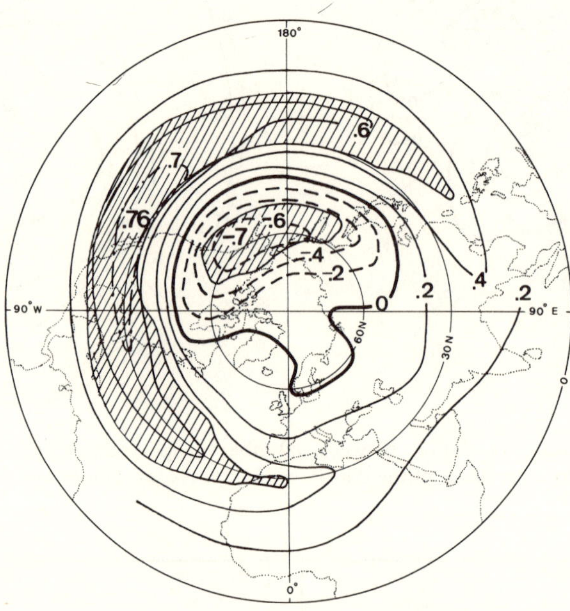

FIGURE 6. As figure 4a, but for March–April in west years of the QBO. The areas where the confidence level is above 95% are shaded. (Labitzke & van Loon 1989.)

place tend to be followed by a late final (spring) warming, whereas a winter with a cold polar vortex tends to precede an early final warming. As noted above, the major midwinter warmings in the west years take place in the maxima of the 11-year solar cycle; the final warmings are then late, and the spring vortex is cold. In solar minima the vortex is cold in west years in midwinter, and the spring vortex is warm. The correlations are therefore negative in the Arctic in spring during west years.

It is also of interest to see if there is an association between the temperature variations in the stratosphere on the Southern Hemisphere and the 11-year solar cycle. Unfortunately, there are no long map series of the stratosphere on the Southern Hemisphere, and one must turn to station data to work with a reasonably long series. It is futile to search for a solar signal in the Antarctic stratosphere in midwinter because the vortex is then intense, cold and stable, and its interannual changes are small. The interannual standard deviation of the 50 mbar temperature in midwinter at the South Pole is, for instance, only 1 °C (Labitzke 1987b). We therefore use the month of October when the stratospheric polar vortex becomes unstable and the interannual standard deviation is five times larger than in winter. The highest level in the stratosphere with a fairly complete record at the South Pole is 50 mbar (ca. 18 km). The correlation between the solar flux and the 50 mbar temperature (figure 7a) is only 0.29; but if one divides the temperatures into the east and west years of the QBO, the correlation is 0.79 in the east years (figure 7c) and insignificantly small in the west years (figure 7b). Both

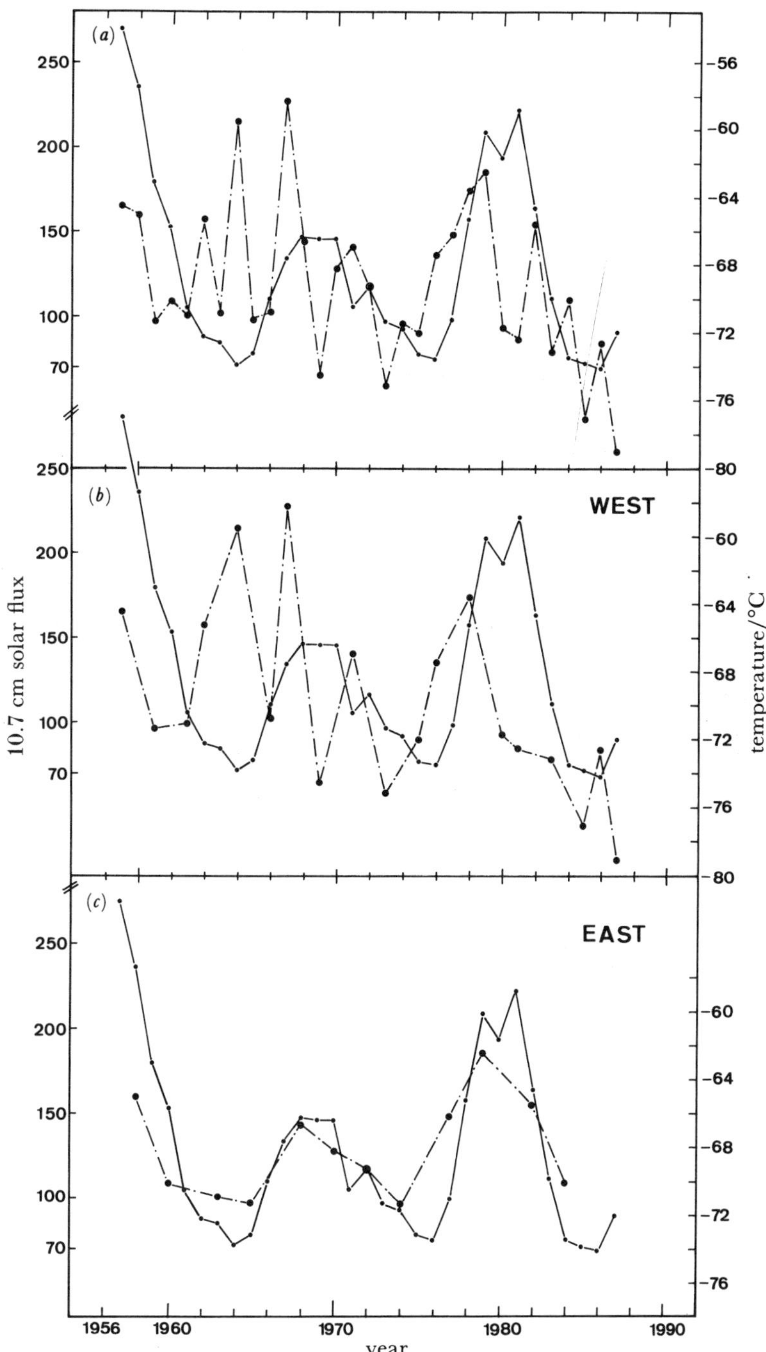

FIGURE 7. (a) Time series of the 10.7 cm solar flux, solid line, and the 50 mbar temperature at the South Pole in October ($r = 0.29$, $n = 31$). (b) As (a), but the temperatures are only in the west years of the QBO ($r = 0.15$, $n = 19$). (c) As (a), but the temperatures are only in the east years. East and west according to the equatorial wind at 50 mbar in September–October ($r = 0.79$, $n = 12$, 95 % interval: 0.45–0.91). (Labitzke & van Loon 1989.)

conventional and Monte Carlo methods indicate that the correlation coefficient of 0.79 is unlikely to be accidental.

3. THE SOLAR SIGNAL IN THE TROPOSPHERE

The marked response to the solar flux in the stratosphere extends through the troposphere to the Earth's surface. In this section we show first the correlations between solar flux and sea level pressure in 19 winters in the west phase (figure 8a), and the correlations in 16 winters in the east phase (figure 8b). In the west phase the most distinct pattern is over North America and the Atlantic Ocean whereas in the east phase it is in the Pacific region. The two patterns resemble closely two well-known teleconnection patterns: the North Atlantic oscillation in the west years, and the North Pacific oscillation in the east years (Walker & Bliss 1932).

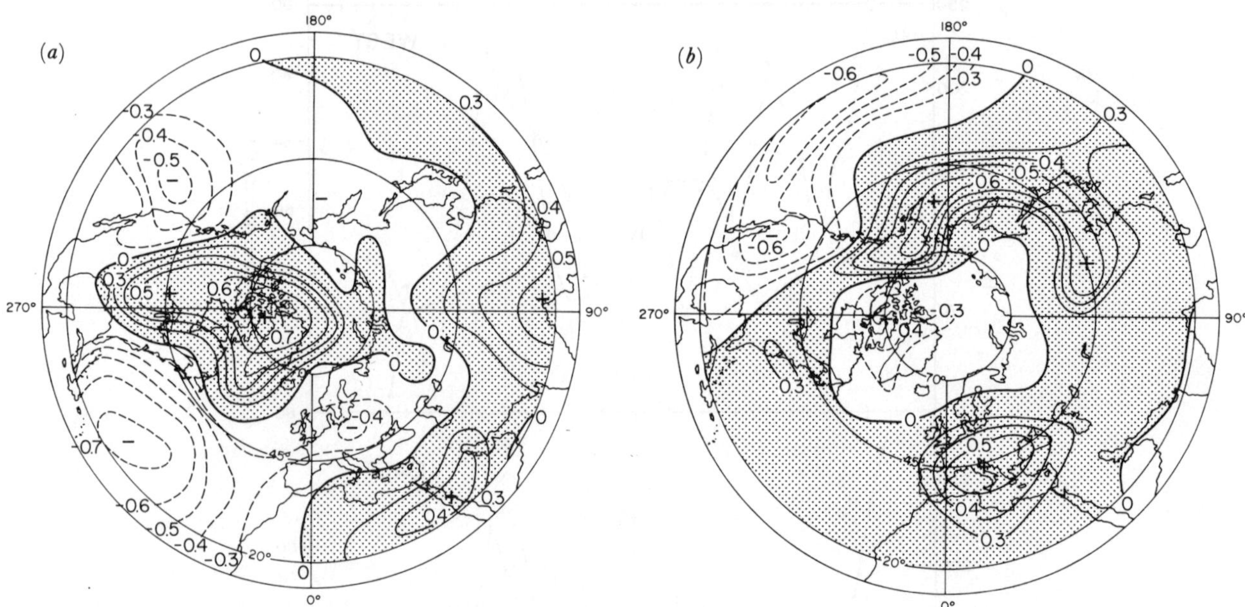

FIGURE 8. (a) Lines of equal correlation coefficient: sea level pressure in January–February at grid points correlated with the 10.7 cm solar flux for 19 winters when the QBO was westerly. (b) As (a), but for 16 winters when the QBO was easterly. Areas with positive correlations are shaded. (van Loon & Labitzke 1988.)

We have chosen three stations (figure 9) in the area of positive correlation over North America in the west phase to illustrate the relation between the solar flux and sea level pressure. Two of them, Des Moines and The Pas, lie where the correlation coefficient is between 0.5 and 0.6, and the third, Resolute, where the correlation is 0.7. In all three instances the association between the solar and pressure curves is conspicuous because the extremes of pressure tend to occur in extremes of the 11-year solar cycle. With a difference of 15 mbar between extremes at Resolute and The Pas and 9 mbar at Des Moines, the solar effect covers the whole range of interannual variability of the mean January–February sea level pressure. One finds the same large response in the atmospheric elements at all levels on both hemispheres. Other examples are the 30 mbar temperature in figure 2 and the curves in figure 10.

Because the correlations change sign on the scale of planetary waves, correlations between

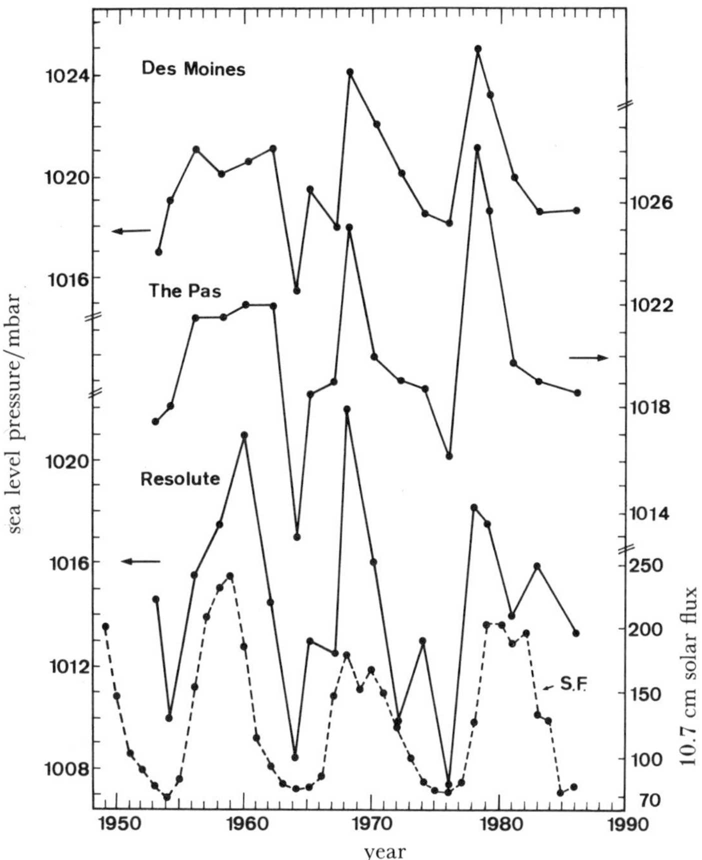

FIGURE 9. Time series of the 10.7 cm solar flux for all years (dashed line), and of sea level pressure at three North American stations for the years when the QBO was westerly in January–February. (van Loon & Labitzke 1988.)

the solar flux and zonal averages of pressure, or averages over large regions, mostly lead to insignificantly small numbers.

The correlation pattern in figure 8a indicates that there is a tendency for an anticyclonic anomalous circulation over North America in the west years in solar maxima that is flanked by anomalous cyclonic circulations, and conversely in solar minima. Therefore, there is a solar signal in the pressure difference between continent and ocean that is especially strong in the western Atlantic Ocean (figure 10a). Correlation between this pressure difference and the solar flux yields a coefficient of 0.77.

The change of geostrophic wind at sea level in winter from solar maximum to solar minimum in the west years can be deduced from the map in figure 8a. In solar maximum the anticyclonic anomalous circulation over North America is associated with northeasterly anomalous winds on its east side. In solar minima the circulation would be the opposite. Because the temperature level in winter is closely associated with the direction of the wind, the southeastern U.S. would thus tend to experience lower temperatures in solar maxima than in solar minima in the west years in the QBO. We illustrate this by the time series of surface air temperature at Charleston, South Carolina, in figure 10.

The correlation between solar flux and surface air temperature in winter over most of the Northern Hemisphere in the west years is shown in figure 11. If one compares this map with

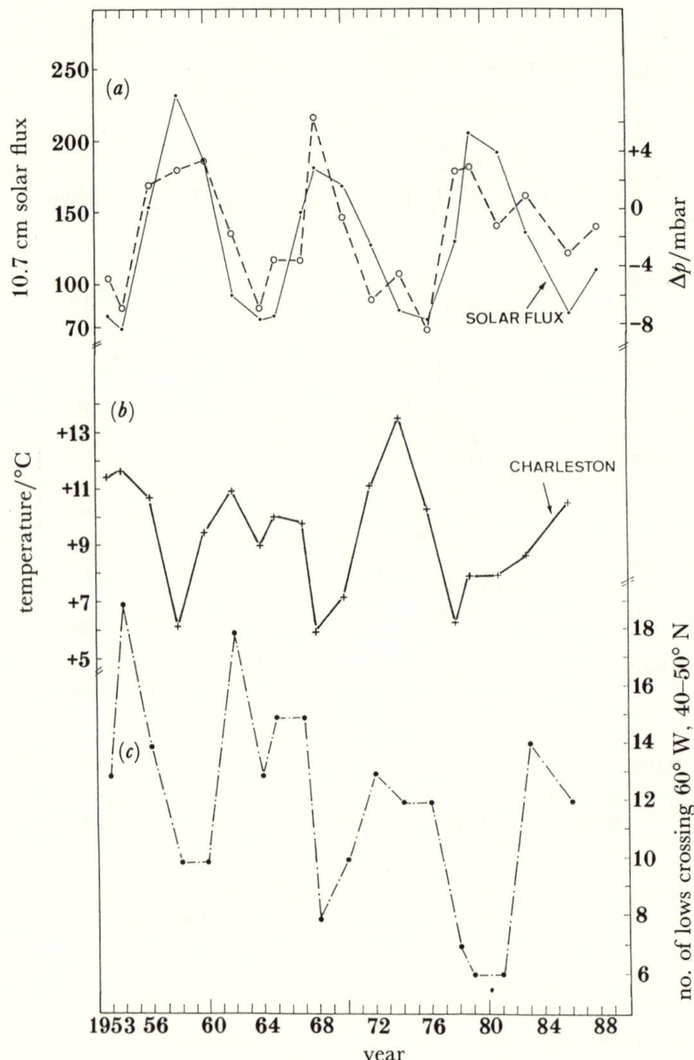

FIGURE 10. (a) Time series of 10.7 cm solar flux and of the pressure difference (70° N, 100° W) minus (20° N, 60° W) in the west years of the QBO, January–February. (b) Time series of the surface air temperature at Charleston, South Carolina, in January–February of the west years. (c) Times series of the number of lows crossing the 60th meridian west between the latitudes of 40° N and 50° N, in January–February of the west years.

the sea level pressure correlations in figure 8 *a*, it is easy to see how well the two fit together if one considers the wind anomalies implicit in figure 8 *a* and the temperature anomalies implicit in figure 11. To a large extent figure 11 also bears the impress of the North Atlantic Seesaw, which is a component of the North Atlantic oscillation. The Seesaw consists of a tendency for the temperature in northern Europe to be in-phase with the temperature in the central and eastern U.S., but out-of-phase with the temperature in southeastern Europe and Labrador–Greenland (van Loon & Rogers 1978).

The correlation between the 700 mbar height (*ca.* 3 km) and the 10.7 cm solar flux is shown in figure 12 for January–February in the 19 west years. The pattern implies that in these years the westerlies across North America and the adjacent ocean areas quite often must be appreciably weaker in solar maxima than in solar minima, whereas the opposite holds from North Africa to central Russia. The pattern in figure 12 resembles a well-known teleconnection pattern: we show as an example an update of a teleconnection map from Namias (1981) in figure 13; the 700 mbar height at 70° N, 90° W is here correlated with the heights everywhere else on the hemisphere in all 35 winters from 1952 to 1986. The pattern of the point correlations in figure 13 is nearly the same as that in figure 12 in which all points on the map are correlated with the solar flux in the 19 west years. Such similarities appear whenever a point that is highly correlated with the solar cycle is correlated with all other points.

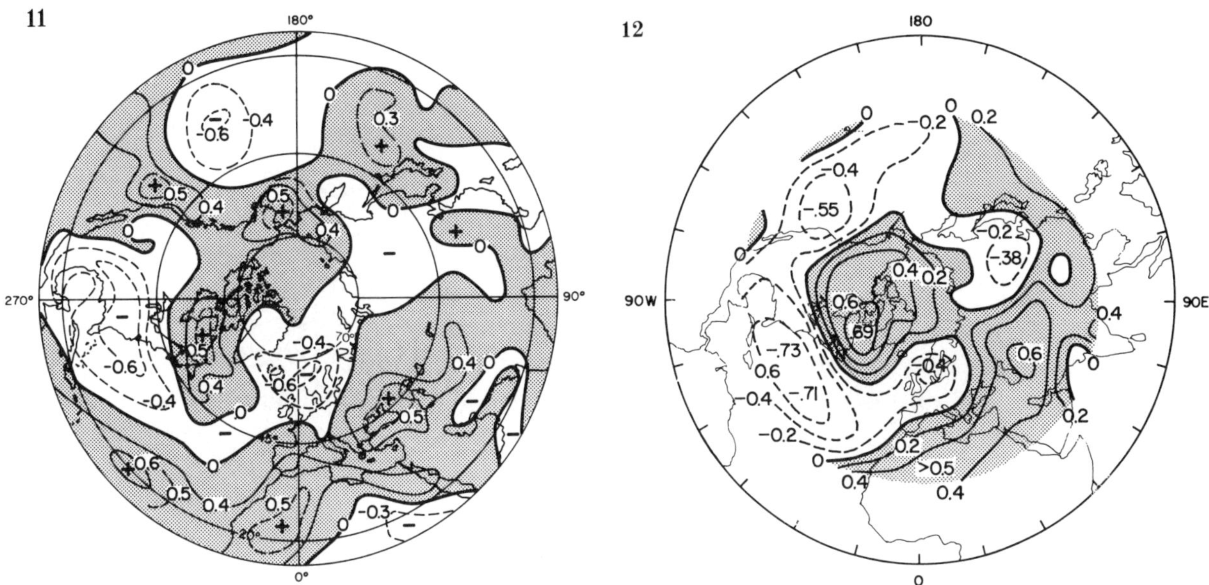

FIGURE 11. Lines of equal correlation coefficient: surface air temperature in January–February in west years at stations and in ocean blocks correlated with the 10.7 cm solar flux. Areas with positive correlations are shaded. (van Loon & Labitzke 1988.)

FIGURE 12. Lines of equal correlation between 700 mbar geopotential height at grid points and the 10.7 cm solar flux, in January–February and for years of westerly QBO. Areas with positive correlations are shaded. (van Loon & Labitzke 1988.)

The fact that a map of the correlations at all points with the solar flux is similar to a map of the correlations of one point with all other points may contain a key to how and where the solar variability affects the atmosphere, but it does not immediately lead to an explanation. Most likely, such a pattern is a property of the atmosphere's internal dynamics, a favoured mode that can be evoked within the atmosphere itself or by extraneous influences. Because the atmosphere responds so readily in the pattern, the forcing may not have to be large.

The correlations of solar flux with sea level pressure and 700 mbar height imply that pressure systems – lows and highs – are also affected by the solar variability. We show in the following an especially marked example of the changes in the frequency of low-pressure systems associated with the 11-year solar cycle.

We have counted the number of lows that crossed the meridian of 60° W between the latitudes of 40° N and 50° N, where the gradient of correlation is the steepest in the hemisphere (figure 12). The westerlies and the horizontal cyclonic shear in this area are weaker in the solar maxima in west years (above normal pressure to the north; below normal to the south) than in the solar minima (below normal pressure to the north; above normal pressure to the south). The number of lows crossing 60° W is therefore lower in the solar maxima than in the minima by about 40%. The correlation between cyclone frequency and the solar flux in the 19 winters is -0.63.

Finally, we stress the strong vertical coherence of the solar signal, from the layers near the surface to the middle atmosphere. We pointed out in the discussion of figure 5a, which shows the difference in 30 mbar heights between west and east years in solar minima, that the circumpolar cyclonic vortex is by far stronger in the west years. The same difference appears in figure 14, but for the 700 mbar level, i.e. 20 km below the 30 mbar level. The two maps are very similar, allowing for more waves in the troposphere and slope with height.

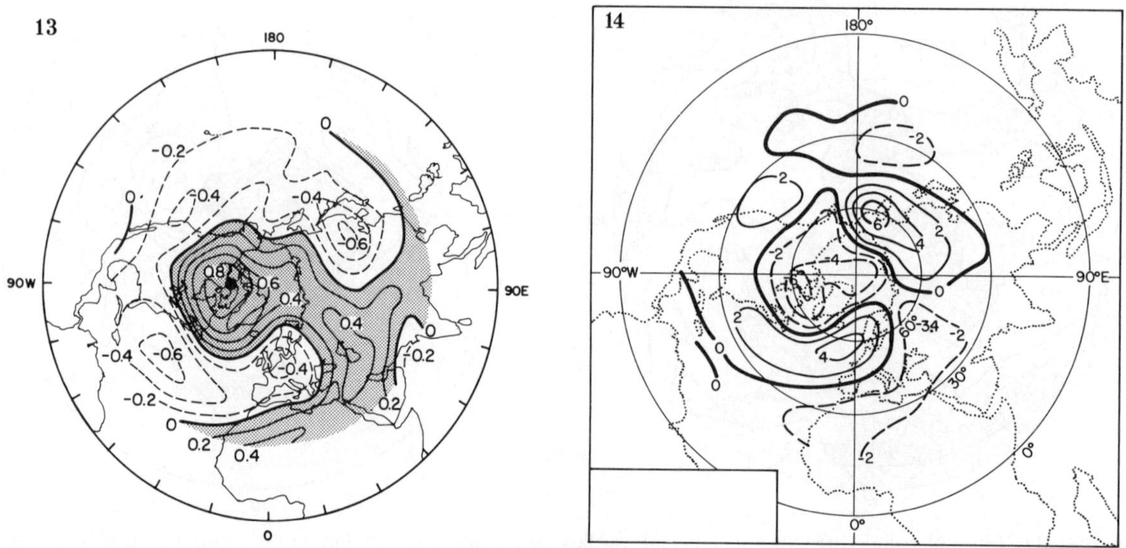

Figure 13. Lines of equal correlation between the 700 mbar geopotential height at 70° N, 90° W and the 700 mbar height elsewhere. For January–February and all years between 1952 and 1986. (van Loon & Labitzke 1988.)

Figure 14. As figure 5a, but for the 700 mbar level. The years in the west phase are: 1953, 1954, 1964, 1965, 1974, 1976, 1986; and the east years: 1952, 1966, 1973, 1977, 1987.

4. Conclusions

The observation (Labitzke 1982) that when the QBO is westerly, major midwinter warmings of the stratospheric cyclonic vortex on the Northern Hemisphere tend to occur only in maxima of the 11-year solar cycle, led to the discovery that one obtains statistically significant correlations between the solar variability and atmospheric elements when one groups the data according to the state of the QBO. As the correlations are often of opposite sign in the west and east phase of the QBO, the correlation coefficients are mostly small when one uses a full time series of an atmospheric element. The spatial patterns of the correlations resemble well-known teleconnection patterns that are inherent in the internal dynamics of the atmosphere and readily evoked within the atmosphere itself or by extraneous effects. The range of an atmospheric element between solar extremes covers the full interannual variability in both east and west years at all levels investigated.

We have tested the correlations both by conventional statistical methods and Monte Carlo techniques, and the tests suggest that our results are unlikely to have occurred by chance. We can, however, examine only three solar cycles for we do not know the state of the QBO before 1952. Added to this limitation is the fact that we cannot explain how the observed small solar variability could produce such large responses in the atmosphere. Therefore, one cannot yet exclude the possibility that the promising results stem from sample variations.

We have by now studied not only the Northern Hemisphere in winter, but also sea level pressure in the Southern Hemisphere in winter and in the Northern Hemisphere in summer. Our results are published in Labitzke & van Loon (1988); van Loon & Labitzke (1988); Labitzke & Chanin (1988) and Labitzke & van Loon (1989).

The National Center for Atmospheric Research is sponsored by the National Science Foundation, K.L. is an Affiliate Scientist at NCAR. We express our gratitude towards Dennis J. Shea of NCAR, who has applied his considerable skills to the analysis and testing.

References

Holton, J. R. & Tan, H.-C. 1980 *J. atmos. Sci.* **37**, 2200–2208.
Labitzke, K. 1981 *J. geophys. Res.* **86**, 9665–9678.
Labitzke, K. 1982 *J. met. Soc., Japan* **60**, 124–138.
Labitzke, K. 1987a *Geophys. Res. Lett.* **14**, 535–537.
Labitzke, K. 1987b *Annls Geophysicae* A5, 95–102.
Labitzke, K. & Chanin, M.-L. 1988 *Annls Geophysicae* **6**, 643–644.
Labitzke, K. & van Loon, H. 1988 *J. atmos. terr. Phys.* **50**, 197–206.
Labitzke, K. & van Loon, H. 1989 *J. Climate* **2**, 554–565.
Namias, J. 1981 *CALCOFI atlas no. 29.* La Jolla, California: Scripps Institution of Oceanography. (vii + 265 pages.)
Naujokat, B. 1986 *J. atmos. Sci.* **43**, 1873–1877.
van Loon, H. & Labitzke, K. 1988 *J. Climate* **1**, 905–920.
van Loon, H. & Rogers, J. C. 1978 *Mon. Weather Rev.* **106**, 295–310.
Walker, G. T. & Bliss, E. W. 1932 *Mem. R. met. Soc.* **4**, 53–84.

Discussion

K. Shine (*Department of Meteorology, University of Reading, U.K.*). I know that most of Professor Labitzke's work has been concerned with the Northern Hemisphere. However, there is much circumstantial evidence linking the QBO with the variability in the Antarctic ozone hole; I wonder if Professor Labitzke could comment on the possible role of solar variability in modulating the ozone hole?

Karin Labitzke. According to the data given in figure 7 and the current development of the QBO, the October 1989 will belong in the east years (figure 7c); as we will be in solar maximum, the correlation shows that it is probable that the temperature at 50 mbar in October 1989 will be above the mean for east years.

Phil. Trans. R. Soc. Lond. A **330**, 591–599 (1990)

Printed in Great Britain

Solar luminosity variations over timescales of days to the past few solar cycles

By P. Foukal

Cambridge Research and Instrumentation Inc., 21 Erie Street, Cambridge, Massachusetts 02139, U.S.A.

[Plate 1]

Dips in the total irradiance of up to 0.2 % and lasting typically 10 days are now well known to be caused by the transit of dark sunspots across the photospheric disc. The large bright magnetic faculae usually associated with spots cause irradiance increases of comparable magnitude although the form of their signal is more subtle. Radiometry from five satellites beginning in late 1978 indicates a minimum in irradiance at the epoch of lowest magnetic activity between solar cycles 21 and 22. Analysis of these radiometric measurements indicates that this irradiance decline between about 1981 and 1986 was caused mainly by decay in the excess radiation of bright faculae in the magnetic network outside of active regions. Empirical models of irradiance modulation extending back to 1874 indicate that the Sun is typically about 0.05 % brighter at activity maximum than at minimum.

1. Introduction

Scarcely 10 years ago, it was still unclear whether the total irradiance, S (the 'solar constant'), varied over timescales accessible to direct measurement. Since then, radiometers flown on the *Nimbus*-7 and *Solar Maximum Mission* (*SMM*) satellites have provided us with essentially continuous daily coverage beginning in late 1978. The similarity of the main features in these two concurrent databases gives us confidence that we now have access to the behaviour of the total solar irradiance over all timescales from a few minutes to essentially one complete solar activity cycle.

The results are of interest to solar and stellar astronomers, as well as to climate physicists. For instance, the variations on timescales of a few minutes contain a contribution from modes of approximately 5 min period associated with acoustic oscillations of the solar convection zone and interior. Analysis of these data has proven to be of interest in helioseismological studies of the rotation of the solar interior, of the damping of the acoustic modes, and of possible structure variations in the solar interior over the 11-year cycle that is discussed by D. O. Gough (this Symposium).

In this paper, I concentrate on the longer-term irradiance variations that seem to be of most direct interest in studies of climate. I first discuss the direct irradiance measurements made from satellites over the past decade, and what they seem to tell us about the main sources of solar luminosity variation. The physical interpretation of these variations has been reviewed recently (Foukal 1987, 1988; Spruit 1988) and I only mention the most recent developments here. I then present a simple empirical model of the irradiance variations developed together with J. Lean (Foukal & Lean 1986; Lean & Foukal 1988). This model is based on the understanding gained from cycle 21, and I apply it to reconstructing the variations expected to have been

592 P. FOUKAL

caused by solar activity during cycles 19 and 20. An adaptation of this model is also described that enables us to estimate what the solar cycle variation might have been back to 1874.

Photometric studies of younger Sun-like stars provide us with the most direct information on the behaviour of the solar luminosity and its variability over longer palaeoclimatological timescales. An excellent review of this topic has been given recently by Baliunas (1988).

2. Radiometry of the total solar irradiance 1978–88

The best existing data on solar irradiance variation are shown in figure 1. The longest time series is from the Earth Radiation Budget (ERB) radiometer on the *Nimbus*-7 satellite. These daily data begin in November 1978 and continue to the present. The experiment and reduction procedures have been described by Hickey *et al.* (1988). The data shown here extend through October 1988.

The most precise measurements are those made with the Active Cavity Radiometer Irradiance Monitor (ACRIM) radiometer on the *SMM* spacecraft beginning in early 1980 and also continuing to the present, with a short interruption between late 1983 and early 1984 (Willson & Hudson 1988). The ACRIM data shown in figure 1 extend to late 1987. Three additional sets of radiometric measurements, from the *NOAA*-9, *NOAA*-10 and *ERBS* satellites, have been available since late 1984 (Mecherikunnel *et al.* 1988). The *ERBS* data to early 1988 are shown here. These measurements are only obtained a few times per month, but they provide useful checks of slow irradiance trends.

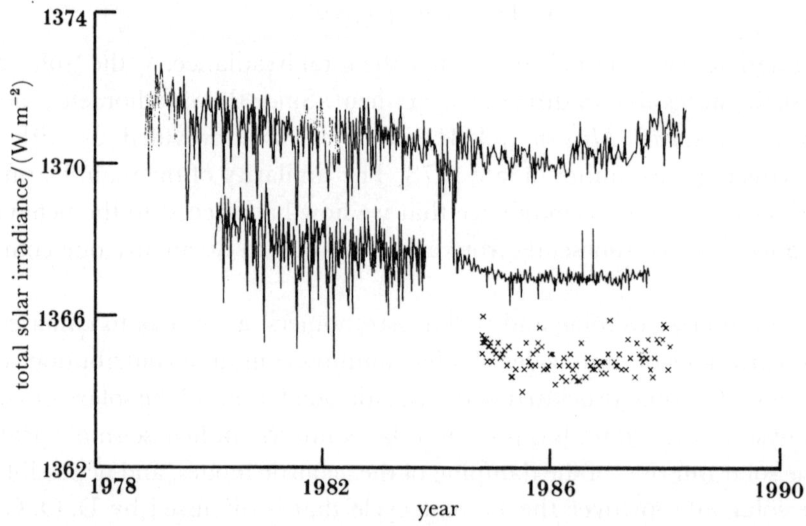

Figure 1. Total solar irradiance measurements from the *Nimbus*-7 (top), *SMM* (middle), and ERBS (bottom) satellites. The vertical offsets correspond to real differences of about 0.2 % in absolute calibration between each of the three radiometers.

The vertical offset of roughly 0.2 % between each of the three data-sets is caused by a real disagreement of absolute calibration between the three radiometers. However, the precision of each is much higher, as shown by the agreement in the short-term variations seen by the ERB and ACRIM.

A high-frequency variation on timescales of about 10 days, with peak-to-peak amplitude reaching about 0.3 % is one prominent feature of the irradiance data shown in figure 1. This

[194]

is caused by dark spots and bright magnetic faculae rotating across the disc, as discussed below. The second is the gradual decline of the irradiance until early 1986, and then its gradual recovery. The decline observed by both the ERB and ACRIM between 1980 and 1986 amounts to somewhat less than 0.1 %. This variation is of greatest interest to climate studies, and its source is currently the object of intensive study.

3. Influence of spots and faculae

The large dips in total irradiance seen in figure 1 are easy to identify with the influence of spots. The irradiance variation expected from changes in the area of dark spots on the solar disc can be calculated from their photometric contrasts and from daily data on their coordinates and areas (Willson *et al.* 1981). As seen in figure 2 the calculated function accounts for the timing and depth of the larger dips rather well. The missing energy not radiated by spots is most probably stored in a small increase in the thermal and potential energy of the convecting layers outside the spot (Spruit 1982; Foukal *et al.* 1983). Thus it seems to represent a true luminosity decrease, not merely a reradiation into a different solid angle.

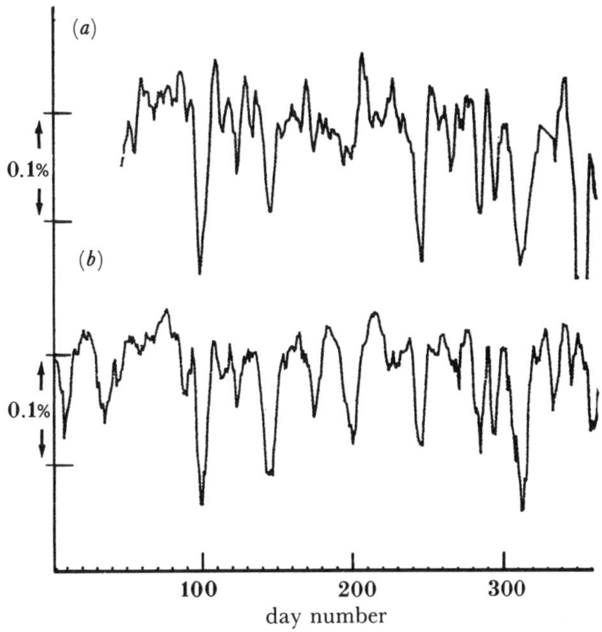

FIGURE 2. Comparison of the variations in measured solar irradiance for 1980 from (*a*) the *SMM* and the variations calculated from an empirical model (*b*) based on measured photometric contrasts of spots and their daily areas and disc coordinates. (From Foukal 1987.)

But this is not the only effect of solar magnetic activity on the total irradiance. If we subtract the calculated sunspot blocking function, P_s, from the measured irradiance, the residuals are still comparable in magnitude with the variations in P_s. Figure 3 illustrates such a subtraction for ACRIM data in 1980. We see that the residuals correlate well with variations in 205 nm ultraviolet (UV) flux. These UV variations are known to be caused primarily by rotation across the solar disc of bright magnetic faculae (see, for example, Lean 1987).

Figure 4, plate 1, shows the appearance of the solar photosphere and chromosphere at times of low and high magnetic activity. The photosphere is the level at which over 99 % of the solar

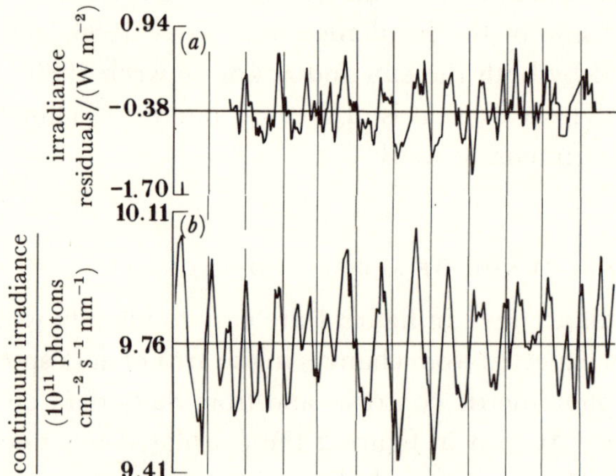

FIGURE 3. (a) Irradiance residuals found after removing the sunspot blocking effect from the ACRIM radiometry for 1980; (b) 205 nm UV continuum irradiance measured from the *Nimbus*-7 spacecraft. (From Foukal & Lean 1986.)

radiation originates. It is seen that as magnetic activity increases, more dark spots and also more bright magnetic faculae appear. The largest faculae are found around the active regions. But outside of these is a network of comparably bright, although smaller structures, that cover the whole disc. These are seen better in the chromosphere where they are brighter.

There is no real distinction between the bright faculae and the network except for the size. They both correspond to regions of intense, radial, small-scale magnetic fields. Their excess brightness in photospheric radiations is most likely caused by the relatively low plasma pressure in these intense magnetic flux tubes. This lowers the opacity and enables radiation to escape from deeper and hotter layers (Spruit 1976; Deinzer *et al.* 1984).

The contrast of faculae is quite low in broadband visible light, where most of their energy is emitted. This makes it relatively difficult to determine whether their signal can account for the amplitude of the irradiance residuals seen in figure 3. However, photometry of active regions (see, for example, Chapman & Meyer 1987) indicates that the relatively large areas covered by faculae in active regions enable them to make an irradiance contribution comparable with that of spots.

A clean test of faculae as the cause of the irradiance residuals is provided by their cross-correlation with the sunspot blocking functions P_s, and with the 205 nm flux (Foukal & Lean 1986). This cross-correlation shows that the residuals S-P_s are caused by solar structures that last typically at least three solar rotations, thus about two rotations longer than do spots. This is consistent with the well-known long lifetime of faculae, and argues against errors in the P_s function as the main cause of the residuals.

4. DETECTION AND UNDERSTANDING OF SLOWER IRRADIANCE VARIATIONS

Identification of the facular signal in the irradiance record seems to be helpful in understanding the nature of the slower variations over timescales from months to the full activity cycle. Thus Foukal & Lean (1988) found a good correlation between slow changes in the irradiance residuals, and in indices of facular area, for the four years between 1981 and 1984. The CaK plage index, one of the two used in that study, is based on the areas of only

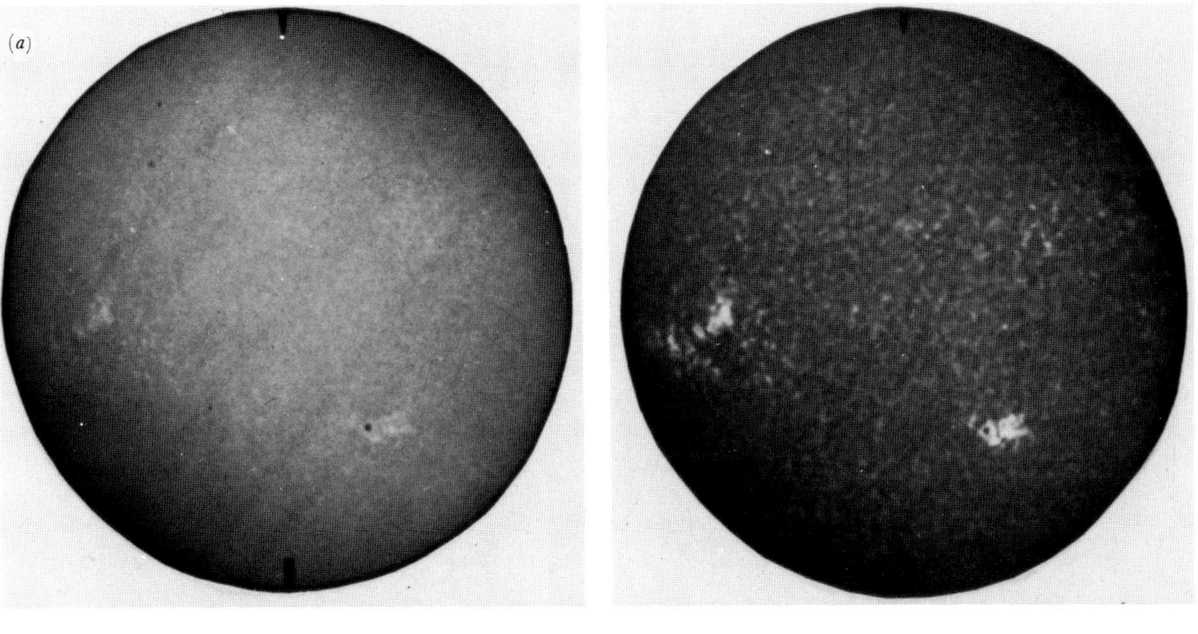

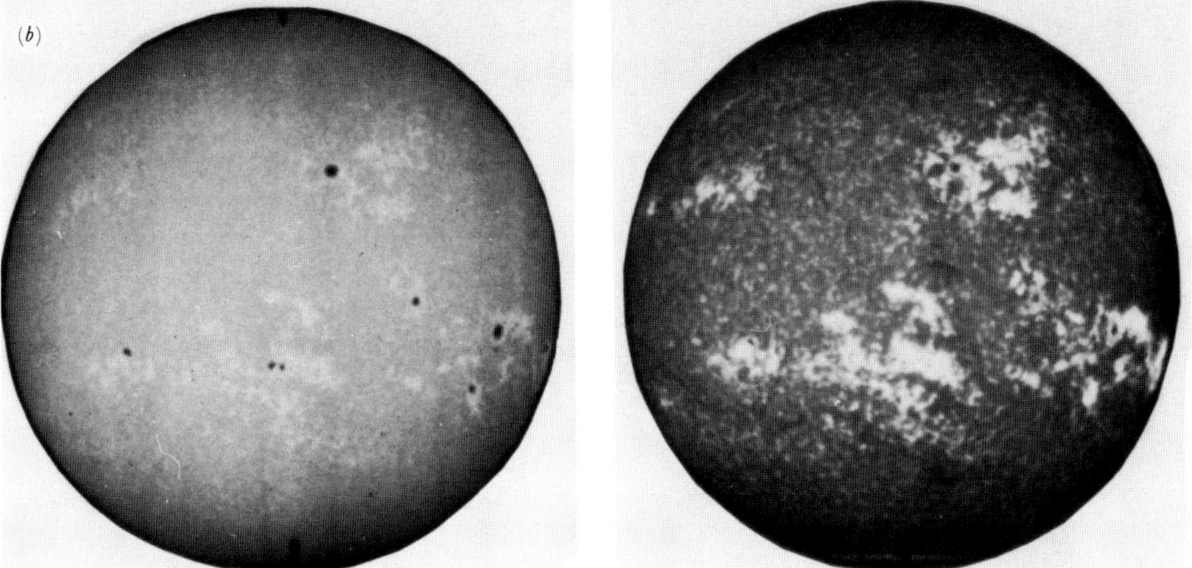

FIGURE 4. Images of the Sun at (*a*) low (10 May 1975) and (*b*) high (7 March 1979) magnetic activity levels. The images on the left are taken in the wing of the Ca II K absorption line, and refer to the highest photospheric levels. Those on the right are in the centre of that line, and show the overlying chromospheric layers. (From Lean 1987.)

the large faculae in active regions. The second is the He I index that is global in the sense that it includes contributions from all the faculae, including those in the network.

The regressions constructed in that study were then used with daily values of the CaK and He I indices between 1981 and 1984, to reconstruct the irradiance residuals obtained for that period. Figure 5 shows the comparison of the reconstruction with the original data. It is seen that when the global index, He I, is used one can reconstruct not only the slow downtrend, but also the variations of 4–9 month timescale, remarkably well. When the CaK plage index is used as a basis for the reconstruction, the 4–9 month variations are reproduced about equally well, but the ability to reproduce the slow downtrend is lost.

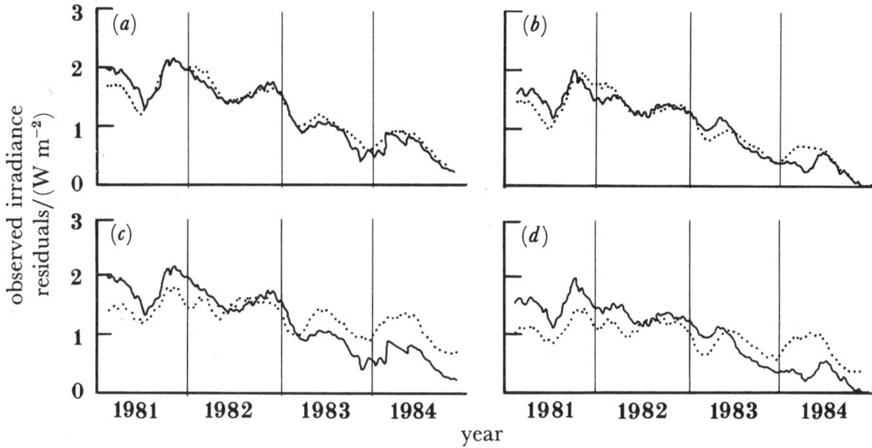

FIGURE 5. Comparison of the observed irradiance residuals (solid lines) for (a) and (c) the ACRIM data and for (b) and (d) the ERB data, with the irradiance residuals (dotted lines) calculated by using (a), (b) the He I and (c), (d) CaK indices. (From Foukal & Lean 1986.)

This result indicates that the 4–9 month irradiance variations are caused by something that is common to both the global index and the active region index, i.e. the large active regions. This timescale seems to be set by the tendency of extended complexes of activity to erupt and subside (Harvey 1984). The slow decline between 1980 and 1984 must be caused by a mechanism residing outside the active regions.

In principle, this mechanism could be a deep-seated alteration of solar convection or even of processes in the core as some have suggested (Kuhn *et al.* 1988; Gough 1988) and a contribution from such effects cannot be dismissed. But the finding that a single linear relation between facular area and total irradiance can account for both the 4–9 month variations and the slow downtrend indicates that most of the decline is caused by a solar-cycle variation in the surface density of the network faculae outside active regions.

A test that could distinguish between the two kinds of mechanism might be based on comparison of the solar cycle variations in effective temperature at relatively deep photospheric layers (where the enhancement of the network is smallest and effects due to changes in underlying convection should be largest) and at higher layers where the relative importance of these two effects should be reversed.

Data that might be used for such a test have been obtained (Livingston *et al.* 1988) by measurement of small changes in the strength of solar absorption lines in disc-integrated solar light between 1978 and 1988. The results (figure 6) show that lines formed around the

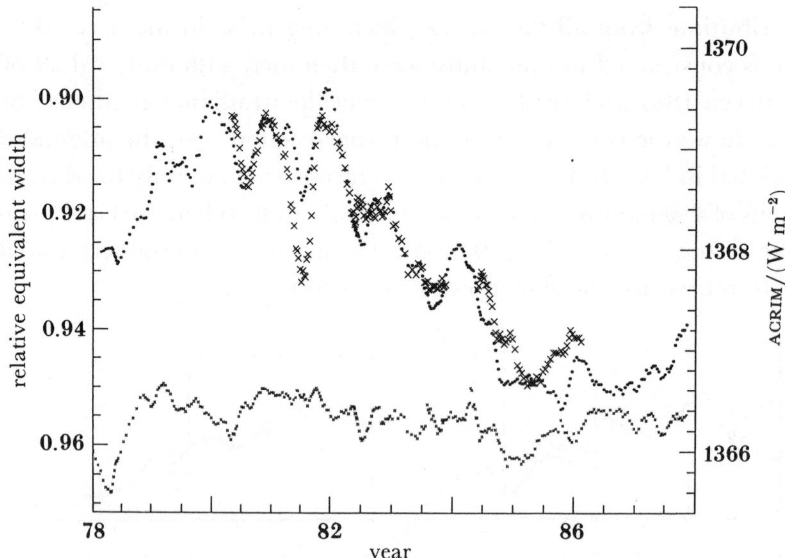

FIGURE 6. Irradiance residuals (crosses), and He I 1083 nm equivalent width (dots). At the bottom is the time series for C I 5380 treated in the same way. (From Livingston *et al.* 1988.)

temperature minimum and in the chromosphere follow the irradiance residuals well. The line C I 5380, formed deeper in the photosphere, shows no significant signal. This result would seem to rule out the deep-seated variations, but one must inquire into the sensitivity of the C I 5380 data to changes in T_{eff}. According to Livingston (personal communication) it turns out to be somewhat inadequate to rule out the 0.07 % change in S found by the radiometry between 1980 and 1986.

An argument for the network as the source of total irradiance change over the solar cycle is that lines such as He I 10830 and Lyman α are formed in the non-thermally heated chromosphere and corona. The finding that they show the same relative amplitudes of the 4–9 month variations and of the slow decline between 1980 and 1986 as does the total irradiance, supports the dominant effect of the magnetic network, in which these lines are well known to be enhanced. It is difficult to understand why changes in deep convection or in nuclear processes should affect these radiations in such a similar way.

5. EMPIRICAL MODELLING OF TOTAL IRRADIANCE VARIATIONS CAUSED BY SOLAR ACTIVITY 1874–1988

If I suppose that the bright faculae in active regions and in the network, together with the spots, are the main cause of the irradiance variations observed so far, I can use our model to calculate these variations back as far as indices of spot and facular area are available. Daily values of spot coordinates and areas are on record from 1874. Good global indices of faculae are more difficult to obtain. The daily He I 10830 index begins in 1975. The 10.7 cm microwave emission, which arises primarily from enhanced bremmstrahlung radiation from the denser coronal atmosphere over the magnetic faculae, is available back to 1947.

The variation of S during cycle 21 as predicted from the daily He I values is shown in figure 7, along with the values of the sunspot and facular contributions plotted separately. The

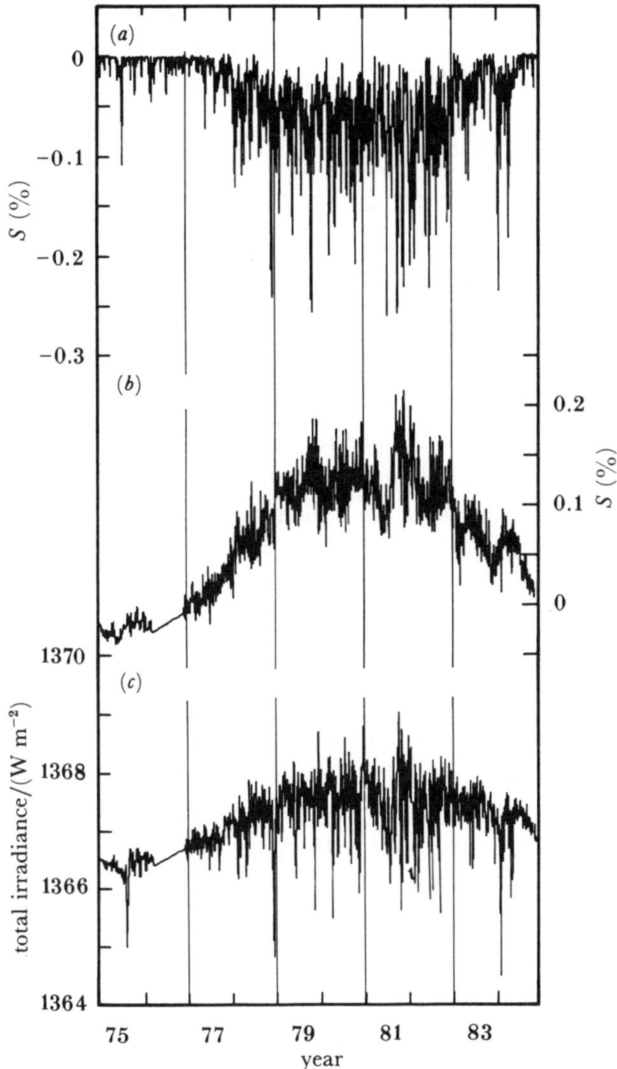

FIGURE 7. Solar cycle 21 behaviour of (a) the daily sunspot blocking function P_s, (b) the daily facular irradiance excess P_f from the He I index, and (c) the reconstructed total solar irradiance. (From Foukal & Lean 1986.)

main finding is that the expected irradiance increases from activity minimum in 1976 to maximum near 1980 by about 0.12 % (Foukal & Lean 1988).

A similar model, but based on the 10.7 cm microwave emissions has been used to estimate the irradiance variation between 1954 and 1984 (Lean & Foukal 1988). The results are shown in figure 8. It is seen that the Sun is consistently brighter at high activity levels. Of most interest is the finding that the calculated irradiance increase of about 0.1 % in cycle 21 was significantly larger than that calculated for cycle 19, which was the largest cycle in the history of reliable spot records (i.e. since about 1850).

This large brightening of the Sun during cycle 21 is caused by the relatively small amplitude of that cycle as measured in spot area compared with its amplitude in other indices, such as the Zurich sunspot number, R_Z, or the microwave flux. That is, during cycle 21 the mean area of a spot tended to be significantly smaller compared with the mean area for all

[199]

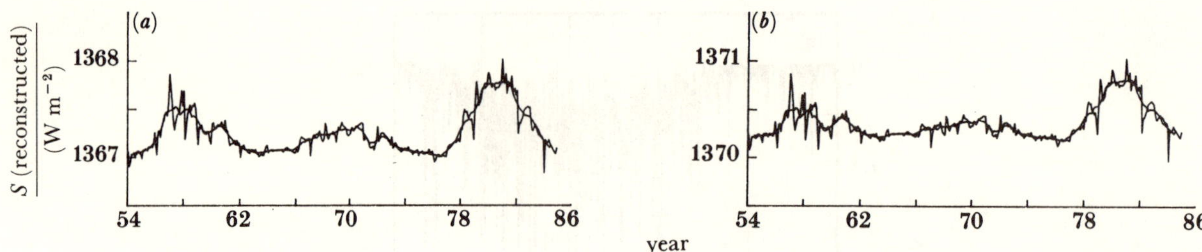

FIGURE 8. Behaviour of the total solar irradiance between 1954 and 1985 reconstructed with (a) the ACRIM data and (b) ERB data. The heavy line represents a 12-month running mean. The difference in the absolute scales in the two panels arises from a difference in the ACRIM and ERB absolute calibrations. (From Foukal & Lean 1988.)

cycles back to at least 1874. The reason for this aberrant behaviour, which has been recognized only recently (Foukal & Lean 1990) is not understood.

There is no reliable direct index of facular activity for calculation of the irradiance further back than cycle 19. The Greenwich photoheliograph data on faculae exhibit marked anticorrelations with other indices that call into question the reliability of that record, probably because of systematic errors caused by the difficulty of detecting white-light faculae near the limb.

However, our work (Foukal & Lean 1990) indicates that irradiance variations back to 1874 can be calculated by using the fact that variations in the sunspot number, R_z, are highly correlated with microwave flux (i.e. facular area) variations. If we are interested only in the relative magnitudes of the cycles in facular activity, the spot number seems to give us a reasonable proxy index for faculae. By contrast, the sunspot area, which determines the sunspot irradiance effect, correlates much less well with $F_{10.7}$ since 1947.

A preliminary calculation of irradiance variation extending to 1874, based on P_s and R_z, indicates a marked brightness increase for the peaks of the four cycles since 1950. This increase can be attributed in part to the general rise of solar activity beginning with cycle 18. But the most remarkable feature is the anomalously large irradiance increase around 1980 caused by the peculiar behaviour of spot areas during cycle 21.

6. SUMMARY AND CONCLUSIONS

Information on the Sun's luminosity behaviour, and our understanding of the mechanisms that seem to cause it, has progressed rapidly in the past 10 years. Relatively short-term, but large-amplitude, variations of total irradiance are caused by sunspots and faculae. Reasonable explanations have been put forward for the irradiance and luminosity variations caused by these large- and small-diameter flux tubes and these are found to be consistent with photometric observations, including recent near-infrared imaging of active regions at the deepest observable photospheric layers (Foukal et al. 1989). Still, the fundamental reasons for spot darkness and facular brightness cannot be considered to be well understood, and in particular the possibility of an energetic connection between these active region structures by non-thermal transports (Schatten et al. 1987) needs closer attention.

The most significant observational result is the finding that the Sun is somewhat brighter, rather than darker, at high magnetic activity levels. The peak-to-peak variation in cycle 21 seems to have been roughly 0.1 %. The most likely source of this variation is the predominant irradiance contribution of bright faculae over that of dark spots. Although the time-integrated irradiance contributions of spots and faculae in active regions roughly balance, the enhancement and subsequent decay of radiation from the magnetic network seems to cause most of the general irradiance variation over the 11-year cycle. Deeper-seated influences on photospheric heat and non-thermal energy flow are certainly not ruled out, but more accurate observations are required to establish their importance.

It is interesting that this predominance of faculae in controlling solar luminosity variation may not have held for the early Sun. Radick *et al.* (1990) point out from stellar photometry that young, late-type dwarf stars of ages less than 10^9 years exhibit an anticorrelation of brightness with magnetic activity. It would seem that in such stars, the propensity to form large dark spots (rather than bright faculae) is greater than in the present Sun, leading to a net darkening at high activity levels.

Regarding slow trends in solar irradiance, the most significant new evidence is that the irradiance increase measured in cycle 21 is likely to have been exceptionally large, exceeding even that of cycle 19, the largest sunspot cycle in the history of reliable records. The anomalously low area coverage of spots during cycle 21, compared to that of faculae, seems to have been an unusual occurrence with no precedent back to at least 1874.

More generally, the rise in irradiance signals associated with the last few cycles appears to produce a significant upward trend in solar irradiance beginning about 1950, and culminating in the exceptionally high irradiance levels associated with cycle 21, which peaked around 1980 (Foukal & Lean 1989). Although the mean irradiance increase averaged over these cycles is small, its duration for almost 40 years now may have begun to produce climatic consequences, that could be contributing to the general trend of global warming.

References

Baliunas, S. 1988 *Solar radiative output variation* (ed. P. Foukal), p. 230. (Available from the editor.)
Chapman, G. & Meyer, A. 1986 *Sol. Phys.* **103**, 21.
Deinzer, G., Henzler, G., Schussler, M. & Weisshaar, E. 1984 *Astr. Appl.* **139**, 435.
Foukal, P. 1987 *J. geophys. Res.* **92**, 801.
Foukal, P. & Lean, J. 1986 *Astrophys. J.* **302**, 826.
Foukal, P. & Lean, J. 1988 *Astrophys. J.* **328**, 347.
Foukal, P. & Lean, J. 1990 *Science, Wash.* (In the press.)
Foukal, P., Fowler, L. & Livshits, M. 1983 *Astrophys. J.* **267**, 863.
Foukal, P., Little, R. & Mooney, J. 1989 *Astrophys. J. Lett.* **336**, 33.
Gilliland, R. 1988 *Solar radiative output variation* (ed. P. Foukal), p. 289. (Available from the editor.)
Harvey, J. 1984 In *Solar irradiance variations on active region time scales* (ed. B. Labonte *et al.*), p. 197. NASA CP2310.
Hickey, J., Alton, B., Kyle, L. & Hoyt, D. 1988 *Space Sci. Rev.* **48**, 321–342.
Kuhn, J., Libbrecht, K. & Dicke, R. 1988 *Science, Wash.* **242**, 908.
Lean, J. 1987 *J. geophys. Res.* D **92**, 839.
Lean, J. & Foukal, P. 1988 *Science, Wash.* **240**, 906.
Livingston, W., Wallace, L. & White, O. 1988 *Science, Wash.* **240**, 1765.
Mecherikunnel, A., Lee, R., Kyle, L. & Major, E. 1988 *J. geophys. Res.* D **93**, 9503.
Radick, R., Lockwood, W. & Baliunas, S. 1990 (In preparation.)
Schatten, K., Mayr, H. & Omidvar, D. 1987 *J. geophys. Res.* **92**, 818.
Spruit, H. 1976 *Sol. Phys.* **50**, 269.
Spruit, H. 1982 *Astr. Appl.* **108**, 348, 356.
Spruit, H. 1988 *Solar radiative output variation* (ed. P. Foukal), p. 254. (Available from the editor.)
Willson, R., Gulkis, S., Janssen, M., Hudson, H. & Chapman, G. 1981 *Science, Wash.* **211**, 700.
Willson, R. & Hudson, H. 1988 *Nature, Lond.* **332**, 810.

Phil. Trans. R. Soc. Lond. A **330**, 601–616 (1990)

Printed in Great Britain

Time series analysis of Holocene climate data

By D. J. Thomson

AT&T Bell Laboratories, Murray Hill, New Jersey 07974, U.S.A.

Holocene climate records are imperfect proxies for processes containing complicated mixtures of periodic and random signals. I summarize time series analysis methods for such data with emphasis on the multiple-data-window technique. This method differs from conventional approaches to time series analysis in that a set of data tapers is applied to the data in the time domain before Fourier transforming. The tapers, or data windows, are discrete prolate spheroidal sequences characterized as being the most nearly band-limited functions possible among functions defined on a finite time domain. The multiple-window method is a small-sample theory and essentially an inverse method applied to the finite Fourier transform. For climate data it has the major advantage of providing a narrowband F-test for the presence and significance of periodic components and of being able to separate them from the non-deterministic part of the process. Confidence intervals for the estimated quantities are found by jack-knifing across windows.

Applied to ^{14}C records, this method confirms the presence of the 'Suess wiggles' and give an estimated period of 208.2 years. Analysis of the thickness variations of bristlecone pine growth rings shows a general absence of direct periodic components but a variation in the structure of the time series with a 2360-year period.

1. Introduction

To most scientists attempting to analyse their data, the subject of statistical time series analysis presents a bewildering array of methods and papers, each invariably purported to have or to show some optimal property. The purpose of this paper is to summarize some features of the multiple-window method as they apply to the analysis of Holocene climate data, to show some results of this analysis, and also to caution against some currently popular spectrum estimation methods whose use, I believe, can result in misleading conclusions. The analysis of Holocene climate data merits special attention as it is beset by a combination of problems that occur in other disciplines, but rarely in such concentration. Given current speculations on the 'greenhouse effect' and related global climate problems, it is mandatory to provide an accurate historical base against which changes may be measured. Specific problems affecting the time series part of data analysis, as opposed to the more difficult problems of interpretation and establishing causality, are the following.

1. Most of the available records are proxies and, as such, record a combination of effects. Changes in ^{14}C, for example, record a confounding of solar activity, geomagnetic changes, and various reservoir effects. Reliable analysis requires proper attribution.

2. Geophysical data almost invariably include a combination of periodic, or almost periodic, signals embedded in a non-deterministic, or noise-like, background. The background signal is usually 'red' that is, it has much more power at low frequencies than at high. The Wold decomposition theorem (Doob 1953), requires that the deterministic and random components be treated separately. Statistically, this must be done as well, but with the complication that both components be estimated simultaneously.

3. Sampling times in geophysical records are often unavoidably sporadic and, in addition to being irregularly sampled, the record may contain large gaps. It is usually assumed that cautious interpolations will allow the recovery of low-frequency terms, but the incommensurable periods of many known effects and large power variations at high frequencies exacerbate the inability to do anti-alias filtering.

4. Climate data contain statistical outliers both from the inevitable recording and measurement errors and also from natural causes. The former, as demonstrated in Kleiner *et al.* (1979), are bad enough; the latter may be seen even more insidious, for example, it has been suggested (see, for example, De Mendoça Dias 1962) that volcanic and solar activity are related. Thus long-term climate data may contain solar effects observed through indirect as well as more conventional mechanisms.

5. Radiometric timescales contain random inaccuracies whereas those obtained from dendrochronology may have jumps. Both have the effect of making a spectrum appear unnecessarily complicated.

6. There is much circumstantial evidence, and even more claims in the literature, for numerous low-level and closely spaced periodic signals in the climate and solar activity records (see, for example, the reviews by Pittock (1978), Shapiro (1979) and Sonett (1984)). Most of the suspected periods are unknown, not simply related, and the existence of closely spaced lines makes confirmation difficult.

7. The length of available data records is short in comparison with the frequency separation of the various suspected lines. Perversely, there is usually too much data to allow numerical calculation of 'exact' procedures.

8. Data for the complete Holocene, even if available, cannot be treated as a stationary process both because of the large residual transient from the last Pleistocene glaciation and to continuing Milankovitch processes. Further, in §7, I give evidence for a process with a 'period' of about 2400 years that alters the structure of the process.

In this paper we cannot address all the above in detail, and so concentrate on those where the multiple window approach is advantageous. The outline of the paper is as follows: the next section gives an abbreviated background on stationary processes, followed by the fundamental equation, with §4 outlining an inverse theory solution of it. Section 5 describes procedures for line detection and mixed spectra, §6 covers miscellaneous topics, with §§7 and 8 containing the examples. Finally, §9 discusses some of the problems with other spectrum estimation procedures in current use.

2. Background on stationary processes

The power spectral density, or power spectrum, or simply the spectrum of a random process is firmly established as the most useful description of most time series encountered in the physical sciences; its definition, and the accompanying spectral representation, despite their general description by Cramér in 1940, are unfortunately less familiar. For a discrete-time harmonizable process with unit time step (so the Nyquist frequency is $\frac{1}{2}$) the *Cramér* or *spectral* representation is written as a generalized Fourier transform,

$$x_t = \int_{-\frac{1}{2}}^{\frac{1}{2}} e^{i2\pi\nu t} \, dX(\nu), \tag{1}$$

and describes the process for all time. If the process is stationary, i.e. if its statistics are invariant under changes of time origin, dX has zero mean, energy at different frequencies is uncorrelated, and the expected value of $|dX(f)|^2$ is, by definition, the spectrum $S(f)\,df$. Detailed properties are available in numerous books (see, for example, Priestley 1981; Brillinger 1975). For many problems in the physical sciences, however, a better, if even less well-known, approach is given by the extended, or Munk–Hasselman (Munk & Hasselman 1964), representation where the deterministic component, or first moment, is explicitly given by

$$E\{dX(f)\} = \sum \mu_j \delta(f-f_j)\,df, \tag{2}$$

where δ is the Dirac delta-function, the f_j are the frequencies of periodic, or line, components, and μ_j their amplitudes. The continuous part of the spectrum is defined by

$$S(f)\,df = E\{|dX(f) - E\{dX(f)\}|^2\}, \tag{3}$$

E denoting the statistical expected value operator. That is, the non-deterministic part is given by the second central moment. Such processes are known as centred or conditionally stationary (as opposed to strict or covariance stationary) and are often noted as having mixed spectra. The distinction between first and second moment properties is exceptionally important.

First moments correspond loosely to what scientists usually term harmonic analysis, that is the study of periodic phenomena. Typically, a process will contain a few such lines, each described by their amplitude, frequency, phase, and perhaps frequency drift rate. Consequently, such parameters may be estimated by using the well-established methods of maximum-likelihood. Moreover, super-resolution, i.e. discrimination of frequencies spaced closer than the Rayleigh resolution of $1/T$, T being the total observation time, is possible in principle (and indeed in practice to some extent) and the accuracy with which closely spaced lines may be estimated is a function of the signal:noise ratio, defined as the ratio of power in the first moment to power in the second moment as a function of frequency.

Second moments, on the other hand, are 'noise-like' and are the non-predictable or non-deterministic part of the process. An important property is that energy at different frequencies is uncorrelated, $\mathrm{Cov}\{dX(f), \overline{dX(g)}\} = 0$ for $f \neq g$, Cov the covariance operator. Here, in contrast to the line spectrum of the first moments, the spectrum is typically continuous and often smooth. Because the second-moment spectrum cannot vanish over an interval unless the process is deterministic, it is often referred to as the continuum spectrum. Also in contrast to the first moments, one is now trying to estimate a function of frequency, not just a handful of parameters. Consequently, one is *not* doing maximum-likelihood estimation but a quadratic counterpart of Backus–Gilbert inverse theory. Again, there is a distinction in achievable resolution. It is essentially impossible to resolve details of the continuum spectrum separated by less than $1/(2T)$, half the Rayleigh limit. Typically, useful resolution is between $2/T$ and $50/T$, much poorer than Rayleigh.

Confusing the two moment properties will result in absurdities like 'smoothing' line spectra and, equally unwarranted, applying super-resolution criteria to noise-like processes. Finally, one must remember that, classically, spectra are defined only for stationary processes. However, I make the usual assumption that the structure of the process varies slowly enough to assume local stationarity (see Priestley 1988; Daubechies 1988).

3. The fundamental equation

I assume that there are N observations from a stationary series, $x_0, x_1, ..., x_{N-1}$, at equally spaced sample times $0, 1, ..., N-1$. Because my goal is to estimate the statistics, in particular the first *two* moments of dX, we take the Fourier transform of the available data,

$$\tilde{x}(f) = \sum_{t=0}^{N-1} x_t \, e^{-i2\pi ft}, \tag{4}$$

and combine it with the spectral representation (1): the result is the fundamental equation of spectrum estimation

$$\tilde{x}(f) = \int_{-\frac{1}{2}}^{\frac{1}{2}} K_N(f-\nu) \, dX(\nu), \tag{5}$$

where the kernel is given by

$$K_N(f) = \sum_{t=0}^{N-1} e^{-i2\pi ft} = \exp\left[-i2\pi f(N-1)/2\right] \frac{\sin N\pi f}{\sin \pi f}. \tag{6}$$

Note that $\tilde{x}(f)$ may be inverse transformed to recover the data and is thus a trivially sufficient statistic; the same is not true for $|\tilde{x}(f)|^2$. Equation (5) is traditionally treated as a convolution; the difference here is that we choose to treat it as a Fredholm integral equation of the first kind and, although it does not have a unique solution, attempt to find an approximate solution with 'reasonable' statistical properties.

4. Spectrum estimation as an inverse problem

In multiple-window spectrum estimation one chooses a bandwidth W and constructs a local least-squares eigensolution of the fundamental equation (5) on the interval $(f-W, f+W)$ for a set of frequencies f. The key to this solution is a set of special functions known as Slepian functions, or discrete prolate spheroidal wave functions, that are the time-limited functions whose Fourier transforms are most concentrated in frequency. Details are available in Slepian (1978), and references given therein. Slepian defines the discrete prolate spheroidal sequences $v_n^{(k)}(N, W)$ as an orthonormal solution to a matrix eigenvalue problem. Their Fourier transforms,

$$V_k(f) = \sum_{n=0}^{N-1} v_n^{(k)}(N, W) \, e^{-i2\pi fn}, \tag{7}$$

are solutions of the integral equation,

$$\lambda_k V_k(f) = \int_{-W}^{W} K_N(f-\nu) \, V_k(\nu) \, d\nu, \tag{8}$$

where the kernel is again given by (6). To agree with common definitions of the discrete Fourier transform, I use V in place of Slepian's U with

$$V_k(f) = (1/\epsilon_k) \exp\left[-i\pi f(N-1)\right] U_k(-f),$$

where ϵ_k is 1 for k even, i for k odd, and the dependence of the λs, Us and Vs on N and W have been suppressed. Note carefully that the fraction of energy retained within the band $(-W, W)$ by the kth function is given by λ_k, and that there are about $K = \lfloor 2NW \rfloor$ functions with eigenvalues near 1. These K functions each have most of their energy concentrated within this band, and thus are classified as 'good' windows from the point of view of spectrum estimation.

Although details of the solution are available in Thomson (1982), and, in easier to read form, in Park *et al.* (1987*b*), a summary is given here. Because the Slepian functions form a complete set, I assume that the observable portion of dX has the expansion,

$$dX(f-\nu) = \sum_{k=0}^{\infty} x_k(f) \cdot \overline{V_k(\nu)} \, d\nu, \tag{9}$$

for $|\nu| < W$. By the assumed stationarity, the part of dX in the inner or local domain, $(f-W, f+W)$ is uncorrelated with dX on the rest of the frequency domain, and it is helpful to think of the 'inner' energy as 'signal', the rest as 'noise'. Using the basic integral equation and orthogonality properties one obtains the expansion or eigencoefficients:

$$x_k(f) = \sum_{n=0}^{N-1} e^{-i2\pi f n} \cdot \nu_n^{(k)}(N, W) \cdot x_n. \tag{10}$$

The expansion coefficients are obtained by windowing the data with a Slepian sequence and taking the Fourier transform. This is identical to the better windowed spectrum estimates except, conventionally, only the first term in this expression has been used. Again I emphasize that the $K = \lfloor 2NW \rfloor$ coefficients with eigenvalues $\lambda_k \approx 1$ are used to represent the information in the signal projected onto the local frequency domain for all subsequent inference. For example, a crude multiple-window spectrum estimate is

$$\overline{S}(f) = \frac{1}{K} \sum_{k=0}^{K-1} |x_k(f)|^2. \tag{11}$$

Temporarily restricting ourselves to the continuous part of the spectrum, the dependence on the bandwidth W of the estimate may be deduced from the existence of $K \approx \lfloor 2NW \rfloor$ windows with eigenvalues near 1. If the spectrum is flat within the local domain the coefficients are uncorrelated (because the windows are orthogonal) and each contributes two degrees of freedom; thus estimate (11) has $2K$ degrees of freedom. If W is too small one has poor statistical stability, but if W is too large the estimate has poor frequency resolution. Typically W is chosen between $1/N$ and $20/N$ with a time-bandwidth product NW of 4 or 5 being a common starting point. (Thus $W = 4/N$ or $5/N$ with corresponding K usually taken conservatively as 6 or 8 giving estimates with 12 or 16 degrees of freedom.) The case $NW = 4$ is interesting for comparison: the bandwidth is the same as that obtained when a Parzen (1957) or Papoulis (1973) window is used as a data taper except that one obtains 10–12 degrees of freedom instead of 2. Obviously a fast Fourier transform (FFT) may be used for efficient computation of the expansion coefficients. Also, the Slepian sequences are eigenvectors of a simple symmetric tridiagonal matrix, equation (14) of Slepian (1978), and so are trivial to compute.

In practice, one uses a local least-squares solution for the integral equation instead of simply truncating the series after the first K coefficients. This gives a data adaptive weighting with superior protection against leakage and bias. An approximate solution is

$$\sum_{k=0}^{K-1} \frac{\lambda_k(\hat{S}(f) - \hat{S}_k(f))}{[\lambda_k \hat{S}(f) + (1-\lambda_k)\sigma^2]^2} = 0, \tag{12}$$

where $\hat{S}_k(f) = |x_k(f)|^2$ and σ^2 is the process variance. Details are given in §V of Thomson (1982). Note that this is the estimate usually used: form (11) is useful for an introduction, (12) is more accurate and reliable. Note also that estimate (12) is not a simple quadratic form but the solution of an implicit equation involving a ratio of two such forms. One consequence is that the estimate of autocovariance, $\tilde{R}(\tau)$ obtained by taking the Fourier transform of the spectrum obtained by solving (12),

$$\tilde{R}(\tau) = \int_{-\frac{1}{2}}^{\frac{1}{2}} \hat{S}(f)\, \mathrm{e}^{\mathrm{i}2\pi f \tau}\, \mathrm{d}f,$$

can be non-zero for lags, τ, greater than N.

5. Harmonic F-test for periodic signals

As emphasized earlier, a major problem with most spectrum estimates is that they ignore differences between first and second moments. Here the multiple-window method is advantageous because it allows us to separate line and continuum components and permits an analysis-of-variance test for significance of possible line components. Given a periodic signal at frequency f_0 the expected value of the eigencoefficients is, by (2) and (7)

$$\mathrm{E}\{x_k(f)\} = \mu V_k(f - f_0). \tag{13}$$

The complex amplitude is estimated by minimizing the residual local least-squares error,

$$e^2(\mu, f) = \sum_{k=0}^{K-1} |x_k(f) - \mu(f)\, V_k(0)|^2, \tag{14}$$

with respect to μ, giving

$$\hat{\mu}(f) = \sum_{k=0}^{K-1} \overline{V_k(0)}\, x_k(f) \Big/ \sum_{k=0}^{K-1} |V_k(0)|^2, \tag{15}$$

which is simply a linear regression of the $x_k(f)$s on the $V_k(0)$s. The fit may be tested for significance by the usual regression F-test, the ratio of the energy explained by the assumption of a line component at the given frequency to the residual energy:

$$F(f) = \frac{1}{\nu} |\hat{\mu}(f)|^2 \sum_{k=0}^{K-1} |V_k(0)|^2 \Big/ \frac{1}{2K - \nu} e^2(\hat{\mu}, f), \tag{16}$$

where ν is two degrees of freedom when the line frequency is known, and is three if frequency is estimated. The location of the maximum value of F provides an estimate of line frequency that has resolution within 5–10 % of the Cramér–Rao bound. This test works well if lines are isolated, i.e. if there is only a single line in $(f - W, f + W)$. The total number of lines in the spectrum is not important as long as they occur singly. For lines spaced closer than W one uses a multiple-line test, a similar but algebraically more complicated regression of the eigencoefficients on a matrix of Slepian functions $V_k(f - f_j)$ with the simple F-test replaced by partial F-tests (see, for example, Draper & Smith 1981). Multiple-line tests should be used with caution, because the Cramér–Rao bounds for line parameter estimation degrade rapidly, when the line spacing becomes less than $2/N$ (Rife & Boorstyn 1976). It is important to remember that in typical time-series problems hundreds or thousands of uncorrelated estimates are being dealt with; consequently one will encounter numerous instances of the F-test giving what would normally be considered highly significant test values that, in actuality, will only be sampling fluctuations. A good rule-of-thumb is not to get excited by significance levels less than $1 - 1/N$.

One aspect of the F-test of particular significance for the analysis of climate data is its use in estimating the mean value of the series. It is uncommon to have measurements free from an average offset, so the usual practice is to subtract the average, and possibly slope, from the data, and then treat it as though it were a sample from a process with known zero mean. The problem is that this approach leaves a hole in the spectrum around zero frequency that, combined with the very red spectra typical of geophysical processes, often leads to the appearance of a 'peak' at a period commensurable with the length of the series. A better approach is to include a term in (2) at zero frequency, use (15) at $f = 0$ to estimate $\hat{\mu}(0)$, and use the residuals $\hat{x}_k(f) = x_k(f) - \hat{\mu}(0) V_k(f)$ in place of $x_k(f)$ at frequencies less than W. This is particularly crucial when the periods of interest are nearly as long as the data; here multiple-line estimates are necessary, as is the inclusion of both positive and negative frequency terms. Further information on F-tests is in Thomson $et\ al.$ (1986) and Park $et\ al.$ (1987a).

6. VARIOUS ASPECTS OF MULTIPLE-WINDOW ESTIMATES

A classical problem in spectrum estimation is the choice of the bandwidth, W. Although several methods for making such a choice exist, they typically depend on an unverifiable smoothness assumption. Here again, multiple-window methods have the advantage of allowing a test for unresolved structure. To do this, take the vector of eigencoefficients

$$\boldsymbol{X}(f) = [x_0(f), x_1(f), ..., x_{K-1}(f)]^{\mathrm{T}}$$

and compute its covariance matrix by averaging over different realizations or sections $\hat{\boldsymbol{C}}(f) = \mathrm{ave}\{\boldsymbol{XX}^\dagger\}$. The expected elements of $\boldsymbol{C}$ are $C_{jk}(f) = \mathrm{E}\{x_j(f)\,\overline{x_k(f)}\}$ or, up to terms in $1 - \lambda_j$, approximately

$$C_{jk}(f) \approx \int_{-W}^{W} V_j(\nu)\,\overline{V_k(\nu)}\,S(f-\nu)\,\mathrm{d}\nu.$$

If S is constant in $(f-W, f+W)$ then, by the orthogonality of the Slepian functions, $C_{j,k} \approx S(f)\lambda_j \delta_{j,k}$ or $\boldsymbol{C} \approx S(f)\boldsymbol{I}$. Consequently one may use a complex analogue of the sphericity (Anderson 1984) or uniformity (Watson 1983) tests on $\hat{\boldsymbol{C}}(f)$ for unresolved structure. Note that assuming S is constant across the local band is the same as assuming the spectrum is locally white, or that the bandwidth is such that details are resolved: practically, S is rarely constant, but if the structural components in a spectrum estimate are much smaller than the random components, the estimate is probably reasonable. Use of quadratic inverse theory gives a better, but more complicated, test for unresolved structure that does not require subdivision or replication of the data. Details will be given in a forthcoming paper.

Cross-spectra and coherences may also be computed without the usual phase-averaging problems. Letting $x_k(f)$ and $y_k(f)$ be the eigencoefficients from series x and y respectively, the cross-spectrum estimate corresponding to (11) is

$$S_{xy}(f) = \frac{1}{K} \sum_{k=0}^{K-1} x_k(f)\,\overline{y_k(f)}$$

from which coherences may be computed as usual. If several series are available, for example solar and climate series, linear transfer functions are easily computed; the paper by Lanzerotti $et\ al.$ (1986) contains an example. Generally speaking, multiple-window methods allow the use of all the inferential techniques of multivariate statistics without the loss of resolution implicit in data sectioning methods but, of course, with the unavoidable complications of having almost everything a complex function of frequency.

Tolerances and confidence limits may be found for multiple-window estimates without gaussian assumptions by jack-knife or bootstrap methods (Efron 1982) by treating the eigencoefficients (10) as data. Details are given in a forthcoming paper by Thomson & Chave (1990).

7. An example: thickness variations of growth-rings in bristlecone pine

This data-set consists of 5405 measurements of the thickness of the annual growth-rings of bristlecone pines from Mount Campito (east-central California) starting at 3434 B.C., generously provided by Professor C. P. Sonett and attributed by him to the late Professor V. C. La Marche of the University of Arizona's tree-ring laboratory. The data is of high quality with few outliers.

The time dependence of this data, however, exhibits unusual non-stationary effects. Because of the numerous claims for periodic effects in the climate literature, the harmonic F-test (16) was used to look for these using the complete data record. F-tests at 8.86 and 19.61 years were statistically insignificant. Periods of 5.429, 5.071, 3.276, 2.500, and 2.143 years had F-tests with significances between the 99.9% and 99.99% levels but, unless confirmed by independent data, are best considered sampling fluctuations. (Remember that this set contains 5400 samples, so one should expect about five spurious peaks above the 99.9% level.) At periods longer than 5.5 years nothing of significance above 99.9% was found. The general level of the F-test at long periods is suspiciously low. A possible, although one hopes improbable, explanation is that this band consists of a continuum of unresolved lines. Next, the series was broken into overlapping subseries of various lengths and the tests repeated. Numerous lines with high apparent significances were found in individual blocks but did not persist between blocks. This could result if the time base had several step discontinuities from errors in the dendrochronology, but the otherwise high quality of this data makes this explanation unlikely. Another mechanism to explain this behaviour is that line components do exist, but as unresolvable multiplets. If true, the isolated line F-test would fail, and its behaviour in short blocks would be apparently erratic. In blocks where the singlets are in phase, a signal at the average frequency would be detected; but in blocks where they are out of phase, the singlets would largely cancel. To check this possibility several approaches were tried: first, a robust prewhitening filter of the type described by Kleiner et al. (1979) was used to reduce the dynamic range of the spectrum and the influence of the few outliers. Then a single prolate data window with time-bandwidth product of 1.5 was used to compute the spectrum, part of which is shown in figure 1. The portion shown, from 0.011 to 0.015 cycles per year (or periods between 66 and 91 years) includes the Gleisberg cycle (Sonett 1982). The solid curve shows a roughly symmetric array of peaks, suggestive of a modulation process. To check this, the analysis was repeated on the data listed in table 3 of Ferguson (1970). This data is a longer bristlecone pine series, beginning in 5142 B.C.: because of numerous anticorrelated errors in adjacent years, the higher frequencies are unrealiable; low frequencies are similarly unrealiable because of the filtering. The corresponding estimate (scaled so the largest peak in this band has magnitude 1) is shown as the dashed line. Remembering that the Rayleigh resolution here is 2×10^{-4} cycles per year, only one major peak, at 0.013 cycles per year (77-year period), is in good agreement with the Campito set, despite the presumably high correlation between the two data-sets.

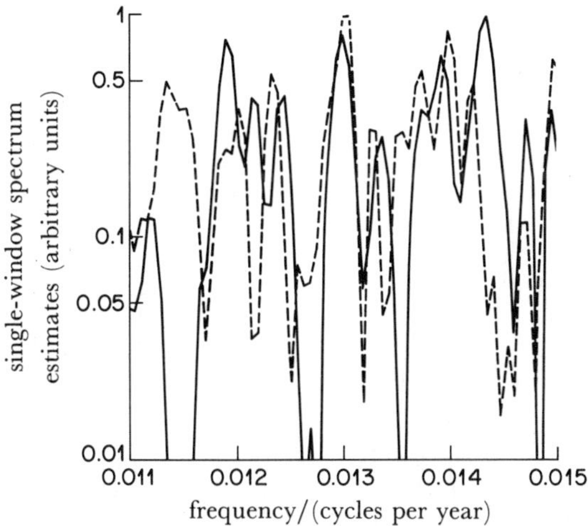

FIGURE 1. Single window (time-bandwidth product of 1.5) unsmoothed spectrum estimates of the Campito (solid line) and Ferguson (dashed line) bristlecone pine series. Note that the only peak with good agreement is the one at 0.013 cycles per year (77-year period). Both estimates have been scaled to have a maximum value of 1 in this band.

Secondly, a narrow-band eigenvalue procedure similar to that described in Thomson (1986) was tried. This, as Pisarenko-style estimators do, produced spectacularly narrow peaks at similar frequencies to those shown. These frequencies were then used in a multiple-line F-test, but the significance of the fit was low. Finally, deleting each line in turn and computing partial F statistics again resulted in mediocre fits. None the less, this band probably should be modelled as a dominant multiplet split into two or three singlets, perhaps by precession, and the whole modulated. The reason for this is that if one writes (9) in the time domain, to give the complex demodulate,

$$x_f(t) = \sum_{k=0}^{K-1} x_k(f)\, v_t^{(k)}(N, W),$$

the magnitudes $|x_f(j/2W)|$ taken at the filter Nyquist rate, have an empirical probability distribution that is a poor match to the expected Rayleigh distribution. Development of such a complicated model, however, is beyond the scope of this paper and the limitations of this data.

Given the failure of the line test procedures to show significant levels even at known astronomical periods that are suspected of being present in climate data, one must attribute most of the variations in the spectrum to the non-deterministic or continuum part of the process. Computing a multiple-window spectrum from the complete record results in an estimate that varies rapidly in frequency, reminiscent of a collection of medium-Q modes. However, when confidence limits are set by a jack-knifing procedure, where individual windows are delected from the estimate, an excessive variance is obtained. As before, the data were subdivided, spectra computed on each block, then tested for stationarity by using the procedure described on p. 1994 of Thomson (1977) with the results shown in figure 2. For this test five sections each of 1800 years duration and offset by 900 years were used. The spectrum for each section was computed by using $K = 5$ windows with a time-bandwidth product $NW = 4$ and the adaptive weighting (12). At each frequency, the stationarity test is a Bartlett M statistic (see Pearson &

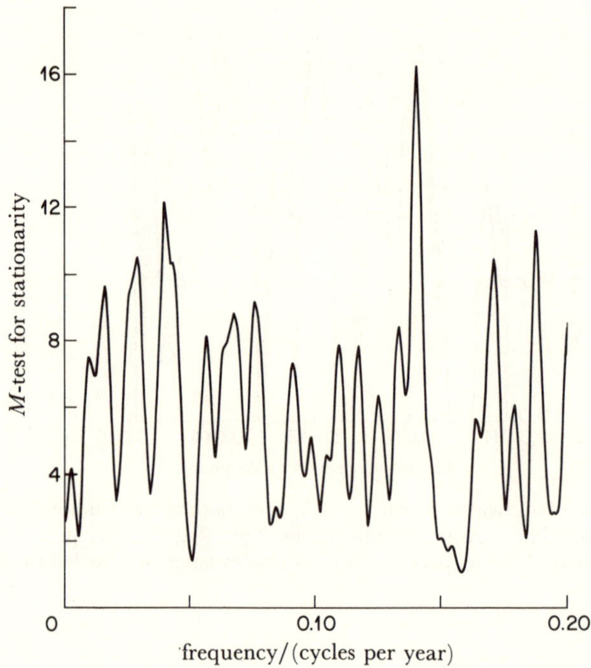

FIGURE 2. Stationarity test for the Campito data. The full data-set was broken into five overlapping segments each 1800 years long and offset from each other by 900 years, starting from an initial date of 3434 B.C. On each segment five tapers with a time-bandwidth of 4 were used, giving spectrum estimates each having 10 degrees of freedom. The expected value of Bartlett's M statistic for homogeneity of variance is 4.16 and the 90 % point is at 8.1. The plotted data have been slightly smoothed.

Hartley 1969), so here would have a distribution which is approximately χ^2 with four degrees of freedom. As the expected value of M is 4.16 and the average, across frequency, is 5.2, the series is only mildly non-stationary.

Figure 3 shows a robust average, a trimmed logarithmic mean, of the five subset estimates. Conventional methods would assign factors of 0.8 and 1.2 for the lower and upper 5 % points, but in view of the non-stationarity, factors of $\frac{1}{2}$ and 2 are more appropriate. Even so, there are many statistically significant features and, at a minimum, one can say that the probability of the spectrum being white is zero.

To explore the non-stationarity and associated excess variability further, spectra were computed on 16 sections each 900 years long and offset by 300 years by using 10 windows with a time-bandwidth product of 6 on each block. Again, to minimize effects of sampling noise, only frequencies below 0.2 cycles per year were kept. Logarithms of the different spectrum estimates were arranged in a rectangular matrix G, the average over blocks subtracted at each frequency, and a singular value decomposition of the result computed. Taking $\delta G = U\Sigma V^{\mathrm{T}}$ the left eigenvectors, the columns of U, describe frequency dependence; the right eigenvectors, the columns of V, the time dependence; and the singular values σ the scale of the effect. The first three eigenvectors are shown in figures 4 and 5. The frequency dependence (figure 4) is complicated: U_1 shows a basic split in behaviour at 0.025 cycles per year; U_2 also has this break but has related peaks near 0.044 cycles per year (22.7 years) and 0.011 cycles per year (9 years), possibly a modulation of low-frequency solar processes by the 8.86-year lunar cycle. U_3 also has a 22-year peak and a related third harmonic at 0.14 cycles per year. The time

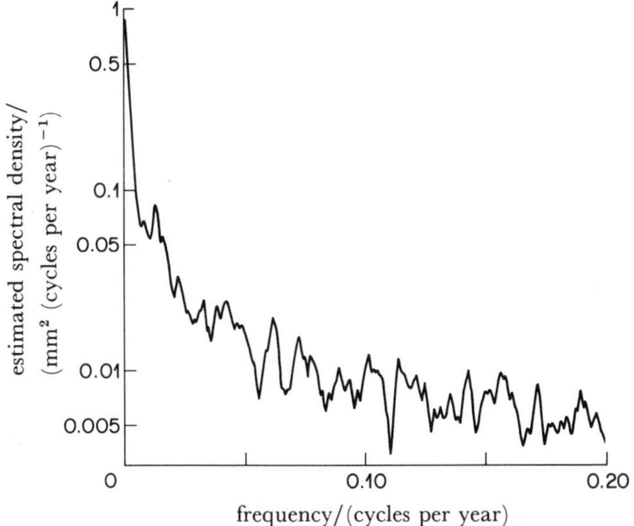

FIGURE 3. An estimate of the spectral density of the Campito data. The estimate is made by averaging the logarithms of the central three of the five sorted section estimates at each frequency. The data segmentation and windowing is the same as in figure 2. Because the sorting introduces discontinuities the estimate shown has been slightly smoothed.

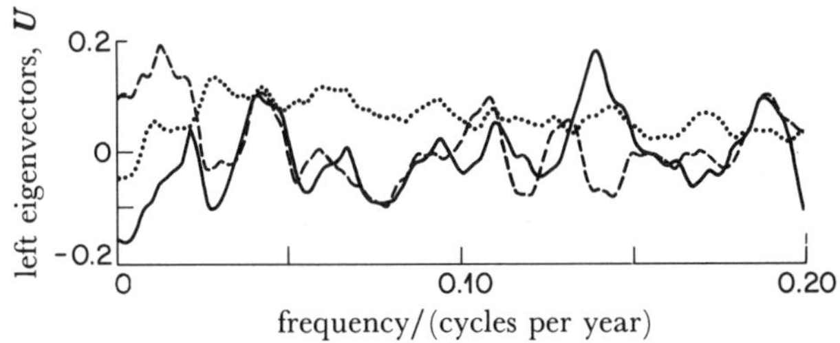

FIGURE 4. Frequency components, the left eigenvectors U, produced by a singular value decomposition of a time-dependent spectrum estimate of the Campito data. The first eigenvector is shown dotted; the second, dashed; and the third, solid, with relative eigenvalues of 259.4, 95.3 and 71.9 respectively. The corresponding right eigenvectors are shown in figure 5.

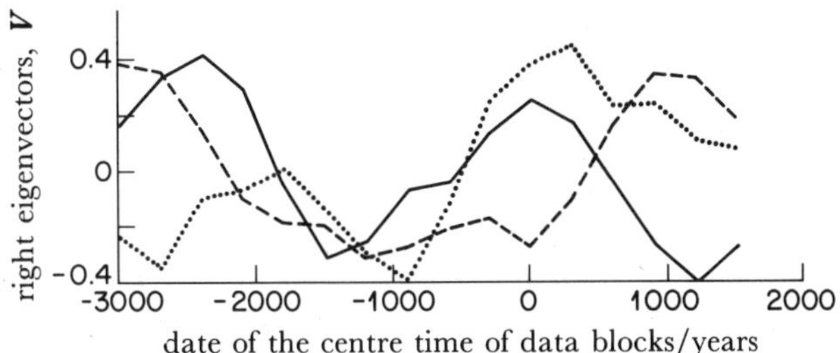

FIGURE 5. The time-dependent part, or right eigenvectors, of the SVD for the Campito data. The line styles are as in figure 4. Note the 2360-year period of the third eigenvector (solid line).

dependence of the second and third components is more interesting: V_2 appears to vary with about a 4000-year period, whereas V_3 has two cycles of a 2360-year period. Applying the harmonic F-test to V_3 gives a significance level of 99.99 %. Note carefully, however, that this effect is not a direct periodic component embedded in the signal but rather the structure of the process varies in a periodic manner.

[213]

612   D. J. THOMSON

Taken together, these results give a different picture of climate variability (as expressed by
the tree-ring record) than usual. Here one sees few, if any, dominant periodic effects in the
direct signal. There are, however, numerous low-level features whose statistical structure varies
with time.

8. Variations in ^{14}C and 'Suess wiggles'

Our second data example consists of 644 values of differences between the ^{14}C and
dendrochronological ages of bristlecone pine as determined by Professor Hans Suess. These
records have long been suspected of having periodic components (Suess 1965; Sonett 1984),
but the point has been contested (Stuiver 1980; Sonett & Suess 1984). The original
measurements, being unequally spaced and sometimes replicated, were interpolated to an
equally spaced grid. Multiple-window estimates with $NW = 4.5$ and $K = 6$ were used, giving
a bandwidth of 5.4×10^{-4} cycles per year. Here periodic effects are detected. In particular a
line with estimated frequency of 104.14 years (one jack-knife standard deviation limits of 104.0
to 104.28 years), an amplitude 0.84 ± 0.07 years, and a significance level above 99.99 % was
found. Possibly related lines with periods of 208.2 years at 95 %, and 34.8 years at 99.6 % were
also present.

The analysis was repeated, starting with 282 samples of differences between the ^{14}C and
dendrochronology ages of samples from Irish oaks taken from G. W. Pearson's thesis, the data
again courtesy of Professor C. P. Sonett. Again, a highly significant line, at 99.7 %, was found
with an estimated period of 103.93 years and one standard deviation limits of 103.72 to 104.15
years, that is essentially identical to the line found in the Suess data, but with a lower amplitude
of 0.48 ± 0.12 years. Again a 208-year component was present. The other lines common between
the two data-sets were 230 years, 60.6 years and 51.6 years. An estimate of the spectrum
showing both continuum and line components is given in figure 6. From this it may be
concluded that the 'Suess wiggles' in ^{14}C data are real and with very high confidence, that they
have a probable fundamental period of 208.15 years, a strong harmonic at 104.07 years, and
probably additional frequencies.

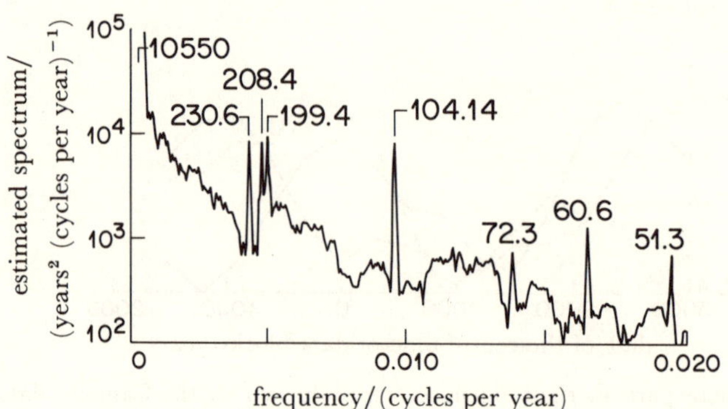

Figure 6. Reshaped multiple-window spectrum estimate of the Suess ^{14}C data. A time-bandwidth product of 4.5
with six windows was used with adaptive weighting resulting in an average stability of 11.6 degrees of freedom.
At frequencies where the harmonic F-test exceeded the 95 % level the spectrum was 'reshaped' to show the line.
The width of each line is proportional to its frequency uncertainty (from the F-test) and the height is scaled
to conserve power. Significant peaks are labelled with their estimated period in years. The estimated low-
frequency spectrum, clipped in the figure, was 2×10^6 but had unresolved components; the true value would
be somewhat higher.

[214]

To conclude this example, note that the interpolation problem was largely ignored; this is not because it is trivial or unimportant, quite the reverse. To proceed rigorously, one develops an integral equation analogous to (6) using the time points (or intervals) observed, then either uses the eigenfunctions of this equation directly, or equivalently, uses them to interpolate to an equally spaced grid. Although details are excessive, we note that the eigenvalues of the random sampling equation are much poorer than for the regular Slepian functions, so that random aliases of high-frequency signals are difficult to exclude on analytic grounds. Bronez (1988) gives an introduction to these problems. Because a variety of interpolation schemes were tried here with similar results, and because of the agreement of the two sets with different sampling times, we conclude that the observed period is real and not an alias.

9. Estimates to be avoided

Having dwelt at length on the multiple-window methods, what of the myraid other spectrum estimation methods? Recall that the multiple-window method occurs as a natural solution of the inverse problem; the better of the single-window estimates, i.e. a Parzen or Papoulis window applied directly to the data, not to the correlations, give the first term in the series. (See Harris (1978) for properties of these, and many other, windows.) They are consequently less efficient and are unable to separate moments. Also, being *ad hoc*, the theory for such windows is much less satisfactory, not to mention algebraically more complicated, than multiple windows. The oldest spectrum estimate (from 1894) still in common use, the unwindowed periodogram,

$$P(f) = \frac{1}{N}\left|\sum_{n=0}^{N-1} x_n\, e^{i2\pi n f}\right|^2, \tag{17}$$

is obsolete, but, because it is computed by simply passing one's data through a discrete Fourier transform (DFT) (usually as an FFT) and squaring, it has had an unfortunate resurgence. It is often disguised, for example: the usual frequency spacing of $\Delta f = 1/N$ gives exactly the Fourier series representation of the periodically extended observations; a least-squares fit to the data by a trigonometric series is again the same thing; and there are innumerable similar variants. As an estimate of the spectrum it is both inconsistent and so badly biased that it is extremely misleading and should not be used. This bias can be exceptionally severe; in Thomson (1977) an example using engineering data is shown where the periodogram is in error by more than a factor of 10^{10} over most of the frequency range.

Indirect or *Blackman–Tukey* (Blackman & Tukey 1958) estimates were developed before the invention of the FFT and traditionally began by computing the sample autocovariances,

$$\hat{R}(\tau) = \frac{1}{N}\sum_{n=0}^{N-1-|\tau|} x_n\, x_{n+|\tau|},$$

as a function of the lag, τ, for lags up to perhaps 40 % of the series length. The spectrum is then estimated by multiplying $\hat{R}(\tau)$ by one of a multitude of lag windows and Fourier transforming. Because the sample autocorrelations may be obtained by Fourier transforming the periodigram,

$$\hat{R}(\tau) = \int_{-\frac{1}{2}}^{\frac{1}{2}} P(f)\, e^{i2\pi f t}\, \mathrm{d}f \tag{18}$$

(and are now usually computed with a second FFT and zero padding), the convolution theorem shows that indirect estimates are simply linear smoothers applied to the periodigram. Thus apart from potentially serious round-off problems, indirect estimates give one a stable and incorrect estimate instead of the unstable and incorrect estimates obtained with the periodigram; again they are obsolete and should not be used.

Autoregressive or 'maximum-entropy' estimates are more insidious; misappropriating authority from statistical mechanics and Shannon theory, such approaches choose the estimate maximizing the Wiener entropy,

$$H\{\hat{S}\} = \int_{-\frac{1}{2}}^{\frac{1}{2}} \ln \hat{S}(f)\, df,$$

subject to the constraints that it reproduces the first few autocorrelations. Because the autocorrelations are given by (18), the 'maximum-entropy' estimate can be no better than the periodogram. Thus in addition to fitting a biased and inconsistent estimate, one imposes what may be totally inappropriate model whose forced fitting can introduce spurious features in the estimated spectrum. The Burg procedure (see Chen & Stegun 1974) avoids use of the autocorrelations but imposes both strong stationarity and time-reversal symmetry constraints. Bretthorst (1987) fits an all-line model by using ordinary least squares and a geophysically unrealistic white noise assumption. For dangers of model misspecification, see Tukey (1982); also Kay & Marple (1982) for some examples.

One positive aspect of Wiener entropy that should not be overlooked is that $\exp\{H\}$ is the variance of the best linear predictor of a stationary process; note that it is a logarithmic mean of the spectrum. Thus a small change in total power can drastically alter predictability so spectra should amost invariably be plotted with a logarithmic ordinate.

Finally, the method of 'linear spectral analysis' recently used in the geological literature, is badly biased, erratic and untrustworthy except when one knows *a priori* that there is a single periodic signal with high signal:noise present (see Stigler & Wagner 1987).

To conclude this section, note that it can be argued that careful prewhitening will reduce the bias problem (and although strongly advocated by Blackman & Tukey, the advice is apparently usually ignored); this is usually the case, but there are examples where it is not possible. Moreover, the same argument can be applied, and with greater success, to multiple-window methods. Even ignoring the overwhelming advantages of the multiple-window methods with mixed spectra, careful analysis of bias and variance invariably show multiple-window methods to have the advantage.

10. CONCLUSIONS

The major point of this paper is that, for the analysis of Holocene climate data, the multiple-data-window method of spectrum estimation is better than the various *ad hoc* methods of time-series analysis. With such methods I have shown that climate variations, as recorded through growth rings of bristlecone pines, have few, if any, direct periodic components but do show evidence for changes in structure with a period about 2360 years. I have also shown that the 'Suess wiggles' in ^{14}C variations have a component with a period of 208.2 years, plus harmonics, with very high certainty.

REFERENCES

Anderson, T. W. 1984 *An introduction to multivariate statistical analysis*, 2nd edn. New York: Wiley.

Blackman, R. B. & Tukey, J. W. 1958 *The measurement of power spectra*. New York: Dover.

Bretthorst, G. L. 1987 Bayesian spectrum analysis and parameter estimation. Ph.D. thesis, Department of Physics, Washington University, St Louis, Missouri, U.S.A.

Brillinger, D. R. 1975 *Time series, data analysis and theory*. New York: Holt, Rinehart and Winston.

Bronez, T. P. 1988 Spectral estimation of irregularly sampled multidimensional proceses by generalized prolate spheroidal sequences. *IEEE Trans. Acoust. Speech, signal Processing* 36, 1862–1873.

Chen, W. Y. & Stegun, G. R. 1974 Experiments with maximum entropy power spectra of sinusoids. *J. geophys. Res.* 79, 3019–3022.

Daubechies, I. 1988 Time–frequency localization operators: a geometric phase space approach. *IEEE Trans. Inf. Theory* IT-34, 605–612.

De Mendoça Dias, A. A. 1962 The volcano of Capelinhos (Azores), the solar activity, and the earth-tide. *Bull. volcan.* 24, 211–221.

Doob, J. L. 1953 *Stochastic processes*. New York: Wiley.

Draper, N. R. & Smith, H. 1981 *Applied regression analysis*. New York: Wiley.

Efron, B. 1982 *The jackknife, the bootstrap, and other resampling plans*. Philadelphia: SIAM.

Ferguson, C. W. 1970 Dendrochronology of bristlecone pine, *Pinus aristata*. Establishment of a 7484-year chronology in the White Mountains of eastern-central California, U.S.A. In *Radiocarbon Variations and Absolute Chronology, Proc. 12th Nobel Symp.* (ed. I. U. Olsson), pp. 237–259. New York: Wiley; Stockholm: Almqvist & Wiksell.

Harris, F. J. 1978 On the use of windows for harmonic analysis with the discrete Fourier transform. *Proc. IEEE* 66, 51–83.

Kay, S. M. & Marple, S. L. Jr 1982 Spectrum analysis – a modern perspective. *Proc. IEEE* 69, 1380–1419.

Kleiner, B., Martin, R. D. & Thomson, D. J. 1979 Robust estimation of power spectra (with discussion). *Jl R. statist. Soc.* B41, 313–351.

Lanzerotti, L. J., Thomson, D. J., Meloni, A., Medford, L. V. & Maclennan, C. G. 1986 Electromagnetic study of the Atlantic Continental Margin using a section of a transatlantic cable. *J. geophys. Res.* B91, 7417–7427.

Munk, W. & Hasselmann, K. 1964 Super-resolution of tides. In *Studies on oceanography* (ed. K. Yoshida), pp. 339–344. Tokyo. (Reprinted 1965 University of Washington Press.)

Papoulis, A. 1973 Minimum bias windows for high-resolution spectral estimates. *IEEE Trans. Inf. Theory* IT-19, 9–12.

Park, J., Lindberg, C. R. & Thomson, D. J. 1987a Multiple-taper spectral analysis of terrestrial free oscillations: part I. *Geophys. Jl R. astr. Soc.* 91, 755–794.

Park, J., Lindberg, C. R. & Vernon, F. L. III 1987b Multitaper spectral analysis of high-frequency seismograms. *J. geophys. Res.* 92, 12675–12684.

Parzen, E. 1957 On consistent estimates of the spectrum of a stationary time series. *Ann. math. Statist.* 28, 329–348.

Pearson, E. S. & Hartley, H. O. 1969 *Biometrika tables for statisticians*, 3rd edn. Cambridge University Press.

Pittock, A. B. 1978 A critical look at long-term Sun–weather relationships. *Rev. Geophys. Space Phys.* 16, 400–420.

Priestley, M. B. 1981 *Spectral analysis and time series*. London: Academic Press.

Priestley, M. B. 1988 *Non-linear and non-stationary time series analysis*. London: Academic Press.

Rife, D. C. & Boorstyn, R. R. 1976 Multiple tone parameter estimation from discrete-time observations. *Bell System Tech. J.* 55, 1389–1410.

Shapiro, R. 1979 An examination of certain proposed Sun–weather connections. *J. atmos. Sci.* 36, 1105–1116.

Slepian, D. 1978 Prolate spheroidal wave functions, Fourier analysis, and uncertainty. V: the discrete case. *Bell System Tech. J.* 57, 1371–1430.

Sonett, C. P. 1982 Sunspot time series: spectrum from square law modulation of the Hale cycle. *Geophys. Res. Lett.* 9, 1313–1316.

Sonett, C. P. 1984 Very long solar periods and the radiocarbon record. *Rev. Geophys. Space Sci.* 22, 239–254.

Sonett, C. P. & Suess, H. E. 1984 Correlation of bristlecone pine ring widths with atmospheric ^{14}C variations: a climate–Sun relation. *Nature, Lond.* 307, 141–143.

Stuiver, M. 1980 Solar variability and climatic change during the current millenium. *Nature, Lond.* 286, 868–871.

Stigler, S. M. & Wagner, M. J. 1987 A substantial bias in nonparametric tests for periodicity in geophysical data. *Science, Wash.* 238, 940–944.

Suess, H. E. 1965 Secular variations of the cosmic ray produced carbon-14 in the atmosphere. *J. Geophys. Res.* 70, 5937–5952.

Thomson, D. J. 1977 Spectrum estimation techniques for characterization and development of WT4 waveguide. *Bell System Tech. J.* 56: part I, 1769–1815; part II, 1983–2005.

Thomson, D. J. 1982 Spectrum estimation and harmonic analysis. *Proc. IEEE* 70, 1055–1096.

Thomson, D. J. 1986 Combined multi-window eigenvalue spectrum estimates. In *Proc. Third IEEE ASSP workshop on spectrum estimation and modelling*, pp. 103–105.

Thomson, D. J. & Chave, A. D. 1990 Jackknifed error estimates for spectra coherences, and transfer functions. In *Advances in spectrum estimation and array processing* (ed. S. Haykin), ch. 2. Englewood Cliffs, New Jersey: Prentice-Hall. (In the press.)

Thomson, D. J., Lanzerotti, L. J., Medford, L. V., Maclennan, C. G., Meloni, A. & Gregori, G. P. 1986 Study of tidal periodicities using a transatlantic telecommunications cable. *Geophys. Res. Lett.* **13**, 525–528.

Tukey, J. W. 1982 Styles of spectrum analysis. In *The collected works of John W. Tukey*, vol. II (ed. D. R. Brillinger), pp. 1143–1153. California: Wadsworth.

Watson, G. S. 1983 *Statistics on spheres*. New York: Wiley–Interscience.

Phil. Trans. R. Soc. Lond. A **330**, 617–625 (1990)

Printed in Great Britain

Periodicity and aperiodicity in solar magnetic activity

By N. O. Weiss

*Department of Applied Mathematics and Theoretical Physics, Silver Street,
University of Cambridge, Cambridge CB3 9EW, U.K.*

Solar activity varies irregularly with an 11-year period whereas the magnetic cycle has a period of 22 years. Similar cycles of activity are seen in other slowly rotating late-type stars. The only plausible theory for their origin ascribes them to a hydromagnetic dynamo operating at, or just below, the base of the convective zone. Linear (kinematic) dynamo models yield strictly periodic solutions with dynamo waves propagating towards or away from the equator. Nonlinear (magneto-hydrodynamic) dynamo models allow transitions from periodic to quasi-periodic to chaotic behaviour, as well as loss of spatial symmetry followed by the development of complex spatial structure. Results from simple models can be compared with the observed sunspot record over the past 380 years and with proxy records extending over 9000 years, which show aperiodic modulation of the 11-year cycle.

1. Introduction

There is evidence of solar variability on timescales ranging from 5 min to 10^{10} years. Variations on an evolutionary timescale are gradual and monotonic; at the other extreme, p-mode oscillations have small amplitudes and periods that are very precisely defined. On intermediate timescales behaviour is typically aperiodic. Magnetic flux erupts irregularly at the solar surface to form active regions within which sunspots may appear. Over the past 275 years the frequency and location of sunspots has varied cyclically with an average period of about 11 years. These aperiodic cycles are interrupted by episodes of reduced activity. Proxy records confirm that grand minima with a characteristic timespan of about 200 years have recurred aperiodically over the past 9000 years and there are suggestions of modulation on even longer timescales still.

This review is concerned with the implications for our understanding of the solar interior and for theories of solar magnetism of variations with timescales of 10–1000 years. I first summarize the relevant observational results. Next I describe the physical processes responsible for solar activity, together with models of the solar dynamo. Then I discuss periodic and chaotic oscillations in nonlinear deterministic systems and contrast chaotic with stochastic models of solar behaviour. Finally I indicate how such considerations affect our interpretation of time series derived from terrestrial data.

2. Observational constraints

The interaction of convection with rotation leads to magnetic activity in lower main-sequence stars like the Sun, with deep outer convective zones. By studying other similar stars we can therefore construct the magnetic history of our Sun (Soderblom & Baliunas 1988; Tayler 1989; Weiss 1989). Magnetic fields have been measured directly in some active stars while indirect evidence of magnetic activity comes from observations of coronal X-ray emission, chromospheric Ca^+ emission and photospheric starspots. These observations show that, for a star of given mass, magnetic activity depends on the rotation rate. Magnetic heating leads to the formation of a hot corona, which drives a stellar wind (Priest 1982) that carries angular momentum away from the star. Thus the star is spun down owing to magnetic braking at a rate that depends upon its angular velocity (Mestel & Spruit 1987). As its rotation rate decreases the star grows less active and decelerates more slowly. The Sun's activity is relatively feeble. Thus the timescale for solar spindown now is around 10^{10} years and its magnetic behaviour should show no significant evolutionary changes over the past 10^4 years (or even since the late Precambrian era).

Cyclic activity has been detected in about a dozen slowly rotating stars like the Sun. The principal features of the solar cycle are well known. Sunspots appear within active regions formed by the eruption of azimuthal flux tubes through the photosphere. Since 1714 the incidence of sunspots has varied cyclically; although the cycles are aperiodic they have a well-defined mean period of 11.1 years. The sense of the magnetic field reverses from one activity cycle to the next, so the underlying magnetic cycle has a 22-year period. Early telescopic observations indicate that the sunspot cycle proceeded much as now during the first few decades of the seventeenth century, but later (throughout the reign of Louis XIV) it was interrupted by the Maunder Minimum. The dearth of sunspots was recognized by many contemporary observers (Eddy 1976) and even referred to by the poet Andrew Marvell (Cohen 1961; Weiss & Weiss 1979).

Fortunately, modulation of the solar cycle can also be detected in terrestrial proxy records. A reduction in solar activity leads to enhanced production of unstable isotopes by galactic cosmic rays, and both the Maunder Minimum and the preceding Spörer minimum are apparent in the ^{14}C and ^{10}Be records (G. M. Raisbeck, this Symposium) and also in thermoluminescence profiles (Cini Castagnoli *et al.* 1988). Moreover, precise calibration of radiocarbon dates from tree rings reveals recurrent anomalies (the 'Suess wiggles') associated with grand minima. Stuiver & Braziunas (1988) have identified a total of 17 such minima over the past 9700 years. These episodes of reduced activity have a characteristic timescale of 150–220 years, but they recur aperiodically with no apparent pattern.

3. ORIGINS OF SOLAR ACTIVITY

It seems likely that solar magnetic fields are generated in a region of weak convective overshoot beneath the base of the convective zone (Weiss 1989). In this region toroidal fields are created from poloidal fields by differential rotation, so it is important to establish how the angular velocity ω varies in the interior of the Sun. The surface rotation rate is a maximum at the Equator and falls by about 30 % towards the poles (Howard 1984). Recent measurements of the rotational splitting of solar p-mode oscillations imply that this latitudinal gradient persists throughout the convective zone and that there is a transition to a uniform angular velocity over a depth of order 50000 km at the top of the radiative zone (Brown & Morrow 1987; Dziembowski *et al.* 1989). Earlier in the Sun's history, when it rotated more rapidly, it is likely that the Proudman–Taylor constraint forced the angular velocity to be constant on cylindrical surfaces in the convective zone, as indicated by experiments (Hart *et al.* 1986). Indeed, it has been suggested that grand minima are associated with changes in the pattern of convection that would affect the distribution of angular momentum (Zel'dovich *et al.* 1983;

Jones & Galloway 1988). There are, however, no indications that the surface rotation changed substantially during the Maunder Minimum.

Two mechanisms for generating the Sun's alternating magnetic field have been proposed. Either the Sun is a magnetic oscillator with a fixed poloidal field and a toroidal velocity that oscillates with a 22-year period or else there is a dynamo, with a poloidal field generated owing to the persistent helicity of convective eddies in a rotating system. The observed poloidal field reverses around sunspot maximum while the only systematic variations in angular velocity are the so-called torsional waves with a period of 11 years at any latitude (Howard & LaBonte 1980; Ulrich *et al.* 1988). That is just what would be expected for torsional oscillations in a nonlinear dynamo, because the Lorentz force does not depend upon the sign of the magnetic field. Thus the observational evidence favours a hydromagnetic dynamo.

We are therefore led to consider a magnetic layer beneath the base of the convective zone (Spiegel & Weiss 1980; van Ballegooijen 1982; Gilman *et al.* 1989), where toroidal flux builds up until the layer is broken up by instabilities driven by magnetic buoyancy (Parker 1979; Hughes & Proctor 1988). There are several timescales associated with this process. Flux escapes through the convective zone to emerge in active regions in a convective turnover time of about a month (comparable with the rotation period). There are fluctuations in activity and luminosity on this timescale. Then there are variations corresponding to the 11-year activity cycle or the 22-year magnetic cycle. (Note that Spiegel & Weiss (1980) erroneously assumed that convective elements retained their excess energy, rather than excess entropy, as they rose through the convective zone; hence their estimate of the luminosity fluctuation should be reduced by the ratio of the temperature at the surface to that at the base of the convective zone, which is around 5×10^{-3}.) Finally, there are modulations associated with grand minima, on a timescale of 200 years.

The systematic properties of the solar cycle suggest that it can be modelled by averaging over small-scale, short-term variations. It is convenient to adopt a three-scale approach (Weiss 1981). By averaging over the smallest scales we obtain eddy diffusivities that can be fed into self-consistent dynamo models with motion extending across the whole convectively unstable region (Gilman 1983; Glatzmaier 1985). These numerical experiments confirm that the dynamo process works, even though details do not match the observations, which is scarcely surprising in view of the simplifications that have to be made to produce a tractable model. Next we can average azimuthally to produce a mean field dynamo (Parker 1979; Zel'dovich *et al.* 1983), where the α-effect describes the generation of poloidal flux through helicity and toroidal fields are produced by differential rotation. There are many examples of linear and nonlinear $\alpha\omega$-dynamos tuned to produce butterfly diagrams that accord with observations. Recently, there has been more interest in studying spatiotemporal structure in idealized nonlinear models (see, for example, Belvedere *et al.* 1990; Brandenburg *et al.* 1989). Even a simple cartesian model with parametrized nonlinearities can possess a complicated bifurcation structure: figure 1 illustrates the bifurcation pattern, involving steady and oscillatory modes with quadrupole and dipole symmetry, as well as mixed-mode solutions lacking any symmetry about the Equator, for a model investigated by Jennings & Weiss (1989). These bifurcations involve changes in spatial structure for solutions with straightforward time dependence. When such models are extended to allow more complicated dynamical interactions further bifurcations may lead to a rich variety of temporal behaviour, including chaos (Weiss *et al.* 1984).

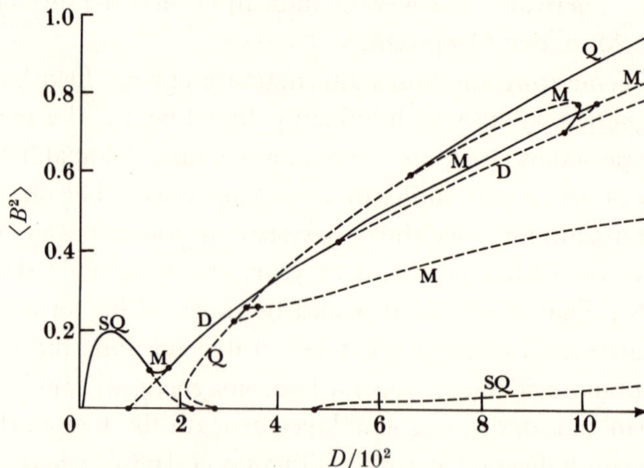

Figure 1. Bifurcation structure for a nonlinear dynamo showing transitions between quadrupole and dipole symmetries (after Jennings & Weiss 1989). The mean square toroidal field $\langle B^2 \rangle$ is plotted against the dynamo number D (proportional to ω^2) for a truncated 14-mode model. Stable solutions are indicated by full lines and unstable solutions by broken lines. There are branches of steady quadrupole (sq) solutions and of oscillatory dipole (d) quadrupole (q) and mixed-mode (m) solutions.

4. Deterministic chaos

To interpret time series derived from natural phenomena it is necessary to understand some theoretical aspects of the behaviour of nonlinear dissipative systems. We need to adopt a geometrical approach: instead of studying the periodic displacement of a nonlinear oscillator we follow trajectories in phase space, where they are attracted to a closed limit cycle. Now suppose that the system is subjected to small stochastic disturbances. (The classic example of a pendulum bombarded by small boys with peashooters was introduced by Yule (1927) in the context of the solar cycle.) Then trajectories will be contained within a tube enclosing the limit cycle and the attractor is only slightly distorted. Obviously, the situation changes as the intensity of the noise is increased (for example, if the peashooters are replaced by cannon).

This must be distinguished from chaotic behaviour associated with sensitivity to initial conditions (Schuster 1988). Figure 2 illustrates the development of a chaotic attractor in two-dimensional double-diffusive convection (Knobloch *et al.* 1986). The variation with time of the kinetic energy $E(t)$ can be followed either by projecting trajectories onto the $E\dot{E}$-phase plane or by forming a power spectrum $\hat{E}(\omega)$, where ω is the frequency. Figure 2*a* shows a periodic solution with a convoluted cycle and a line spectrum (limited by the finite data-set). In figure 2*b* the solution is aperiodic, with a chaotic attractor in the vicinity of the unstable periodic orbit; the spectrum is intrinsically noisy but the peaks are still clearly visible. (Such behaviour is called semiperiodic.) In figure 2*c* and *d* the trajectories are more obviously chaotic. Peaks corresponding to the basic cycle frequency and its first harmonic can still be discerned in figure 2*b*, but are submerged by noise in figure 2*d*. Such results indicate how sophisticated frequency analysis can pull convincing line spectra out of intrinsically aperiodic data. Moreover, the frequencies may indeed correspond to those of unstable periodic orbits that are approached by the chaotic trajectory.

More generally we may expect to find not only periodic but also quasi-periodic (i.e. multiply periodic) orbits, with trajectories that lie on tori in phase space. A two-torus can be identified

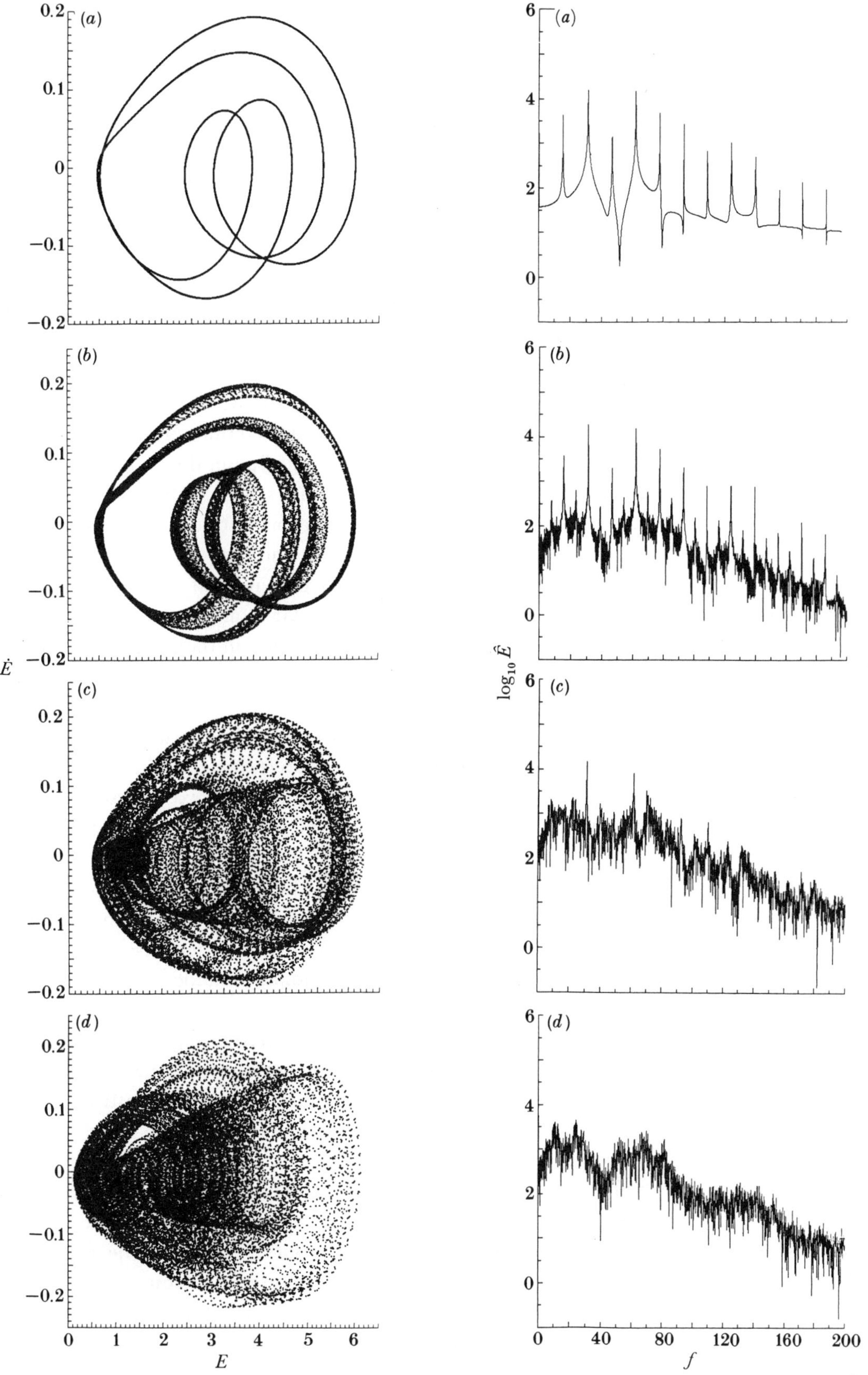

FIGURE 2. Chaos in thermosolutal convection (after Knobloch *et al.* 1986). Trajectories projected onto the $E\dot{E}$-plane and corresponding power spectra $\hat{E}(f)$ for different values of the Rayleigh number R. (*a*) $R = 10450$: periodic solution; (*b*) $R = 10475$: semiperiodic solution; (*c*) $R = 10625$: chaotic solution; (*d*) $R = 11000$: chaos.

by taking a Poincaré section, which yields a closed curve. At every resonance this curve collapses to isolated points and frequency-locking may be followed by a transition to chaos (cf. Jones *et al.* 1985). Once again, the basic form of the attractor is not affected by adding weak stochastic disturbances, though violent disturbances can change its structure (Lorenz 1987).

The prevalence of chaos in complicated nonlinear systems could therefore be anticipated. On a laboratory scale transitions to chaos have been identified in experiments on convection, where temporal chaos is followed by weakly turbulent behaviour as more spatial structure is involved (Libchaber 1987). Meteorology and astrophysics provide examples of chaotic behaviour in natural systems (Lorenz 1984; Spiegel 1985). Short-term climatic changes are aperiodic (Ghil 1987) and the weather is notoriously unpredictable (Lorenz 1984, 1987). Turbulent convection in the Sun is a good example of spatiotemporal chaos and the solar cycle resembles a chaotic oscillator.

To model aperiodic solar activity we once again assume a separation of scales in time. Convective timescales range from 5 min for the photospheric granulation to 25 days for giant cells that span the full depth of the convective zone, while the lifetime of active regions is comparable with the rotation period (also 25 days). Short-term fluctuations in solar activity can be regarded as a dynamic process associated with instabilities of the submerged magnetic layer (Spiegel & Wolf 1987). We shall, however, average over rapid variations with periods less than a year and regard the fluctuations as noise. Can medium and long-term variations, with timescales of 10–500 years be described by a relatively low-dimensional dynamical system? Here it is encouraging that a simple sixth-order model of the solar dynamo can yield aperiodic modulation with episodes of reduced activity resembling those found in the solar record (Weiss *et al.* 1984). These grand minima are associated with the persistence of a 'ghost' torus in the chaotic régime.

There are procedures that allow us to establish whether a given time series corresponds to a low-dimensional attractor (Schuster 1988). First the time series is converted into a trajectory embedded in an m-dimensional phase space by introducing time lags. Next we take N points $x_i (i = 1, ..., N)$ on this trajectory. Let $n(l)$ be the number of pairs (x_i, x_j) with separations $l_{ij} = |x_i - x_j|$ such that $l_{ij} < l$; then the quantity $C(l) = n(l)/N^2$ is an estimate of the correlation integral and $\nu = \mathrm{d}\ln C/\mathrm{d}\ln l$ is an estimate of the correlation dimension. To demonstrate the existence of an attractor with dimension ν we need $m > \nu$ and N sufficiently large. Smith (1988) has shown that the number of (independent) points required is of order $N = (40)^\nu$: so for $\nu \approx 3$ we require more than 10^4 data points and computational demands rapidly become prohibitive as ν increases.

L. A. Smith has compared the sunspot record with two very different models. The first is a well-known third-order system that exhibits chaotic oscillations associated with a heteroclinic bifurcation (Spiegel 1985). For suitably chosen parameters there is a chaotic attractor with dimension $\nu \approx 2.1$. The estimate converges to this value over an acceptable range in l for $N > 2^{12}$ and $m = 3$, but more points are needed as m is increased. The second model is purely stochastic, constructed to mimic solar activity by filtering gaussian noise with a frequency corresponding to that of the magnetic cycle (Barnes *et al.* 1980). This stochastic (ARMA) oscillator yields a remarkably convincing simulation of the solar cycle and the computed dimension ν approaches the embedding dimension m if the time series is adequately long. The actual record of annual sunspot numbers provides a time series with $N \approx 2^8$. Figure 3a shows the estimated dimension ν as a function of l for an embedding dimension $m = 3$, compared with

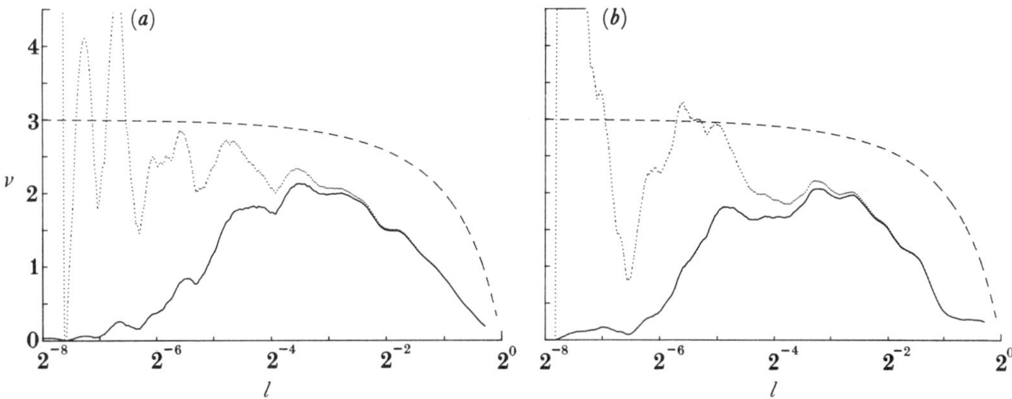

FIGURE 3. Estimating the dimension of an attractor. The correlation dimension ν is plotted as a function of l for an embedding dimension $m = 3$ and a total of $N = 2^8$ points. (a) Annual sunspot numbers and (b) the stochastic oscillator of Barnes et al. (1980). ———, Excluding points with $i = j$; ·····, including points with $i = j$; ———, white noise. The actual record and the artificial series cannot be distinguished.

that for white noise, where $\nu \to m$ as $l \to 0$. Results for the stochastic oscillator, again with $N = 2^8$, are shown in figure 3b and the two figures are virtually indistinguishable. This demonstrates that the sunspot record since 1714 is not long enough to distinguish between stochastic and chaotic models.

We might also attempt to discover whether modulation of the solar cycle is chaotic, by averaging over the 11-year cycles. The bidecadal averages of fluctuations in the ^{14}C production rate provide just such a record, extending over the past 9000 years (Stuiver & Braziunas 1988). This yields a time series with $N \approx 2^9$. Once again, this series is too short for us to establish whether there is a low-dimensional attractor.

5. CONCLUSION

Periodicity is characteristic of linear systems, where periodic solutions can be superposed to produce a complicated signal. Nonlinear systems are likely to be intrinsically aperiodic, though frequency analysis may still yield peaks in the power spectrum of a chaotic record. Historically, attempts to interpret the record of solar activity seem to have favoured aperiodic and periodic descriptions at different epochs. After the Maunder Minimum astronomers apparently regarded variations in the Sun's activity as deviations from its normal spotted state. Thus Herschel (1801), referring to the prolonged minimum of 1795–1800, commented: 'It appears to me, if I may be permitted the metaphor, that our Sun has for some time past been labouring under an indisposition, from which it is now in a fair way of recovering'. Such interruptions were thought to be irregular both in duration and in frequency. It took Schwabe (1844), who ignored previous observations and referred only to his own, to recognize the 11-year activity cycle. Since then it has been fashionable to look for periodic modulations of solar activity. Both the sunspot record since the first telescopic observations and the ^{14}C proxy records are, however, consistent with aperiodic modulation and irregularly spaced grand minima. Unfortunately, the data are insufficient to establish whether this aperiodic modulation is a consequence of stochastic disturbances or deterministic chaos. Thus the latter cause can only be preferred on the grounds that it provides a more economical explanation of the observations.

Although we know that the solar luminosity does vary slightly with the sunspot cycle there is as yet no clear evidence of a direct correlation between solar activity and the weather or between the Maunder Minimum and the Little Ice Age (Stuiver 1980). It may be more profitable to think of the solar dynamo and the Earth's climate as two weakly coupled nonlinear systems, each behaving chaotically with a variety of characteristic timescales. Occasionally oscillations with nearby frequencies will be locked in phase but this locking should be temporary and unlikely to persist.

REFERENCES

van Ballegooijen, A. A. 1982 *Astron. Astrophys.* **113**, 99–112.
Barnes, J. A., Sargent, H. H. & Tryon, P. V. 1980 In *The ancient sun* (ed. R. O. Pepin, J. A. Eddy & R. B. Merrill), pp. 159–164. New York: Pergamon.
Belvedere, G., Pidatella, R. M. & Proctor, M. R. E. 1990 *Geophys. Astrophys. Fluid Dyn.* (In the press.)
Brandenburg, A., Krause, F., Meinel, R., Moss, D. & Tuominen, I. 1989 *Astron. Astrophys.* **213**, 411–422.
Brown, T. M. & Morrow, C. A. 1987 *Astrophys. J.* **314**, L21–L26.
Cini Castagnoli, G., Bonino, G. & Provenzale, A. 1988 *Solar Phys.* **117**, 187–197.
Cohen, I. B. 1961 *The birth of a new physics.* London: Heinemann.
Dziembowski, W. A., Goode, P. R. & Libbrecht, K. G. 1989 *Astrophys. J.* **337**, L53–L57.
Eddy, J. A. 1976 *Science, Wash.* **192**, 1189–1202.
Ghil, M. 1987 In *Irreversible phenomena and dynamical systems analysis in geosciences* (ed. C. Nicolis & G. Nicolis), pp. 313–320. Dordrecht: Reidel.
Gilman, P. A. 1983 *Astrophys. J. Suppl.* **53**, 243–268.
Gilman, P. A., Morrow, C. A. & DeLuca, E. E. 1989 *Astrophys. J.* **338**, 528–537.
Glatzmaier, G. A. 1985 *Geophys. Astrophys. Fluid Dyn.* **31**, 137–150.
Hart, J. E., Toomre, J., Deane, A. E., Hurlburt, N. E., Glatzmaier, G. A., Fichtl, G. H., Leslie, F., Fowlis, W. W. & Gilman, P. A. 1986 *Science, Wash.* **234**, 61–64.
Herschel, W. 1801 *Phil. Trans. R. Soc. Lond.* **91**, 265–318.
Howard, R. 1984 *A. Rev. Astr. Astrophys.* **22**, 131–155.
Howard, R. & LaBonte, B. J. 1980 *Astrophys. J.* **239**, L33–L36.
Hughes, D. W. & Proctor, M. R. E. 1988 *A. Rev. Fluid Mech.* **20**, 187–223.
Jennings, R. L. & Weiss, N. O. 1989 In *Solar photosphere: structure, convection, magnetic fields* (ed. E. A. Gurtovenko & J. O. Stenflo). Dordrecht: Kluwer.
Jones, C. A. & Galloway, D. J. 1988 In *Secular solar and geomagnetic variation over the last 10,000 years* (ed. F. R. Stephenson & A. W. Wolfendale), pp. 97–108. Dordrecht: Kluwer.
Jones, C. A., Weiss, N. O. & Cattaneo, F. 1985 *Physica* D14, 161–176.
Knobloch, E., Moore, D. R., Toomre, J. & Weiss, N. O. 1986 *J. Fluid Mech.* **166**, 409–448.
Libchaber, A. 1987 *Proc. R. Soc. Lond.* A **413**, 63–69.
Lorenz, E. N. 1984 *Tellus* A **36**, 98–110.
Lorenz, E. N. 1987 In *Irreversible phenomena and dynamical systems analysis in geosciences* (ed. C. Nicolis & G. Nicolis), pp. 159–179. Dordrecht: Reidel.
Mestel, L. & Spruit, H. C. 1987 *Mon. Not. R. astr. Soc.* **226**, 57–66.
Parker, E. N. 1979 *Cosmical magnetic fields.* Oxford: Clarendon Press.
Priest, E. R. 1982 *Solar magnetohydrodynamics.* Dordrecht: Reidel.
Schuster, H. G. 1988 *Deterministic chaos*, 2nd edn. Weinheim: VCH.
Schwabe, S. H. 1844 *Astr. Nachr.* **21**, 233–236.
Smith, L. A. 1988 *Phys. Lett.* A **133**, 283–288.
Soderblom, D. R. & Baliunas, S. L. 1988 In *Secular solar and geomagnetic variations over the last 10,000 years* (ed. F. R. Stephenson & A. W. Wolfendale), pp. 25–48. Dordrecht: Kluwer.
Spiegel, E. A. 1985 In *Chaos in astrophysics* (ed. J. R. Buchler, J. M. Perdang & E. A. Spiegel), pp. 91–135. Dordrecht: Reidel.
Spiegel, E. A. & Weiss, N. O. 1980 *Nature, Lond.* **287**, 616–617.
Spiegel, E. A. & Wolf, A. 1987 *Ann. N. Y. Acad. Sci.* **497**, 55–60.
Stuiver, M. 1980 *Nature, Lond.* **286**, 868–871.
Stuiver, M. & Braziunas, T. F. 1988 In *Secular solar and geomagnetic variations over the last 10,000 years* (ed. F. R. Stephenson & A. W. Wolfendale), pp. 245–266. Dordrecht: Kluwer.
Tayler, R. J. 1989 *Q. Jl R. astr. Soc.* **30**, 125–161.
Ulrich, R. K., Boyden, J. E., Webster, L., Snodgrass, H. B., Padilla, S. P., Gilman, P. & Shieber, T. 1988 *Solar Phys.* **117**, 291–328.

Weiss, J. E. & Weiss, N. O. 1979 *Q. Jl R. astr. Soc.* **20**, 115–118.
Weiss, N. O. 1981 In *Solar activity* (ed. C. Jordan), pp. 35–50. Oxford: EPS.
Weiss, N. O. 1989 In *Accretion discs and magnetic fields in astrophysics* (ed. G. Belvedere), pp. 11–29. Dordrecht: Kluwer.
Weiss, N. O., Cattaneo, F. & Jones, C. A. 1984 *Geophys. Astrophys. Fluid Dyn.* **30**, 305–341.
Yule, G. U. 1927 *Phil. Trans. R. Soc. Lond.* A**226**, 267–298.
Zel'dovich, Y. B., Ruzmaikin, A. A., & Sokoloff, D. D. 1983 *Magnetic fields in astrophysics.* London: Gordon & Breach.

Discussion

P. FOUKAL (*CRI, Cambridge, Massachusetts, U.S.A.*). To return to the subject of solar irradiance variations, Richard Radick, Wes Lockwood and Sallie Baliunas have reported recently on the behaviour of flux variations in stars similar to the Sun, but much younger (about 3.5×10^9 years younger). They find that in such stars high levels of magnetic activity tend to anticorrelate with stellar brightness. That is, unlike the present Sun, these young stars get darker at high activity periods of their sunspot cycles. Presumably the propensity to form dark spots in these younger stars dominates over the formation of faculae, the opposite of what we now see in the Sun.

N. O. WEISS. That is a significant observation. It confirms my view that one has to be cautious before extrapolating from the Sun to young, rapidly rotating and more active stars. The pattern of convection, the angular velocity distribution and the behaviour of the dynamo are likely to be very different. In particular, stars with spots that produce variations of up to 30% in luminosity must have far more flux in their poloidal fields than the Sun.

Phil. Trans. R. Soc. Lond. A **330**, 627–640 (1990)
Printed in Great Britain

On possible origins of relatively short-term variations in the solar structure

By D. O. Gough

Institute of Astronomy, and Department of Applied Mathematics and Theoretical Physics, University of Cambridge, Silver Street, Cambridge CB3 9EW, U.K., and Joint Institute for Laboratory Astrophysics, University of Colorado and National Institute of Standards and Technology, Boulder, Colorado 80309-0440, U.S.A.

The general large-scale redistribution of magnetic field over the solar cycle is possibly associated with an overall variation of thermal structure of the convection zone, which modulates not only the total luminosity but also the latitudinal distribution of radiative flux, thereby modifying the irradiance of the Earth. Whether the cause of this variation lies within the convection zone or is more deeply seated is still an open question.

1. Introduction

There are three obvious natural timescales on which one would expect the Sun to vary: the nuclear evolution time, about 10^{10} years, the thermal diffusion time characteristic of the entire Sun, about 3×10^7 years, and the acoustic travel time from centre to surface, about 1 h. The last of these timescales is a dynamical time; it is of the order of the free-fall time from the surface to the centre of the Sun, and it is approximately equal to the period of the gravest mode of acoustic oscillation.

There is strong astronomical evidence that the Sun varies on the nuclear timescale. Moreover, it is also an inevitable consequence of hydrostatic balance under newtonian gravity, granted that the fusion reactions cause the mean molecular mass (the stellar physicists' jargon for the mean mass of the particles that constitute stellar material, measured in units of the mass of the hydrogen atom; the interior of the Sun is too hot for there to be any molecules) in the core of the Sun to increase with time. Moreover, it has been observed that the Sun varies dynamically; in particular, acoustic oscillations have been well studied. Their amplitude is extremely low, and it is unlikely that the oscillations that have been observed have a significant influence either on the hydrostatic structure of the Sun or on its temporal variation on timescales substantially greater than hours. However, the oscillations have important diagnostic power, and I shall therefore return to them later.

There is no direct evidence that the Sun varies on the thermal timescale. Linearized stability analyses have found the Sun to be thermally stable, so one would expect variation on the thermal scale only as a relaxation to thermal equilibrium if the structure of the Sun were to have been perturbed by some other process. I shall not discuss such possibilities here, because the timescale is not of interest to this meeting. Indeed, nor are the nuclear and the dynamical timescales. I have introduced them simply to point out that the phenomena discussed in this meeting do not find a ready explanation as a simple natural solar variation.

There are two other, somewhat less natural timescales that come to mind: the characteristic turnover time of turbulent eddies in the Sun's convective envelope, and the thermal relaxation time of the convective envelope. The former varies from about 15 min at the top of the convection zone to about a month in the almost adiabatically stratified layers beneath. The characteristic rotation period of the Sun is also about a month (the Sun's angular velocity is not uniform), which means that rotation neither dominates the dynamics of the convection zone nor acts as a small perturbation. Therefore, the interaction is quite complicated, and probably leads to a complicated flow field. The motion can twist and stretch magnetic fields, producing the rich variety of magnetic activity observed in the surface layers that is normally attributed to a dynamo. The thermal relaxation time of the convection zone is about 10^5 years, and may very well be of some significance to climatic issues discussed in this meeting, even though the timescale is greater than those of direct concern. As is the case with the thermal relaxation of the entire Sun, however, it can be manifest only as a transient response to some other stimulus, or via an interaction with some other process. I should perhaps also mention at this juncture the thermal relaxation timescale of the core of the Sun within which energy is being generated: it is *ca.* 10^6 years.

I do not include the rotation period in the same category as the five natural timescales I have just enumerated, even though it is a very obvious property of the Sun. The reason is that at our present state of understanding it is hardly natural. It is a consequence of the particular value of the angular momentum the Sun possesses at the moment. To be sure, a systematic variation of angular velocity Ω of main-sequence stars has been observed: Ω depends both on the mass and age of a star. Moreover, there are theoretical rationalizations of this variation. However, the theories have been tuned, and even the most modern ideas would have been unlikely to have successfully predicted the observations had the observations not preceded them.

Another obvious solar timescale is the characteristic period of the solar cycle, *ca.* 22 years. This is very relevant to this meeting. But it is not an obviously natural timescale of the Sun. It can, however, be constructed from natural timescales; it is the geometric mean of the characteristic dynamical time of the Sun and the thermal relaxation time of the energy-generating core. Whether this coincidence has any physical significance is not known.

It is evident from this prologue that any theoretical discussion of temporal solar variations on timescales relevant to this meeting must necessarily be either subtle or uncertain. It is probably true to say that the subtle discussions always rest on severe idealizations, and although they may be correct within the restricted context defined by those idealizations, their application to the Sun lies outside the realm of validity of the arguments, and therefore they too are uncertain. I have dynamo models particularly in mind now. Therefore, notwithstanding the intricacies of the mathematical models and the many hours of computer time that have been devoted to studying those models, either of which can sometimes leave unwarranted impressions of reliability in the mind of an unwary bystander, I feel obliged to regard all theories that address solar variation on the timescales relevant to this meeting as being quite uncertain. I hasten to add that this is not to say that those theories are of no value. I am quite sure that some of their components well describe processes that do go on, and without an understanding of those processes we would never understand the Sun.

With this uncertainty in mind, I shall now discuss, quite briefly, some of the ideas that have been aired. The purpose is to provide some idea of the kinds of processes that may be taking place inside the Sun, and which cry for further study. Most of my discussion will be directed towards the solar cycle, for that is the only solar temporal variation directly relevant to this meeting that we are really sure takes place.

2. The standard solar model

The so-called standard theoretical solar model is important because it provides a basis from which to discuss the processes that might lead to interesting temporal variations. The model is a highly idealized representation of the Sun. Nevertheless, astrophysicists believe it contains the most important ingredients of a more realistic description.

The model assumes the Sun to be hydrostatic and essentially spherically symmetric; rotation and magnetic fields are ignored, which is certainly a good first approximation. There is no motion other than a gradual expansion or contraction, except where the stratification is convectively unstable. This occurs only in the outer 30%, by radius, of the envelope, where the only role the turbulent fluctuations are assumed to play is in providing contributions to the flux of energy and, sometimes, momentum, via a turbulent pressure tensor. Elsewhere, energy is presumed to be transported solely by radiation. There is no significant loss of mass from the surface (the mass-loss timescale due to the present solar wind is about 10^{13} years) nor accretion of mass onto the surface.

The star changes its structure because nuclear fusion reactions in the core increase the mean molecular mass, which leads to contraction and a rise in temperature in the radiative interior. According to the simple mixing-length prescriptions used to model the outer boundary layer of the convection zone, the upper layers of the Sun expand. Concomitant with there being no motion in the radiative interior, the products of the nuclear reactions are assumed to remain *in situ*. Notice that by assuming no motion outside the convection zone the nonlinear development of any instability to which the radiative interior might be subject is ignored.

The rise in the density and temperature of the core leads to a rise in the rate of thermonuclear energy generation, and an increase of entropy. Because evolution is slow compared with the rate at which energy is transported from the core to the surface, there is a close balance between the energy generation rate and the solar luminosity L, the net rate at which energy is radiated from the surface. According to the theory, L rose gradually and monotonically from the initial value of about 70% of its value today, the initial value being the luminosity after the Sun had settled down to a state of thermal balance. Roughly speaking, that was the epoch at which the earth was formed, about 4.6×10^9 years before present (BP). The rise in L is a robust conclusion, and is not dependent on any subtle details of the theory.

Temporal variation on the nuclear timescale is not the subject of this meeting. Nevertheless, I have taken the space to mention it because, so far as I am aware, all self-contained quantitative climatic models, which are relevant to our discussions, cannot reconcile the past low luminosity of the Sun with the Earth not having been almost entirely glaciated for most if not all of its existence. Therefore a healthy scepticism must always be maintained when discussing long-term solar–terrestrial relations.

Some scepticism must also be maintained when discussing the standard solar model. Aside from the adoption of what might seem highly idealized assumptions, we are faced with the fact that the model that otherwise best fits the astronomical observations yields a neutrino flux on Earth some three times greater than the value measured by Davis and his colleagues. This too appears to be an astrophysically robust result, provided the physics used to determine the nuclear reaction rates is correct. I shall mention this discrepancy again later.

Unfortunately, despite their name, standard solar models have not yet been standardized. The models are continually being updated as new modifications to the microscopic physics determining the opacity, the nuclear reaction rates and the equation of state are suggested. Fortunately, the differences between the models are much too small to be of any concern to my discussion here.

3. Competing hypotheses about the solar cycle

An early suggestion for explaining the solar cycle of magnetic activity was that of Walèn (1946). The solar interior was thought to be undergoing oscillations for which magnetic stresses provided the major component of the restoring force; the oscillation is essentially a resonant Alfvén mode. Evidently the motion must be essentially horizontal, to avoid substantial generation of buoyancy forces that would lead to periods of hours rather than 22 years. The oscillations are modes that become the zero-frequency torsional modes as the magnetic field strength approaches zero. They could therefore be called magnetic torsional oscillations. Because the Sun rotates with a period much less than the cycle period, the motion must also be such that the Lorentz force largely balances the Coriolis force, otherwise the oscillation period would once again be too short.

The oscillation period depends on the strength of the magnetic field. A value of a few hundred milliteslas yields 22 years. That is typical of the fields seen in sunspots. Therefore, if one believes that there is a global field in the interior of the Sun, loops of which sometimes rise through the convection zone and break through the photosphere to form sunspot pairs, 22 years does not seem too unnatural a value to expect for the period of the cycle. At least the hypothesis is economical in assumptions. The detailed implications of Walèn's idea appear not to have been worked out.

There is another reason for not being too surprised at a large-scale interior magnetic field with magnitude of a few hundred milliteslas. It is natural to presume that the Sun condensed from a typical interstellar gas cloud, as have most other stars. The cloud was presumably intrinsically rotating, thereby having vorticity with respect to an inertial frame, and was pervaded by a magnetic field. Both vorticity and magnetic field in the interstellar turbulence are random; for want of better data I shall adopt typical values of 2×10^{-8} a^{-1}, corresponding to the galactic rotation, and 3 μG†, respectively. We know that neither angular momentum nor magnetic flux were conserved by the material that eventually formed the stars. For that to have been so, the collapse would have had to be restricted to parts of the gas cloud that happened to have essentially no vorticity and no magnetic field; how would similar neighbouring parts of the cloud, with somewhat greater vorticity and magnetic field, know not to collapse because it would present them with serious dynamical problems in the distant future? It is more likely that these problems were faced by all protostars, and solved. A plausible and commonly held opinion is that much of the solution occurred in a turbulent phase of contraction in which magnetic field was expelled. In an ideal homentropic fluid, both potential vorticity ω and $b = B/\rho$, where B is the magnetic field and ρ is the fluid velocity, behave like a material line element: a streak of dye that has been placed in the fluid. Even if dissipation is considered, the equations for ω and b are similar; so one might expect ω and b to behave similarly, particularly in a turbulent fluid where any differences in initial conditions have been forgotten. This argument was first used by Batchelor (1950) in a discussion of hydromagnetic turbulence in an incompressible fluid, and although particular cases are known

---

† 1 G = 10^{-4} T.

where the analogy between ω and b breaks down, it is an open issue whether it holds under the circumstances I am discussing here. Nevertheless, if one accepts it, even though the collapsing gas is not strictly homentropic, and even though one does not understand the mechanism of vorticity and magnetic field expulsion well enough to predict how much is lost, one can deduce that the ratios of the final to the initial values of both the mean vorticity (and therefore intrinsic angular velocity) and the mean magnetic field are the same. The present solar rotation period is about 27 d, corresponding to a mean vorticity of 160 a^{-1}, an increase over the initial value by a factor of about 10^{10}. Thus a magnetic field of $10^{10} \times 3 \times 10^{-6} = 30$ kG seems not unreasonable. Some further credence has been given to a value of this order by Mestel & Weiss (1987), who point out that fields of about 500 G have been observed at the surfaces of T Tauri stars. These stars are at the end of the extensively turbulent phase of their evolution, and the surface fields may be in the process of being expelled. If they are representative of the interior field, and flux is conserved throughout the subsequent evolution to the main sequence, an increase in field strength by a factor of 10 occurs, yielding a zero-age main-sequence field of 5 kG. The difference by a factor 2π between these two estimates is hardly material for such a rough argument. Of course, in the absence of dynamo action in the radiative interior the field will decay on the main sequence; moreover, there is also strong evidence for substantial angular momentum loss during the main-sequence lifetime. There is no reason to believe that the different processes causing these decays would keep in step, so the angular velocity to field ratio might have changed by a factor 10 or so since the Sun arrived on the main sequence. Nevertheless, this factor is quite insignificant in the light of the wide range of values (1 μG–10 MG) proffered by various authors for the present field in the solar interior.

The major competing hypothesis to explain the solar cycle is that the convection zone of the Sun is a turbulent dynamo. Theories of dynamos in rotating, roughly spherical electrically conducting fluid bodies were developed initially to explain the existence of the Earth's magnetic field. The natural decay time of the field is much less than the age of the Earth. Therefore, if the terrestrial field were not regenerated, any primeval field would by now have decayed to an imperceptible level; a dynamo is therefore essential. As in the Sun, the dipole component of the terrestrial field reverses sign, roughly but apparently not exactly, in a periodic manner. Thus it is perhaps no surprise that having laboured long to develop the mathematical armoury to attack the terrestrial problem, the temptation to turn it towards the Sun was irresistible, even though the natural decay time of the largest-scale component of the solar magnetic field is greater than the age of the Sun.

The characteristic cycle period of the supposed dynamo is not an easy quantity to predict. It depends on subtle properties of the structure of the turbulence in the convection zone, which are difficult to ascertain. It is the case, however, that the observed period is not in obvious conflict with the theoretical estimates.

The earliest dynamo theories assumed the dynamics to be confined entirely within the convection zone. The major difficulty encountered by even the most elaborate studies was that consistent dynamical theories could not reproduce simultaneously the two most pertinent observations: that the photospheric angular velocity is greater at the equator than it is at the poles, and that immediately after field reversal the predominant magnetic activity occurs at mid-latitudes and subsequently migrates towards the equator. To reproduce the more rapidly rotating equator the uncertain parameters of the theory must be adjusted in such a way as to tend to an angular velocity that increases with depth, at least in the equatorial regions; whereas to reproduce the equatorial migration of magnetic field the opposite radial rotation gradient is required.

In recent years, consideration of the dynamics has been extended a little beneath the base of the convection zone. It has been suggested that the essence of the dynamo action is confined to a relatively thin interfacial layer between the true convection zone and the radiative interior (Spiegel & Weiss 1980). Basically, magnetic field is forced downwards through the convection zone as a result of a topological asymmetry in the flow (Drobyshevski & Yuferev 1974). It accumulates as a thin toroidal belt near the base of the convection zone until it becomes buoyant enough to overcome the convective downthrust. Then it surges to the surface to start a new cycle. It was hoped that with such an approach the apparently insuperable difficulty encountered by previous theories of reproducing both the rapid equatorial rotation of the photosphere and the equatorial flux migration at low latitudes might be surmounted. The attention of dynamo theorists has not yet penetrated fully into the radiative interior of the Sun.

How might one decide between the two hypotheses? Before discussing that it would be wise to make it quite plain what the fundamental difference between them is. To be sure, if dynamo calculations were extended deep into the radiative interior, some response would be found, but because some 98% of the inertia of the Sun resides in the radiative region, the amplitude of that response would be relatively small. The existence of the radiative region is not essential to the operation of the dynamo, and presumably its presence modifies the cycle only slightly. The basic mechanism is a loss of magnetic field either by dissipation in the convection zone or loss of flux through the photosphere, followed by amplification of the residual field due to stretching by shearing motion; field amplification to cancel losses is an essential feature of a dynamo. The torsional oscillation, on the other hand, does not require the dissipated field to be replaced. The dominant controlling dynamics of the cycle is in the radiative interior; it is there that the characteristic cycle period is determined. Of course, the processes that are presumed to take place in the turbulent dynamo would be operative in the convection zone, and no doubt would have a very strong influence on the observable features of the oscillation in the photosphere. Therefore, there may be similarities in some of the superficial consequences of the two hypotheses.

Perhaps the principal superficial observed property of the magnetic field is the polarity reversal of the dipole component. This feature is often considered to be an overwhelming objection to torsional oscillations, it being implicitly assumed that superficial behaviour would also be required to take place throughout the Sun, and that therefore the entire radiative interior would have to turn over. That is not the case. If the oscillation were one predominantly in rotational shear, for example, the interior field would be stretched azimuthally, presenting the convection zone with a field that is hardly discernible from that of the interfacial dynamo. The difference between the two hypotheses is not essentially in the field configuration at the base of the convection zone, but in how it is produced.

Another argument that is often tendered in favour of the solar dynamo is that some of the more subtle features of the photospheric flow field have been reproduced by suitably tuned dynamo models, whereas that is not so of a global torsional oscillation. The torsional oscillations discovered by LaBonte & Howard (1982) constitute an example of such flow. When considering this argument one should remember that, as I have just pointed out, many of the processes in the convection zone that might constitute an integral part of a turbulent dynamo are also likely to be operative, albeit incidently, if the cycle is controlled from within the radiative interior. More importantly, it should be realized that one reason certain phenomena have not been reproduced by a theory of global torsional oscillations is that nobody has attempted to do so.

4. TEMPORAL COHERENCE OF THE SOLAR CYCLE

Because the convection zone is turbulent, any model of the cycle that depends in any way on the dynamics of the convection must be subject to some stochastic element. There may be other sources of variation, such as the deterministic chaos discussed by Professor Weiss (this Symposium). It is not out of the question that some information could be gained about the dynamics controlling the cycle by studying the properties of the variations from cycle to cycle. This is quite an old idea, and there have been many studies since the pioneering paper by Yule (1927) on the behaviour of a randomly perturbed pendulum.

The issue I address here is that of phase maintenance. My interest was raised in this subject by Dicke (1970), who remarked that because the early arrivals of the sunspot maxima of 1778 and 1788 were followed immediately by a compensating long cycle, thereby restoring the phase, the cycle must be controlled by some mechanism that keeps time. This could be so of a coherent laminar torsional oscillation of the relatively quiescent radiative interior of the Sun (whose signature at the surface is delayed by some random process), but it would hardly be true of a process that is controlled largely by the random motions in the turbulent convection zone, unless, of course, the dominant motion in the convection zone were not really random.

Of course the occurrence of but a single pair of exceptional cycles whose mean period is similar to that of the entire record is only weak evidence for phase maintenance. The entire record should be examined. Thus with the aim of distinguishing between the two hypotheses discussed in the previous section in mind, two simple stochastic models have been tested against the sunspot record. The first, which is a simple yet extreme representation of a turbulent dynamo, is a process with no long-term memory. Cycle n is considered to be of duration $P_n = P + \sigma_n$, where P is the characteristic period of the process that controls the dynamics of the oscillation and σ_n is a random variable resulting from the stochastic turbulent perturbation which is considered to be uncorrelated with the perturbations $\sigma_m (m \neq n)$ to the periods of the other cycles. The time of occurrence of some well defined epoch (say sunspot maximum or sunspot minimum) signifying the 'beginning' of cycle $N+1$, measured from the same epoch in cycle 1, is thus $T_N = NP + \sum \sigma_n$, where the sum is over all cycles from 1 to N. I am assuming that there is nothing special about cycle 1; it is simply the first cycle of the record available for analysis. The second model is intended to represent a periodic torsional oscillation of the radiative interior. The oscillation keeps perfect time, but because information must pass through the turbulent convection zone to produce a visible change in the photosphere there is again a stochastic element to the observable signal. In this case the time of some epoch signifying the 'beginning' of cycle $n+1$ relative to the same epoch in cycle 1 is $T_n = nT + \tau_n - \tau_1$, where T is the period of the torsional oscillation and τ_n is a random variable representing the turbulent perturbation which, as in the first model, is uncorrelated with the perturbations to the other cycles. Thus in this model the period of cycle n is $T + \tau_n - \tau_{n-1}$. The idea is to compare the properties of the two models with the sunspot record.

No statement has yet been made about how close to being periodic the sunspot record actually is. That would depend on the standard deviations σ and τ of σ_n and τ_n, which depend on how much the turbulence in the convection zone influences the sunspot signal, and about which it is assumed we have no *a priori* information. What must therefore be tested is a statistic that is independent of σ or τ. The obvious choice is the variance σ_T^2 of T_n from the perfectly regular time sequence whose period and phase best match the data, measured in units of the variance σ_P^2 of P_n^2. This is then compared with the value of the ratio $\Phi = E(\sigma_T^2)/E(\sigma_P^2)$ predicted by the two models, where E denotes expectation value. When N is large, Φ increases linearly with N for the turbulent dynamo model, whereas it tends to $\frac{1}{2}$ for the torsional oscillation model. For $N = 2$, the smallest value for which the analysis can formally be carried out, Φ is the same for both models. The crucial question is whether the sunspot record is long enough for the predictions of the two models to be distinguishable.

The analysis was first carried out by using separately the epochs of sunspot maxima and sunspot minima, dating from immediately after the end of the Maunder Minimum (Gough 1978). The result was inconclusive: the time series of sunspot maxima is slightly closer to the prediction of the periodic oscillation model, whereas the series of sunspot minima is slightly closer to the model representing the turbulent dynamo. It appeared that the sunspot record is too short. Shortly afterwards, Dicke (1978) analysed a modified time series based on the epochs of sunspot maxima alone, and concluded that phase has been maintained. The analyses have been rediscussed, and extended by dividing the record into shorter segments and comparing the dependence of Φ on N with the predictions of the models (see, for example, Gough 1988). The results depend on whether 11-year records or 22-year (derectified) cycles are used. In the former case, Φ follows the trend of the stochastic dynamo model except for the greatest value of N, at which, as I have already mentioned, it lies inconclusively between the two models. Because it should be the longest record that is the most reliable, one is bound to be left in an uncertain state. The derectified cycles are similar, except that the undivided record is closer to the dynamo model than it is for the 11-year rectified cycle.

An obvious drawback to taking the epochs of either sunspot maxima or sunspot minima is that in a noisy record they are subject to error due to local fluctuations, particularly because the first derivative is zero at extrema. It is surely preferable to use all the data in the record. Recently, Bracewell (1985) has carried out a Hilbert transform of derectified sunspot data, representing the record as a mildly nonlinear function of a sinusoid that is displaced from zero by a variable amount, having varying amplitude and phase offset. The variance of the phase offset, which represents the deviation of the oscillator from a perfect clock, is very much less than the value of σ_T^2 computed from the raw sunspot data. Indeed, Bracewell (1988) has regarded so small a variance as evidence that phase is maintained, and proposed an explanation in terms of wave motion that is not very dissimilar from the torsional oscillator that I have discussed here. However, as I have already argued, a small value of σ_T^2 does not by itself distinguish between the two models; for that a statistic must be used that does not depend on the magnitude of the stochastic fluctuations. Accordingly, I have defined two time sequences of a given epoch in the 22-year magnetic cycle (representing odd- and even-numbered sunspot maxima) from the phase offset in Bracewell's Hilbert transform, and have subjected them to the same analysis as the raw sunspot data. The result is illustrated in figure 1, together with the values of Φ predicted by the two statistical models. Once again the results are inconclusive: although the N dependence of the ratios obtained from the subsequences looks more similar to the predictions of the turbulent dynamo model, the standard deviations are large enough for

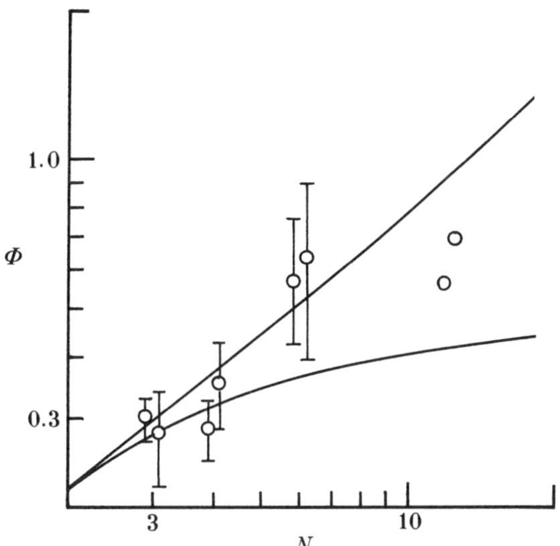

FIGURE 1. Ratios Φ of phase deviation to period deviation from a perfect clock of the solar cycle for two time sequences of constant cycle phase computed from Bracewell's (1985) Hilbert transform of the $12\frac{1}{2}$ cycles of derectified sunspot record between 1705 and 1979. The sequences start at 1705 and 1716, and represent approximate sunspot maxima. (To compute the final date of the second sequence it was necessary to extrapolate Bracewell's phase to 1980.) They have each been divided into contiguous subsequences of length N, from each of which Φ has been computed separately. The points plotted are the means of the values from each subsequence, and the error bars denote $\pm$ one standard deviation. To avoid overlapping, points computed from the sequence starting in 1705 are displaced slightly to the right; those from the sequence starting in 1716 to the left. The upper continuous curve is $N(N+3)/15(N+1)$ and represents the predictions of the turbulent dynamo model; the lower curve is $N^2/2(N+1)^2$ and represents the prediction of the periodic internal oscillator model.

either model to be possible; and the longest records yield ratios lying between the two theoretical curves.

There thus seems to be little evidence in the sunspot record to support the idea that the cycle is controlled by a periodic oscillator. But that does not imply that the Sun is not undergoing torsional oscillations in its radiative interior. If the oscillations are nonlinear, they could be chaotic, in a manner similar to the oscillators discussed by Professor Weiss (this Symposium) in connection with deterministic dynamo models. In that case phase would not be well maintained, and might be consistent with the data plotted in figure 1.

5. THE SEAT OF THE CYCLE

There are other diagnostics that can shed light on the mechanism of the cycle. To interpret them, the structure of the entire Sun must be taken into account.

Any perturbation to the balance of forces in a star causes a global readjustment of its structure. Just how that adjustment takes place depends on the nature of the perturbation and the timescale on which it occurs. Here, interest is on a timescale of order 10 years, which is much less than the thermal relaxation times of either the solar core or the convection zone. However, it is much greater than the internal thermal readjustment timescale of the convection zone, which is the time it takes for a convective wave to propagate through the extent of the zone: about one month for a spherically symmetrical readjustment. Therefore any perturbation on the timescale of the solar cycle would cause an adiabatic hydrostatic readjustment of the radiative interior, leaving the convection zone in internal thermal balance. Of course, the convection zone would not be in balance with the radiative interior; a thermal boundary layer would develop, of characteristic thickness 0.1 % of the solar radius (1 % of the pressure scale height), if intermittent convective overshooting were unimportant.

The hydrostatic readjustment of the convection generally changes the radius R of the photosphere and the photospheric temperature. Consequently, there is a change in the luminosity L. This changes the solar irradiance S on Earth by the same relative amount. Broadly speaking, the magnitude of the ratio W of the relative perturbations to L and R tends to increase with the depth of the perturbation (Gough 1981; Däppen 1983), and so provides a diagnostic for the location of the perturbation. A recent paper by Willson & Hudson (1988) reports that S varies with the rectified 11-year cycle (figure 2), with a relative amplitude of 0.04 %, maximum S corresponding to maximum sunspot number. Unfortunately, unambiguous variations in the photospheric radius have not yet been measured. There are many problems associated with the determination of R, among which is the variation of the limb-darkening function at the very limb resulting from magnetic activity in the solar atmosphere, which varies with the solar cycle. Therefore, this potential diagnostic has not yet given us any secure information.

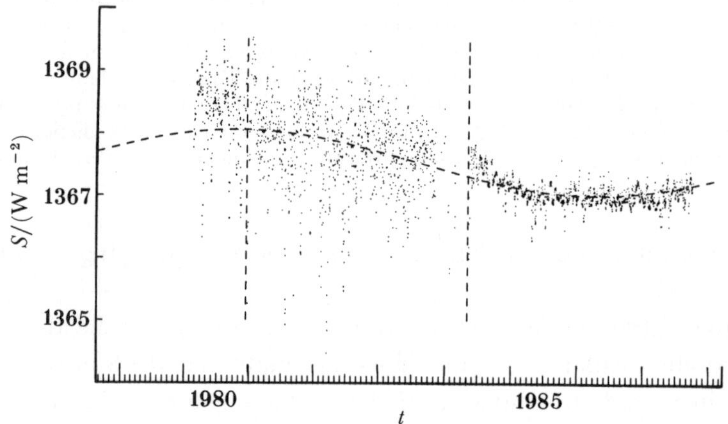

FIGURE 2. Solar irradiance S from the *Solar Maximum Mission* Active Cavity Radiometer Irradiance Monitor reported by Hudson & Willson (1988). Data in the interval from 1980 day 346, when the satellite's attitude control system failed, until the repair in March 1984 (delimited by the vertical dashed lines) are of lower quality than the rest. The dashed curve is the least-squares cosine function
$$S = S_0\{1 + 0.00039 \cos [2\pi(t - 1980.82)/10.95]\},$$
where S_0 is constant and t is time in years from the beginning of the century. Fluctuations are lower at sunspot minimum because there are fewer sunspots appearing and disappearing to modulate the flux.

If one does accept at face value the semi-diameter measurements plotted in figure 3, combines them with the irradiance measurements of figure 2, and assumes that $\delta S/S \approx \delta L/L$, one obtains $W \approx -0.7$. This is larger in magnitude than the theoretical values that have been reported for even the most deeply seated perturbations that occur on a solar-cycle timescale. It is interesting to note, however, that this value agrees almost exactly with the value one can infer from the calculations reported by Gough & Thompson (1990), who considered the response of the Sun's structure due to a broad equatorial belt of toroidal magnetic field at the base of the convection zone. The vertical extent of the belt was about 10 % of the solar radius; and the characteristic field intensity required to obtain variations of L and R of the observed

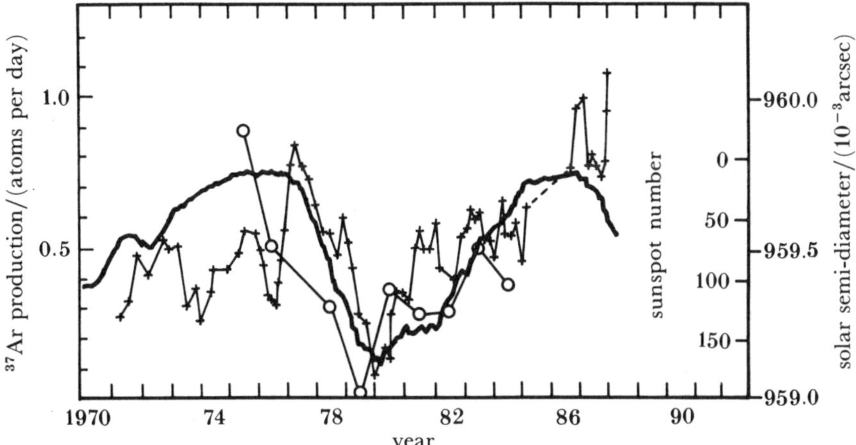

FIGURE 3. Five-monthly averages of sunspot number (thick line), solar neutrino flux recorded by Davis and his collaborators (crosses) and apparent semi-diameter of the Sun from astrolabe measurements (circles), perhaps suggesting a causal connection (after Davis *et al.* 1988).

magnitude was about 10 MG, which is rather greater than the values I have been discussing explicitly in this paper. The computations were not strictly applicable to the solar cycle, however, because the radiative interior was maintained in thermal balance rather than being perturbed adiabatically as would essentially be the case for a variation on a timescale of 10 years. (The purpose of the computations was not to describe solar-cycle variations.) Therefore, the numerical coincidence might just be fortuitous.

Magnetic fields cannot be everywhere spherically symmetrical. Therefore, any magnetic field that perturbs the solar structure must distort the Sun from its otherwise almost spherical configuration. This distortion results, in general, in an aspherical variation of any property of the Sun. In particular, it causes a variation of the photospheric temperature T_e, leading to a redistribution of radiant flux over the surface. This provides another contribution to the variation in the irradiance S on Earth. Recently, Kuhn *et al.* (1988) have shown that the latitudinal variation δT_e of T_e is correlated with the sunspot distribution, a result that is hardly surprising. But what is surprising, however, is Kuhn's (1988) observation that if δT_e is assumed to translate directly into a variation δc of sound speed c, whose relative value $\delta c/c$ is then assumed to be independent of depth in the convection zone, then a certain degeneracy splitting observed in the frequencies of acoustic oscillations of the Sun could be reproduced theoretically. This would imply that the latitudinal relative variation $\delta T/T$ of temperature is independent of depth throughout the convection zone. No consistent calculation of the Sun's convective envelope, perturbed in some plausible way by a magnetic field, has yet yielded such a property. Kuhn's remarkable result therefore offers an interesting challenge to theoretical heliophysicists, which is bound eventually to yield a better appreciation of solar-cycle dynamics.

It should be appreciated that a naive identification of acoustic wave propagation speed with the sound speed of the solar gas may not be correct. Small-scale fibril magnetic fields, temperature inhomogeneities and convective velocities all modify the speed of propagation of large-scale acoustic waves, and thereby modify the resonant frequencies of global oscillations. Therefore it does not necessarily follow from the observations that $\delta T/T$ is independent of depth. It must also be borne in mind that the measurements of Kuhn *et al.* (1988) might themselves have been influenced by fibril fields, and therefore do not reflect true temperature variations.

Solar oscillations have not yet provided a clear indication of whether the asphericity extends into the radiative interior. Degeneracy splitting in low-degree acoustic modes, which penetrate to the energy-generating core of the Sun, is difficult to discern; no convincing measurement has yet been made. Absolute frequency shifts, correlated with the solar cycle, of groups of nearly degenerate modes have been reported by several observers. The shifts are of such a magnitude as to require a substantial contribution to come from an aspherical component of the structure of the radiative interior (Gough 1988), should they be real. However, one must always be wary of measurements of groups of unresolved modes, because one runs the risk of the data being biased by interference phenomena.

There is another direct measurement I wish to mention that might be pertinent to the issue: the solar neutrino flux. In figure 3 is plotted the smoothed neutrino counts of Davis and his collaborators (1988) on the same diagram as smoothed sunspot numbers and measurements of the apparent solar diameter by Laclare and his collaborators (cf. Delache *et al.* 1985). There is not an obviously convincing correlation. I know that because I submitted some of the data to a critical test: I took the solar diameter measurements (sequence A) and the neutrino data over the same interval (sequence B) and supplemented them with a series of random numbers (sequence C). I took the liberty of removing just one (outlying) point from each sequence, and offered then to a well-known expert in correlative science for his advice. Sequences B and C were unmarked, but sequence A was labelled in Laclare's handwriting, and therefore identifiable. My adviser found an interesting correlation between sequences A and C (subsequently admitting that he had believed sequence C to be T. M. Brown's unpublished solar-diameter measurements), and recommended that I devote some time to work more on the subject. Notwithstanding the outcome of this isolated experiment, it may seem for some that there is at least a hint of a causal connection, and indeed there have been serious claims of a significant correlation (Davis 1988), in addition to claims to the contrary (Bahcall *et al.* 1987). If future data support that hint, then something very important will be learned; indeed, it is because it is so important that I have taken the trouble to discuss what at present is such weak evidence.

A detectable flux of solar neutrinos can come only from the solar core. Therefore, if a variation in that flux is convincingly found, and if it is associated with the solar cycle, one can conclude only that either the source of neutrinos is modulated or that the neutrinos are changed as they pass through the Sun. If the former were the case, then the very core of the Sun would be partaking in the cycle; the cycle would not be caused by a dynamo confined to the outer convective envelope and its immediate environs. In this context it is perhaps worth mentioning that the cycle could result from a 22-year modulation of a relatively short-period nonlinear oscillator, such as an internal gravity mode; as Roxburgh (1986) has pointed out, because of its spatial structure such an oscillator is likely to have the added attraction of reducing the neutrino flux of theoretical models to the observed value without having recourse to modifications to nuclear or particle physics. If the latter were the case, it would follow that neutrino transitions take place, the most likely cause being a dipole interaction with the magnetic field inducing neutrino helicity flipping. This would require the neutrino to have a magnetic moment. Current bounds on the value of that moment, coupled with the uncertainty in the interior solar magnetic field, certainly do not rule out this interesting possibility (see, for example, Leurer & Liu 1989), particularly if the transition were a resonance phenomenon (Akhmedov 1988; Minakata & Nunokawa 1989).

Finally, I return to the problems encountered by dynamo theorists in simultaneously explaining both the decrease in the photospheric angular velocity with latitude and the equatorward migration of low-latitude magnetic activity. As I have already mentioned, dynamical calculations require parameters to be adjusted such that the angular velocity Ω increases with depth to produce the observed latitudinal dependence of Ω; yet Ω needs to decrease with depth to reproduce the observed migration of magnetic activity. Recent analyses of rotational degeneracy-splitting of the Sun's acoustic oscillations (Christensen-Dalsgaard & Schou 1988; Brown *et al.* 1989; Dziembowski *et al.* 1989) have revealed in some detail how Ω really varies in the convection zone. The result is illustrated in figure 4. Throughout most of the convection zone Ω increases with depth. But in a possibly thin transition layer at the base of the convection zone Ω decreases somewhat with depth at low latitudes. (The separation of the contours in figure 4 is too great for this small decrease to be visible.) In a discussion of the

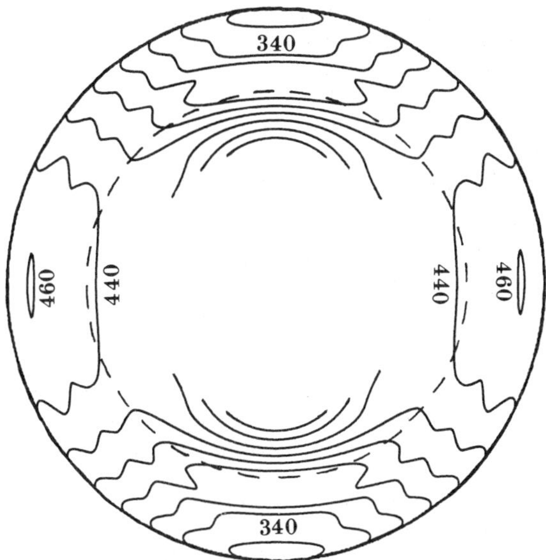

FIGURE 4. Contours of constant angular velocity Ω in the Sun, obtained by averaging the results of the inversions by Christensen-Dalsgaard & Schou (1988), Brown *et al.* (1989) and by Dziembowski *et al.* (1989) of measurements of rotational splitting of solar five-minute acoustic modes. The contours are drawn in a meridional plane through the Sun. They are labelled in nanohertz, and are separated by 20 nHz. They are drawn only in the outer half, by radius, of the Sun; the data are unreliable nearer the centre. The dashed circle indicates the base of the convection zone. The axis of rotation is vertical.

dynamical implications of this discovery, Morrow *et al.* (1988) have recently pointed out that this leads the way to a dynamo theory that might explain the observations: the negative value of $d\Omega/dr$ throughout most of the convection zone is consistent with the overall dynamics of the convection zone in which there is a decline of Ω with latitude, whereas the dynamo is situated in the thin transition zone in which the possibly time-dependent value of $d\Omega/dr$ is positive at low latitudes, leading to low-latitude dynamo waves that propagate equatorward. The ideas also suggest explanations to other properties of the solar cycle, which I have not discussed in this paper.

6. Conclusion

After carefully weighing the evidence, and discounting the numbers of scientists who would vote for each of the principal types of proposed explanation of the solar cycle, the conclusion must be that we do not yet know whether the Sun is an oscillating dynamo or whether the solar magnetic field is decaying from its primeval state. Nor do we know whether the mechanisms that cause the polarity of the dipole component of the surface field to reverse are confined to the convection zone or whether they are controlled by a deep-seated oscillator in the solar core.

I am grateful to R. N. Bracewell for giving me the phases of his Hilbert transform of the sunspot record, and to W. J. Merryfield for his assistance in producing figure 4. I gratefully acknowledge support from NASA grant NSG-7511.

References

Akhmedov, E. K. 1988 *Phys. Lett.* B **213**, 64–68.

Bahcall, J. N., Field, G. B. & Press, W. H. 1987 *Astrophys. J.* **320**, L69–L73.

Batchelor, G. K. 1950 *Proc. R. Soc. Lond.* A **201**, 405–416.

Bracewell, R. N. 1985 *Aust. J. Phys.* **38**, 1009–1025.

Bracewell, R. N. 1988 *Q. Jl R. astr. Soc.* **29**, 119–128.

Brown, T. M., Christensen-Dalsgaard, J., Dziembowski, W. A., Goode, P., Gough, D. O. & Marrow, C. A. 1989 *Astrophys. J.* **343**, 526–546.

Christensen-Dalsgaard, J. Schou, J. 1988 *Seismology of the Sun and Sun-like stars* (ed. E. J. Rolfe), pp. 149–153. ESA SP-286, Noordwijk.

Däppen, W. 1983 *Astron. Astrophys.* **124**, 11–22.

Davis, R. 1988 Seventh workshop on grand unification/Icoban '86 (ed. J. Arafune), pp. 237–276. World Scientific.

Delache, P., Laclare, F. & Sadsaoud, H. 1985 *Nature, Lond.* **317**, 416–418.

Dicke, R. H. 1970 *Stellar rotation* (ed. A. Stetteback), pp. 289–317. Dordrecht: Reidel.

Dicke, R. H. 1978 *Nature, Lond.* **279**, 676–680.

Drobyshevski, E. M. & Yuferev, V. S. 1974 *J. Fluid Mech.* **65**, 33–44.

Dziembowski, W. A., Goode, P. & Libbrecht, K. G. 1989 *Astrophys. J. Lett.* **337**, L53–L57.

Gough, D. O. 1978 *Pleins feux sur la physique solaire* (ed. J. Rösch), pp. 81–103. Paris: CNRS.

Gough, D. O. 1981 *Variations in the solar constant* (ed. S. Sofia), pp. 185–206, NASA Conf. Publ. 2191. Washington, D.C.: U.S. Govt. Printing Office.

Gough, D. O. 1988 *Solar–terrestrial relationships and the Earth environment in the last millenia* (ed. G. C. Castagnoli), pp. 90–132. Bologna: Soc. Italiana di Fisica.

Gough, D. O. 1988 *Seismology of the Sun and Sun-like stars* (ed. E. J. Rolfe), pp. 679–683. ESA SP-286, Noordwijk.

Gough, D. O. & Thompson, M. J. 1990 *Mon. Not. R. astr. Soc.* (In the press.)

Kuhn, J. R. 1988 *Astrophys. J. Lett.* **331**, L131–L134.

Kuhn, J. R., Libbrecht, K. G. & Dicke, R. H. 1988 *Science, Wash.* **242**, 908–911.

LaBonte, B. & Howard, R. F. 1982 *Sol. Phys.* **80**, 373–378.

Leurer, M. & Liu, J. 1989 *Phys. Lett.* B **219**, 304–308.

Mestel, L. & Weiss, N. O. 1987 *Mon. Not. R. astr. Soc.* **226**, 123–135.

Minakata, H. & Nunokawa, H. 1989 *Phys. Rev. Lett.* **63**, 121–124.

Morrow, C. A., Gilman, P. A. & DeLuca, E. E. 1988 *Seismology of the Sun and Sun-like stars* (ed. E. J. Rolfe), pp. 109–115. ESA SP-286, Noordwijk.

Roxburgh, I. W. 1986 *The internal solar angular velocity* (ed. B. R. Durney & S. Sofia), pp. 1–5. Dordrecht: Reidel.

Spiegel, E. A. & Weiss, N. O. 1980 *Nature, Lond.* **287**, 616–617.

Walèn, C. 1946 *Ark. Mat. astr. Fys.* A **33** (18).

Willson, R. C. & Hudson, H. S. 1988 *Nature, Lond.* **332**, 810–812.

Yule, G. V. 1927 *Phil. Trans. R. Soc. Lond.* A **226**, 267–298.

Phil. Trans. R. Soc. Lond. A **330**, 641–643 (1990)
Printed in Great Britain

641

Variability in the solar output

BY I. W. ROXBURGH

*Astronomy Unit, Queen Mary and Westfield College, University of London, Mile End Road,
London E1 4NS, U.K.*

Evidence for variability in the solar output is briefly discussed. If the solar neutrino
flux and the solar oscillation frequencies vary over a solar cycle this could indicate
that the solar cycle has its origin in the solar core rather than be due to dynamo action
in the solar convective zone.

Direct observations of solar variability reveal little information about solar variability other
than the 11- or 22-year solar cycle as determined from sunspot numbers, reliable data exists for
less than 200 years. Thus most inferences about solar variability come either from proxy data,
or from unreliable historic data on sunspots. There is more recent data from the Active Cavity
Radiometer Irradiance Monitor (ACRIM) experiment on the *Solar Maximum Mission* (see P.
Foukal, this Symposium), which shows a variation in solar output with sunspots and possibly
a 0.1 % variation in total irradiance over the last solar cycle, attempts to reconstruct recent
cycle variations suggest that typically the solar luminosity varies by somewhat less than this,
about 0.05 %. There are also claims that the solar radius also varies over a solar cycle, and over
longer timescales (see E. Ribes, this Symposium). As is well known the sunspot cycle itself
appears to be modulated by a 200-year cycle giving rise to the Maunder and other minima
and, it is conjectured, to 'little ice ages' on Earth. More startling is the claim that the solar
neutrino flux measured on Earth also varies with the solar cycle, and that the properties of the
p-modes of oscillation of the Sun may show a similar cyclical variation.

It is premature to take these claims seriously, but my task is to be the 'agent provocateur',
so I must challenge both the accepted wisdoms and put forward provocative suggestions!

The standard explanation of the solar sunspot (and magnetic) cycle is that it results from a
dynamo acting in, or immediately below, the solar convective zone (see N. O. Weiss, this
Symposium). Why should we believe this explanation? It is true that detailed calculations of
dynamo models have been carried out, but these are not based on the correct physics, in spite
of their numerical intricacy they are really no better than back of the envelope calculations
using incorrect physics. Cray time is no substitute for understanding. The models are based on
the following foundations: convection in a rotating shell produces non-uniform rotation, non-
uniform rotation of a magnetic field stretches out the field to produce enhanced toroidal field
from an initial poloidal field, a convective cyclonic eddy rising under gravity can generate
poloidal field from toroidal field. But the solar convective zone is not a simple laminar fluid,
it is as far as we can see, and as far as we can predict from terrestrial physics, in a state of
highly developed turbulence. Thus any model of the solar dynamo has to model the small-
scale turbulence by some eddy transport coefficients, the resulting model is not based on
the equations of magnetohydrodynamics, or plasma physics, but on a set of macro-model
equations, which contain unknown quantities representing smaller-scale behaviour. These

[243]

48-2

equations are solved and the unknown eddy coefficients and transport processes adjusted until one finds a result that bears some resemblance to the observed solar cycle. It is not obvious to me that this is a valid procedure, why should there be any cycle at all in a fully developed turbulent medium? We anyway know from observations that magnetic flux is concentrated into small-scale intense flux tubes, so the microscale behaviour is important and may be dominant. At best all one can infer from the detailed macro-models is that there are solutions of the macro-equations that can be adjusted to give something like the observed magnetic cycle.

The 200-year modulation is more difficult to explain. As has been shown by Weiss, one can construct sets of nonlinear equations that produce 'chaotic' solutions with something like a short period modulated by a longer period. Because the equations governing the behaviour of the convective zone are nonlinear then perhaps one could construct a set of macro-model equations for the solar dynamo that exhibited the same behaviour. But whether this is really what is going on in the Sun I cannot say.

An alternative hypothesis discussed by D. O. Gough (this Symposium), but with a long history, is that the 11- or 22-year period is determined by an internal clock. The obvious candidate for such a clock being the Alfvén travel time round an internal magnetic field of order 0.3 T. This could be a torsional oscillation of the solar interior. In this picture, the details of which remain unspecified, the solar convective zone responds to the oscillation, giving rise to the observed phenomena of the solar cycle. This hypothesis of an internal toroidal oscillation was raised by myself (Roxburgh, unpublished work) as a possible explanation of the non-uniform rotation of the solar interior deduced from solar oscillations; the idea was that because the Alfvén time varies throughout the interior different parts of the Sun will oscillate at different rates so that the differential rotation is time dependent, all we see, or rather deduce from oscillations, is a 'snap-shot', which may well have unusual properties.

But this model is difficult to sustain. If the oscillation of neighbouring parts of the solar interior are out of phase, large gradients will be established on a short timescale (22 years), this will lead to enhanced dissipation which will tend to destroy the differential rotation. For a given angular momentum the lowest energy state is uniform rotation so that one should expect either uniform rotation or disordered differential rotation, in neither case would one have a nice clock.

Although it is still premature to reach conclusions there is some hint that the p-modes detected as solar oscillations show some variation with the solar cycle. The high-order modes are reflected in the surface convective zone and might therefore be expected to show some dependence on the solar cycle; but if, as seems to be suggested, the low-order modes show differential effects, then this may be some hint that the interior of the Sun also varies with the solar cycle. This would be more difficult to understand, although a change in the properties of the base of the convective zone could possibly effect the interior structure.

An even more puzzling phenomenon, if real, is the possible variation of the measured solar neutrino flux with the solar cycle. This suggests that the central temperature undergoes an 11-year variation of the order of a few percent. This is large and were it true would presumably give rise to variations of the internal structure that would in turn effect the properties of low-order p-modes of oscillation. But what could cause such a variation? One suggestion that has not been explored is that the nuclear reaction network plus internal structure is unstable to long-period (11-year) oscillations, driven by overstability in the reaction network (Roxburgh 1985). Another is that the outer convective zone generates gravity waves that propagate into the interior and are focused into the centre (Press 1981); because the outer convective zone varies on a solar cycle so too would the generation of gravity waves and hence their contribution in the very central regions where the neutrinos are generated (Roxburgh & Schatzman 1989). In this scenario the basic 22-year period, or quasi-period, is due to dynamo action in the convective zone.

And what about changes over a longer timescale? Models of the solar interior are really still in their infancy. A great deal of effort and computer time is spent on calculating spherically symmetric hydrostatic models with the latest equation of state, nuclear reaction cross sections, opacity; comparatively little is spent on studying time-dependent three-dimensional models. A good example is the ^{3}He instability discovered by Dilke & Gough (1972), nothing is really known about the development of this instability; these authors suggested it could drive episodic mixing in the solar core leading to variations in solar output over a timescale of 10^8 years. Perhaps there are other instabilities waiting to be uncovered by the questioning scientist. Then our models of the solar convective zone are crude, based on a steady-state mixing-length analysis. It is not even obvious to this author that this bears any relation to reality. Perhaps the convective zone does not settle down to some quasi-steady-state but spends some time in one state then switches to another, as suggested some years ago by Tavakol (1978). There are plenty of examples of nonlinear systems that can behave in this way. Since the thermal relaxation lifetime of the whole solar convective zone is of order 10^5 years it is possible to imagine that somehow this timescale enters into the nonlinear dynamo leading to variations in solar output on this timescale (Roxburgh 1980). Then if the internal magnetic field is only of the order of 1 mT the Alfvén travel time is reduced to 10^6 years and this timescale could enter into the variations in solar flux. Perhaps there are other instabilities on long timescales in the solar interior waiting to be uncovered.

Finally, what about the energetics of the variations of luminosity over a solar cycle? The *SMM* observations imply a variation of up to 0.1 % over the last cycle, and perhaps somewhat less during previous cycles, this is some 10^{-9} of the internal energy of the Sun over a cycle. If this were stored in the equatorial regions of the convective zone its contribution to the distortion of the Sun would be small, but if a larger amount is stored over longer periods during 'Maunder Minima' then it is of the order of magnitude where the distortion would produce an external gravitational field comparable with the solar gravitational quadrupole moment and therefore sufficient to have some effect on the orbits of the interior planets.

I do not claim that any of the above thoughts are correct. My brief is not to provide answers but to stimulate discussion!

References

Davis, R. 1989 In *The solar interior and atmosphere* (ed. A. N. Cox, W. C. Livingstone & M. S. Mathews). Tucson: Arizona Press.
Dilke, F. & Gough, D. O. 1972 *Nature, Lond.* **240**, 262.
Press, W. H. 1981 *Astrophys J.* **245**, 286.
Roxburgh, I. W. 1980 In *Soleil et climat*, p. 261. CNES, Toulouse.
Roxburgh, I. W. 1985 *Sol. Phys.* **100**, 21.
Roxburgh, I. W. & Schatzman, E. 1989 In *The solar interior and atmosphere* (ed. A. N. Cox, W. C. Livingstone & M. S. Mathews). Tucson: Arizona Press.
Tavakol, R. K. 1978 *Nature, Lond.* **276**, 805.

Phil. Trans. R. Soc. Lond. A **330**, 645–655 (1990)

Printed in Great Britain

Archaeological evidence and non-evidence for climatic change

By P. I. Kuniholm

Department of the History of Art and Archaeology, Cornell University, G-35 Goldwin Smith Hall, Ithaca, New York 14853-3201, U.S.A.

'Climate' is often used by historians to explain phenomena for which they cannot otherwise account. Accordingly, much of what has been written about climatic effects and climatic change must be read with extreme scepticism. Even though a disturbance may be obvious in the archaeological record, and it may be synchronous with a climatic event, a cause and effect relation should be demonstrated before one can say with any degree of confidence that the evidence is secure. Only when a number of separate lines of investigation agree on the same thing are we safe in positing true climatic 'effect' or 'change'. This paper focuses on several instances in Mediterranean and Aegean archaeology where more or less satisfactory evidence for climatic change may be sought among a number of disciplines.

The title of this paper as originally assigned was 'Archaeological evidence for climatic change'. The qualifying phrase 'and non-evidence' was my addition, reflecting not just routine academic sophistry but a deep-rooted suspicion that we often attribute to 'climate' phenomena in the archaeological record that we cannot otherwise explain. In layman's language climate becomes the historian's cop-out. One example of the lengths to which some writers will go should be sufficient: 'Climate in Greco–Roman history,' (Eddy 1980) in which the author sets out to match two graphs, each one an inverted V, one of the thickness of sequoia tree rings in California, the second of dedications and building starts in Africa between A.D. 138 and 244. The logic seems to be, after the two Vs are superimposed, that because the latitudes of California and the Mediterranean are the same, so should be the climate. Six data points in 2000 years allegedly prove this. So much for non-evidence, but where might we look for something more satisfactory?

First, I have no doubt that climate does indeed change, both over the long and short term, although, after recent reports announcing that there is less than total agreement among the climatologists that the 'Little Ice Age' ever existed and that there is doubt as to whether there is really a so-called greenhouse effect after all, I wonder how the archaeologist is supposed to provide information about events whose very existence has been called into question.

Secondly, I have no doubt that we are in one way or another affected by climatic changes, particularly from single catastrophic events. The severe drought in the mid-western United States last summer and the billions of dollars' worth of damage that resulted, all duly recorded by the press, is a dramatic example of the degree to which we are at the mercy of the climate, and the people who lost their livestock and had the banks foreclose on their mortgages do not need to be informed by a Royal Society Discussion Meeting that indeed 1988 was a bad year. But again I ask whether an archaeologist digging 2000 years from now in the remains of a mid-western U.S. town and coming down to the depressed 1988 level would be able (*a*) to realize that there had indeed been a horrendous drought, and (*b*) to discriminate between the effects of the drought and some adverse aspect of, let us say, Reagan economics. Other less-dramatic climate changes have less obvious effects. Here the problem is simply one of recognition.

There is also the question of cause and effect. The same hypothetical archaeologist continues to dig down to the 1930s level and finds the great 'dust bowl.' Does this have anything to do with the stock-market crash of 1929 or were the two events simply coincident?

At the first International Conference on Climate and History at the Climatic Research Unit, University of East Anglia, Norwich, in 1979, the point was made repeatedly in both the printed review papers and the discussions that went on all week that it is not so much climate that is important but rather man's *perceptions* of it. Colleagues who work with the English Manor House records used as an example a society where conventional father-to-son wisdom is that you have to save one-third of your harvest to survive until the next harvest and still have enough grain for the next year's planting. Then a series of mild years comes along with abundant rain and sun, and very quickly the old verities are forgotten. One-fourth of the harvest, let us say, is saved, and the rest is sold for immediate profit. Then the climate reverts to normal (except that people have forgotten what 'normal' is: the unofficial Norwich after-dinner estimate of how long our memories work was about four years maximum), and people starve. Has the climate really 'changed?' Not at all. Human memory has once again played us false. How, I wonder, is the archaeologist, measuring the human record as it bumbles along in this fashion, going to be able to match it in any meaningful way with the estimates of the palynologists (who, by the way, should have been represented at this Discussion Meeting), the geomorphologists, and the meteorologists, even presuming that nobody has made any faulty calculations?

One final example follows (here I speak as a member of our local volunteer fire brigade). In October 1981 five inches (13 cm) of rain in five hours sent the Ithaca (New York) creeks into full flood, and for the next three days we pumped out basements of houses along the banks of Fall Creek in which an average of six feet (2 m) of water had collected. As we pumped, we noticed that there was not a single nineteenth-century house in the lot. The old houses were up on top of the bluff. Obviously, in the last century the creek had flooded often enough so that any right-thinking house-builder built on high and dry ground. But somewhere around World War I enough time went by without a flood to entice would-be money makers into subdividing the land along the stream banks into house-lots and then building the houses that the fire brigade had to pump. A glance at the flood-scoured walls of the gorge through which the creek flows would have told any moderately competent geologist that this was no place to build houses, either then or now, but avarice won out over common sense.

Antoine Meillet's great comment on his study of Indo–European linguistics was 'La linguistique est une système ou tout se tient,' that is to say, linguistics are a system in which everything hangs together, and so it should be with archaeology and climate. Because much of my own work is with tree rings (Kuniholm & Striker 1983, 1987), I am continually reminded that there are indeed periodicities of one kind or another in all the records with which we work, but all too often our conclusions about climatic forcing mechanisms or stimuli are drawn based on a single line of evidence without independent corroboration and should therefore be treated with extreme scepticism. Moreover, I find the vast majority of speculations on the relation between archaeology and climate to be precisely that: speculation based often on inadequate or unverifiable data. Pollen diagrams down to the end of the Upper Palaeolithic, for example, may be significant indicators for climatic change, but after intensive agricultural exploitation begins in the Neolithic there is always the danger that a certain element of anthropogenic 'noise' has been introduced into the palynological record. Moreover, how should we expect various bodies of evidence to interact in the archaeological record? There will be lags in some cases, parallel change in others, divergent change in yet others. At best the possibilities are precarious. A model might be the careful, cautious, exhaustive work of Emmanuel Le Roy Ladurie whose *Times of feast, times of famine: a history of climate since the year 1000*, first published in 1971, is now again in print and which makes even better reading the second time around. Nothing is too humble for Le Roy's attention: wine harvest dates, advances and retreats of glaciers, grain prices, documentary evidence for crop yields, taxation rates, tree-ring widths, etc., and his conclusions are equally humble: something indeed seems to be going on with the climate, but he refrains from forcing conclusions upon us.

Twenty years ago the British palaeoclimatologist H. H. Lamb, while commenting on the climatic changes that may have put an abrupt end to Bronze Age civilization in Greece and in other areas of marginal agricultural productivity (Carpenter 1966, and see below), issued a challenge to researchers in the Mediterranean and in the Near East (Lamb 1968).

Our knowledge of this 3000-year climatic sequence in the middle and northern latitudes of Europe has been built up partly from pollen analysis of numerous bog and lake site deposits and partly from analysis of documentary records, especially those which reveal the incidence of extreme warm or cold, wet or dry seasons. By comparison, the Mediterranean is a strangely neglected region despite the many ancient cultures there and the wealth of literature that has survived. Surely, we need not remain for ever ignorant of which years in classical times had unusually long, or short, dry seasons or of which winters were striking for warmth or coldness. What seems most important, if we are to get at the truth of these matters, is to encourage many more historians and archaeologists to dig out the documentary and other sorts of evidence. Every report which is specific as to year and place, giving or implying the general character of a particular summer or winter, or the dates of notable rains or floods at any season, is important. One would imagine that Greek or Roman horticulturists must have recorded the actual dates of the first and last rains. Any compilation of such reports from anywhere and any period in the Mediterranean and Near East is badly needed.

Let us look around the Mediterranean now to see what, if anything, may be said about archaeological evidence for climatic change.

Iran

A comprehensive and cautious survey of a wide variety of evidence is by Butzer (1958). He finds it difficult to correlate evidence for short, minor rainfall ameliorations with anything cultural. He laments the lack of systematic study and systematic observations in both Anatolia and Iran. His final conclusion is that although he believes small-scale variations of climate occurred continually during historical times, he nevertheless strongly negates any overall climatic change within the past 2500 years. More recent, and indeed systematic, work (Van Zeist & Wright 1963; Van Zeist 1967; Van Zeist *et al.* 1968) bears this out. Van Zeist concludes that after about 5500 before present (BP) the climate may have shown minor fluctuations but no major changes.

Mesopotamia

Despite millennia of written records, Mesopotamia is a frustrating area with which to deal, partly because the evidence is very unevenly distributed, partly because there has been hardly any research done on its ecological microstructures (Nissen 1988). Plant cultivation must have been always precarious at best with the bulk of the water coming through irrigation channels rather than from the wretched 200 mm a^{-1}† rainfall of southern Mesopotamia (Van Zeist 1969). The most entertaining single item for a climatic event is Sir Leonard Woolley's finding of a water-borne layer, 8–11 ft (2–3 m) thick, of clean mud and silt at Ur between 'Ubaid I and II levels. Mrs Woolley took one look at it and said, 'Well, of course, it's the Flood!' (Woolley 1954). Spoilsports (Lees & Falcon 1952) have cast doubt on this, saying that the date is not contemporary with other early floods, and that the major cause is tectonic subsidence rather than the simultaneous rising of the Euphrates and the Tigris that can set southern Mesopotamia awash (and see now Nützel 1976). Whatever the cause or combination of causes (downward sea-level change, tectonic uplift, silting-in of the floodplain, or climatic amelioration), H. J. Nissen's new (1988) book on Mesopotamian prehistory concludes that southern Mesopotamia at last became inhabitable in the 4th millennium B.C. only because the land was finally dry enough to farm and raise animals. Another excellent but cautious recent paper (Neumann & Parpola 1987) lays out climatological, meteorological, and Assyriological data for a significant climatic change at the end of the Bronze Age (see also Brinkman 1984).

Egypt

The major review of Nile flooding and its causes is by Lyons (1905), updated by Bell (1975) with an historical commentary particularly regarding the low flood levels that might have helped bring on the so-called Little Dark Age at the end of the Middle Kingdom sometime after *ca.* 1768 B.C. Bell notes that this took place at a time when there was unusual uncertainty about the proper order of succession to the Kingship, a general increase in governmental instability, and numerous short reigns. She does not claim that low water caused the concurrent political upheaval, but poor crops certainly did not help matters. What bothers me about the evidence from the Nile is the apparently deliberate falsification of the Nile records for revenue purposes (Lyons 1905), which certainly should cast doubt on their utility as a reliable source for climatic change unless supported by external evidence.

The Levant

The most comprehensive recent study I have seen is by Horowitz (1979). Yet his climatic evidence boils down to pollen analysis from a grand total of two bore holes, one in Lake Hula and one in the Mediterranean, with one uncalibrated radiocarbon date for each. More evidence is needed here. A recent paper by Koucky (1987) on the Roman frontier in central Jordan goes at long length into 570-year cycles of drought and plenty, and indeed one of his graphs is quite impressive, but I am suspicious of his chronology, and his references, albeit somewhat more recent than Eddy's, do not inspire confidence.

Turkey

Careful, competent work (they call their observations on climate 'speculations') is being done by Van Zeist and others in the Gröningen group, but a major portion of their work (1975) concerns only prehistory for which we have no corroborating evidence. In a limited study in southeastern Turkey (1968) their pollen diagrams for three different lakes show that the climate has not changed noticeably during the last 3000–4000 years.

† mm a^{-1} is millimetres per year.

The Roman world

Clive Foss's (1979) explanations for the silting-in of the harbour at Ephesus dodge between deforestation and *possibly* (my emphasis) climatic change. In his own words, 'the evidence is scattered and ambiguous, and subject to much controversy and variety of interpretation…A theory of climatic change, if it could be developed and maintained, could do much to explain the decline of the ancient world and the conditions which prevailed in the early Byzantine period.'

Last month David Whitehouse, whose scepticism about the lack of evidence for climatic change at the end of the Roman period has been in print for some time (Hodges & Whitehouse 1983), was kind enough to show me part of an expanded manuscript on which he is now working concerning the apparent absence of climatic change in Roman and early post-Roman times. After considerable thought he concludes that the evidence is disappointing: the written records of the River Tiber floods neither support nor contradict the hypothesis of a Dark Age drought. A second class of evidence, Vita-Finzi's (1969) so-called Younger Fill, can be shown, says Whitehouse upon reconsideration, to be equally explicable without climatic change.

Greece

A useful new book, *Beyond the Acropolis: a rural Greek past* (Van Andel & Runnels 1987), has just appeared, with commentary on the Argolid, including its climate, from the earliest times to the present. Speculations on droughts in Attica or at least changes in the water levels in the 8th, 4th and 2nd centuries B.C. as attested by the filling in of Attic wells and the digging of other, deeper wells have been published by John Camp (1979, 1982, 1984). One obvious criticism of Camp's work is that wells go dry for a variety of reasons besides drought, including heavy consumption of water by a large population, but Camp is also able to bring to bear both epigraphical and textual evidence for drought, shortage, and famine datable to specific years, plus the need for Athens to import large amounts of grain from abroad.

We have also the Sanctuary of Zeus Ombrios or Rainy Zeus on the top of Mt Hymettus (Langdon 1976), a rural sanctuary to which local farmers brought offerings in inverse proportion and quality to the known prosperity of Athens. When Athens was prosperous and importing goods from all over the Mediterranean world, the Sanctuary of Rainy Zeus languished; but when Athens fell upon hard times and was forced to patronize the local farmers, the sanctuary flourished. In isolation this does not seem like very hard evidence, but taken together with Camp's wells, and the inscriptions, and the literary texts, it seems to me a promising beginning.

Another useful new work, fulfilling a number of Lamb's *desiderata*, is Peter Garnsey's *Famine and food supply in the Graeco–Roman world: responses to risk and crisis* (1988). His commentary on the Eleusis First Fruits inscription (*Inscriptiones Graecae* II^2 1672), concerning the harvest of 329–328 B.C. that yielded enough wheat and barley to feed only some five-eighths of the Attic population is both fascinating and frustrating because we simply do not have that kind of detailed documentation for other years besides 329–328. Would that we had an absolute Mediterranean tree-ring chronology in place for those centuries to test these snippets of evidence, but for the moment we have only disconnected dendrochronological sequences.

Now that the *Thesaurus Linguae Graecae* is largely complete, containing some 60 000 000 words from almost 3000 authors on compact disc, plus the 20 000 or so Attic inscriptions we have been adding at Cornell, we are at last in a position to begin part of the textual searching, at least for the Greek world of which Lamb spoke. Although some of the literary references are no doubt metaphorical, others are probably every bit as literal as the fourth-century lead tablet from Dodona in which a supplicant asks the oracle, 'Is the present severe winter due to the impiety in the city?' (Parke 1967). And of course the cuneiform and hieroglyphic evidence needs to be searched as well.

After all these remarks about other people's work, here are a few speculations of my own, arrived at during the last fifteen years' work of building tree-ring chronologies in the northeast corner of the Mediterranean, which I think might represent evidence rather than non-evidence for climatic change.

Example 1. Short-term events: the Great Drought of 1873–74 in Turkey

The first is a typical short-term event. In 1874 in the Province of Ankara, District of Keskin, a drought occurred of such devastating proportions that 81 % of the cattle and 97 % of the sheep died. Of the population of 52000, some 7000 moved out of the district and 20000 died (Christiansen-Weniger & Tosun 1939). The traveller C. Naumann reported that in the Provinces of Kastamonu, Ankara, and Kayseri 150000 people and 100000 head of livestock (40 % of all herds) died. Hunger and sickness through the 1873–74 winter killed another 100000 people (Naumann 1893).

Plots of tree-ring growth for 11 sites in or near the area of the reported catastrophe show subnormal growth for 1873 and 1874, following four other years of significantly subnormal growth, as described in table 1.

Table 1

site	1873 growth index (%)	1874 growth index (%)
Mihaliççik	25.2	20.8
Çeltikçi	23.0	54.2
Çamlitepe	46.7	87.8
Hamidiyeköy	54.6	85.7
Yozgat	66.1	67.7
Ovacik	66.3	73.9
Güdül	74.4	69.6
Kizilcahamam	70.9	70.0
Bağlum	59.4	94.4
Yarakin	69.1	95.1
Çatacik	84.6	87.1

This is not to say that every time a band of narrow rings appears on Turkish trees one is to judge that a disastrous famine has occurred, but we feel that this is more than just coincidence, and at least the trees are a lot closer to the troubled area than the California sequoias. Eventually, we hope to be able to reconstruct more sophisticated palaeoclimatic anomalies for the Aegean as have been done most recently for Europe by Schweingruber *et al.* (1987) and Briffa *et al.* (1987) to whose work we were able to contribute nine chronologies from the southeast corner of Europe. Another potentially rich source of information is the Ottoman Prime Ministry Archives in Istanbul with their estimated 40000000 documents covering the past 500 years.

Example 2. Longer-term events: the Celali Rebellions of *ca.* 1585–1640

The second kind of event is rather longer term, from a decade to several decades. Again archival and tree-ring evidence, even in raw form, can be combined with intriguing results. Below (figure 1) is a profile of tree-ring growth from 1560 to 1620 in five forests and two archaeological sites in and around west-central Anatolia. The oscillations below and above the base-line represent departures from mean annual growth. Note that only 13 out of 61 years have above-normal growth, and the cumulative depletion of ground water and the subsequent effect on cereal crops must have been much more severe than on mature forest trees that at least have deep root systems and food reserves which enable them to survive a long-term drought.

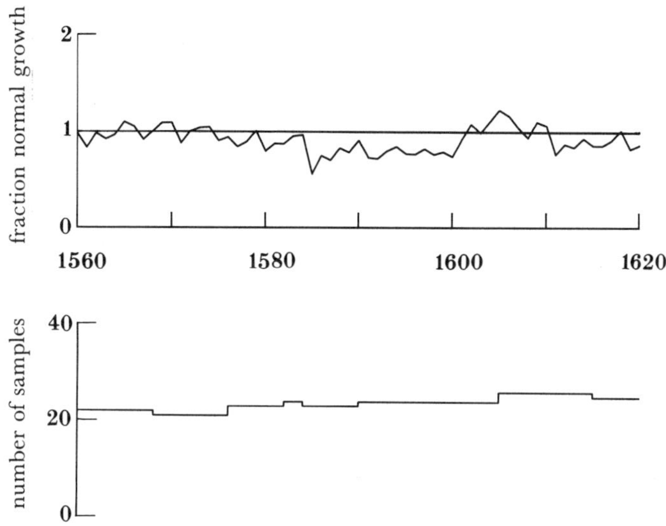

FIGURE 1. Çatacik, Elmadağ, Antalya, Grevena conifers and the archaeological oaks.

Now compare this composite table of archival information and travellers' reports from 1564 to 1612 compiled independently by Professor H. Inalcik and Professor W. Griswold and kindly communicated by them to me as in table 2.

The late sixteenth and early seventeenth centuries in Anatolia were marked by peasant unrest, even revolt (the so-called Celali Rebellions), large-scale changes in land use, and unexpectedly large fluctuations in urban populations (Griswold 1977, 1983, 1989). Obviously, climatic change is not to be blamed for all otherwise inexplicable occurrences in history, especially in a period of great political and economic turmoil, and the correlation between the two sets of information may be merely fortuitous, but it is interesting that there is such a high correlation between years of poor tree-ring growth and years reported to be years of shortage or famine. Remember, too, that the drought effect was undoubtedly cumulative, and a look at the table will show that often the year or years immediately preceding a reported famine are notable for reduced growth. Although the export grain market reopened in 1591 after a year in which tree-ring growth reached only 90%, 1590 was still the best year out of the previous six, and one imagines that the Ottomans must have been sorely pressed for foreign exchange. Of course there are years of subpar tree-ring growth that go unmentioned in the chronicles, and

TABLE 2

year(s)	report of conditions	tree-ring growth (%)
1564	Widespread shortage in Anatolia	96
1565–67	Ottoman prohibitions on grain export	110 105 91
1570–71	More prohibitions. The Venetian ambassador to the Sublime Porte complains that prices have quadrupled	109 87
1574–76	Grain shortage. Famine in Anatolia and Istanbul	105 90 94
1579	Shortage in Archipelago and Syria	100
1580	Shortage in Western Asia Minor and Archipelago	79
1583	Shortage in Archipelago and Aleppo	95
1584	Shortage in Western Anatolia, Syrian coasts, Tripoli	96
1585	No rain in January and February in Istanbul; no rain in summer in Rumeli, Edirne. Shortage in Western Anatolia, Rumeli (Edirne, Berkofca, Temesvar), Lepanto, Zulkadriye in Eastern Asia Minor	55
1586	Famine in Çorum	74
1588	Shortage in Istanbul	82
1589	Great shortage in the Levant	78
1590	Shortage in Damascus. Great shortage in Italy; wheat imports from Northern countries	90
1591	Shortage in Skoplje. Re-opening of the Levant wheat market	72
1592	Fall famine in Damascus; cold January, plague	71
1594	Plague, storms, Ottoman prohibitions. Land 'remains uncultivated because there are no farmers....' Italy makes massive imports from Northern countries	84
1595	Famine	76
1598	Famine. Caspian area hot; rough seas in July on Mediterranean; Sir Anthony Sherely reports exceeding barrenness in Anatolia	76
1599	Unusual contrary winds in the Adriatic. Drought in Zante	78
1610	'Unhusbanded plaines for many miles together,' says traveller Charles Robson. Plague of grasshoppers	106
1611	Famine in Anatolia. Aleppo snow awful	75
1612	French Consul 'killed when snow broke through his house on him'	86

the poor French Consul seems to have survived years in which growth rates reached as high as 122% without succumbing to any snow.

EXAMPLE 3: MAJOR CATASTROPHES

This leads me to my final kind of event, the catastrophic event that occurs only once or twice in a millennium, the hardest kind to document convincingly, and about which most of the nonsense unfortunately gets written. When Rhys Carpenter first published his *Discontinuity in Greek civilization* in 1966, suggesting climatic reasons for the disaster attendant upon the end of the Late Bronze Age in the Mediterranean – specifically a northward shift of the winds that now desiccate the Sahara so that they desiccated instead Crete, the Peloponnese, and Anatolia – he was roundly condemned by the meteorologists with the exception of Professor Reid Bryson, who did the logical thing: assign the topic to a graduate student, David Donley. What Donley found in his dissertation – summarized succinctly and neatly in an apologetic article in *Antiquity*, 'Drought and the decline of Mycenae' (Bryson *et al.* 1974) – was the winter weather pattern of 1954–55 that yielded precisely the conditions posited by Carpenter, i.e. 60% normal rainfall and significantly higher temperatures in Crete and the Peloponnese. The mechanism was not as simple and linear as Carpenter's first scheme. Some areas, Elis and Athens, showed no change whatever. Other areas, particularly in the north, were much wetter than normal.

[254]

Bryson *et al.*, however, in their discussion make the following noteworthy point that echoes Lamb's quoted above:

> It is reasonable to say that if the pattern of the winter of 1954–55 had dominated the climate around 1200 B.C., or even if that pattern had been a good deal more frequent than in modern times, Mycenaean agriculture would have been very precarious indeed. A short, more intense episode might have been disastrous....Clearly more field data are needed, especially in critical areas. It is likely that the yet uncovered record of the past contains the answers we seek – if we are wise enough to ask the right questions.

Donley's climate diagram for 1954–55 did not include Anatolia, but this has now been done by Weiss (1982), and his diagram deserves careful study. Several interesting new additions can be pointed out: the southwest corner of Anatolia had very heavy precipitation, up to 140%. The North Syrian coast received 40% normal precipitation. The Anatolian plateau east of Ankara was dry, with the area around Malatya having as little as 7% of normal precipitation, suggesting that it was not just the Aegean proper that experienced unusual climatic phenomena. This is not to say that this is *exactly* what happened at the end of the Late Bronze Age in the twelfth-century B.C., but it represents what could have happened if Carpenter's modified scenario is correct.

What, in fact, do we know about what happened in the twelfth-century B.C.?

1. A large number (70–90%, depending on whose figures are used) of Mycenaean sites are destroyed or abandoned or both.

2. Troy, Boğazköy on the Anatolian plateau, and Ugarit on the North Syrian coast are destroyed. In one of the so-called 'oven texts' at Ugarit (i.e. the last day's mail before the city was destroyed) mention is made of famine.

3. The Pharaoh Ramesses fights off the so-called Peoples of the Sea (whoever they were), who were 'restless in their isles.'

4. The Greenland ice-core researchers have a big acidity layer, indicating that Hekla 3 erupted in 1100 +50 B.C.

5. M. G. L. Baillie (1988*a*, *b*) finds that the Irish tree rings dwindle to practically nothing in 1159 B.C. and the two decades following, no doubt a direct result of the Hekla eruption.

6. For Mesopotamia Neumann & Parpola (1987), mentioned above, cite both textual data for nomadic incursions and unrest from the middle of the twelfth century onward as well as evidence for notable warming and aridity.

7. Even as far away as China in documents of Chou, the last king of the Shang Dynasty, the effects of Hekla 3(?) were noted: dust and ash rains, a foot of snow in July, all five cereals killed by frost (Pang *et al.* 1988).

8. At Gordion, Turkey, ever since we first measured the rings of our 806-year-long tree-ring chronology that extends from the sixteenth century B.C. to the eighth century B.C., we have known that there was one 20-year period where annual growth was abnormally large, accompanied by abnormal fluctuations both up and down. Now that Bernd Kromer at Heidelberg has successfully wiggle-matched 18 sets of specifically numbered rings from the Gordion chronology with the high-precision oak curve from Europe (Kuniholm & Kromer 1990), we also see that this 20-year anomaly is centred on 1159 B.C. and the two decades following.

9. And finally Professor Donald Sullivan (personal communication) has now found in one of his cores at Gölcük, a small lake above Sardis in Lydia in Western Turkey, a section showing

that the lake deepened drastically between 1200 and 1100 B.C., with peat giving way to lacustrine mud. The date is based on three calibrated radiocarbon determinations. Taken in isolation, the Gölcük core could be blamed on the spread of agriculture in the region, but a period of extraordinary rainfall as is suggested by the Gordion tree rings could also have produced this effect.

In and of themselves these anomalies do not yet prove Carpenter right, and whether there is a convincing cause and effect relation among these various concurrent phenomena has yet to be determined, but Gordion and Gölcük are two more pieces of evidence to show that the middle of the twelfth century B.C. was not a normal time in the Aegean and elsewhere. When similar twelfth-century tree-ring sequences and more carefully studied and dated cores are available from additional sites in the Aegean, and with additional help from a wide variety of specialists such as phytologists, palynologists, ecologists, vegetable crops specialists, microfaunal analysts, and others, we may one day be able to say that Carpenter was right all along about the twelfth-century catastrophe, and, more importantly, that a number of less-dramatic climatic events also had a significant impact upon the human affairs of the past 9000 years. The exercise seems to me well worth the effort.

Endnote

The Aegean Dendrochronology Project is supported by the National Endowment for the Humanities, the National Geographic Society, the Institute for Aegean Prehistory, the Samuel H. Kress Foundation, the David and Lucile Packard Foundation, and a number of private contributors.

References

Baillie, M. G. L. 1988a Irish oaks record volcanic dust veils drama! *Archaeology Ireland* **2**(2), 71–74,

Baillie, M. G. L. 1988b Marker dates – turning prehistory into history. *Archaeology Ireland* **2**(4), 154–155.

Bell, B. 1975 Climate and the history of Egypt. *Am. J. Archaeology* **79**, 223–269.

Briffa, K. R., Wigley, T. M. L., Jones, P. D., Pilcher, J. R. & Hughes, M. K. 1987 Patterns of tree-growth and related pressure variability in Europe. *Dendrochronologia* **5**, 35–57.

Brinkman, J. A. 1984 Settlement surveys and documentary evidence: regional variation and secular trend in Mesopotamian demography. *J. Near Eastern Stud.* **43**(3), 169–180.

Bryson, R. A., Lamb, H. H. & Donley, D. L. 1974 Drought and the decline of Mycenae. *Antiquity* **47**, 46–50.

Butzer, K. W. 1958 Quaternary stratigraphy and climate in the Near East. *Bonner Geographische Abhandlungen* **24**, 103–128.

Camp, J. M. II 1979 A drought in the late eighth century B.C. *Hesperia* **48**, 397–411.

Camp, J. M. II 1982 Drought and famine in the 4th century B.C. *Hesperia Suppl.* **20**, 9–17.

Camp, J. M. II 1984 Water and the Pelargikon. *Studies Presented to Sterling Dow on His Eightieth Birthday. Greek, Roman, and Byzantine Monographs* **10**, 37–41.

Carpenter, R. 1966 *Discontinuity in Greek civilization.* New York: Norton.

Christiansen-Weniger, F. & Tosun, O. 1939 *Die Trockenlandwirtschaft im Sprichwort des anatolischen Bauern.* Ankara.

Eddy, S. K. 1980 Climate in Greco–Roman history. *Syracuse Scholar* **1**, 19–30.

Foss, C. 1979 *Ephesus after Antiquity: a late antique, Byzantine and Turkish city*, appendix III, pp. 185–187. Cambridge University Press.

Garnsey, P. 1988 *Famine and food supply in the Graeco–Roman world: responses to risk and crisis.* Cambridge University Press.

Griswold, W. J. 1977 The Little Ice Age: its effect on Ottoman history, 1585–1625. (Paper presented at the Middle East Studies Association Meeting, New York.)

Griswold, W. J. 1983 *The Great Anatolian Rebellion 1000–1020/1591–1611.* Berlin: Klaus Schwartz Verlag.

Griswold, W. J. 1989 Climatic change: a possible factor in the social unrest of seventeenth century Anatolia. *Studies in honor of Andreas Tietze.* Istanbul: Divit Press.

Hodges, R. & Whitehouse, D. 1983 *Mohammed, Charlemagne & the origins of Europe.* Ithaca: Cornell University Press.

Horowitz, A. 1979 *The quaternary of Israel*, pp. 211–230. London: Academic Press.

International Conference on Climate and History 8–14 July 1979, Review Papers. Climatic Research Unit, University of East Anglia, Norwich, U.K.

Koucky, F. L. 1987 The regional environment. In *The Roman frontier in Central Jordan* (ed. S. T. Parker). *BAR Int. Ser.* **340** (i–ii), 18–25.

Kuniholm, P. I. & Striker, C. L. 1983 Dendrochronological investigations in the Aegean and neighboring regions, 1977–1982. *J. Field Archaeology* **10**, 411–420.

Kuniholm, P. I. & Striker, C. L. 1987 Dendrochronological investigations in the Aegean and neighboring regions, 1983–1986. *J. Field Archaeology* **14**, 385–398.

Lamb, H. H. 1968 Climatic changes during the course of early Greek history. *Antiquity* **42**, 231–233.

Langdon, M. 1976 *A sanctuary of Zeus on Mount Hymettos. Hesperia, Suppl.* **16**.

Le Roy Ladurie, E. 1971 *Times of feast, times of famine* (revised edn 1988). New York: Farrer, Straus & Giroux.

Lees, G. M. & Falcon, N. L. 1952 The geographical history of the Mesopotamian plains. *Geogr. J.* **118**, 24–39.

Lyons, H. G. 1905 On the Nile Flood and its variation. *Geogr. J.* **26**, 249–272; 395–421.

Naumann, C. 1893 *Vom goldenen Horn zu den Quellen des Euphrat.* Munich & Leipzig.

Neumann, J. & Parpola, S. 1987 Climatic change and the eleventh–tenth-century eclipse of Assyria and Babylonia. *J. Near Eastern Stud.* **46** (3), 161–182.

Nissen, H. J. 1988 *The early history of the Ancient Near East 9000–2000 B.C.* University of Chicago Press.

Nützel, W. 1976 The climate changes of Mesopotamia and bordering areas 14000 to 2000 B.C. *Sumer* **32**, 11–25.

Pang, K., Srivastava, S. K. & Chou H.-h. 1988 Climatic impacts of past volcanic eruptions: inferences from ice core, tree ring and historical data (abstract). *1988 Fall Meeting of the American Geophysical Union.*

Parke, H. W. 1967 *The oracles of Zeus: Dodona, Olympia, Ammon*, pp. 261–262. Oxford: Blackwell.

Schweingruber, F. H., Bräker, O. U. & Schär, E. 1987 Temperature information from a European dendro-climatological sampling network. *Dendrochronologia* **5**, 9–33.

Van Andel, T. H. & Runnels, C. 1987 *Beyond the Acropolis: a rural Greek past*. Stanford University Press.

Van Zeist, W. 1967 Late quaternary vegetation history of Western Iran. *Rev. Palaeobot. Palynol.* **2**, 301–311.

Van Zeist, W. 1969 Reflections on prehistoric environments in the Near East. In *The domestication and exploitation of plants and animals* (ed. P. J. Ucko & G. W. Dimbleby), pp. 35–46. Chicago: Aldine.

Van Zeist, W. & Wright, H. E. 1963 Preliminary pollen studies at Lake Zeribar, Zagros Mountains, Southwestern Iran. *Science, Wash.* **140**, 65–67.

Van Zeist, W., Timmers, R. W. & Bottema, S. 1968 Studies on modern and Holocene pollen precipitation in southeastern Turkey. *Palaeohistoria* **14**, 19–39.

Van Zeist, W., Woldring, H. & Stapert, D. 1975 Late quaternary vegetation and climate of southwestern Turkey. *Palaeohistoria* **17**, 53–143.

Vita-Finzi, C. 1969 *The Mediterranean valleys: geological changes in historical times*. Cambridge University Press.

Weiss, B. 1982 The decline of Late Bronze Age civilization as a possible response to climatic change. *Climatic Change* **4** (2), 173–198.

Woolley, L. 1954 *Excavations at Ur: a record of twelve years' work*, pp. 26–36. London: Benn.

Phil. Trans. R. Soc. Lond. A **330**, 657–663 (1990)
Printed in Great Britain

Climate and Holocene culture change: some practical problems

By A. C. Renfrew, F.B.A.

Department of Archaeology, University of Cambridge, Downing Street, Cambridge CB2 3DZ, U.K.

In recent years, not least through tree-ring studies for the Holocene and studies of oxygen isotope ratios in Foramenifera in deep-sea cores for the Pleistocene, both linked with radioactive chronometry, useful and well-dated information has become available for global temperature variations. Yet we seem at present little closer to understanding the climatic influences upon human settlement, or upon such major episodes in human existence as the agricultural revolution or the emergence of pastoral economies.

In making reference to the developments in archaeological survey techniques over the past 20 years, and the increasing collaboration with geomorphologists and settlement geographers, I seek to highlight the gap in the chain of argument between data for global climatic parameters and impact on human communities. Where are the phytologists, the ecologists, the crop plant geographers? Where is the necessary focus upon the crucial themes of changing microclimates and changing agricultural productivity for specific species? An attempt is made to define more closely this deficiency.

Archaeology and chronology

Chronology has always been one of the principal preoccupations of archaeology, and it is probably fair to say that the relation between archaeology and geophysics in this area has traditionally been a symbiotic one. For instance that pioneering work by Zeuner, *Dating the past* (Zeuner 1946), naturally drew upon a wide range of geophysical phenomena, most of them climatically dependent, and also relied to some extent upon calculations of solar radiation. But at the same time, he quite frequently found it convenient, especially when dealing with the palaeolithic period (i.e. the Pleistocene) to 'date' geological deposits (in a broad, relative sense) by the precise typological features of the industries of stone tools that some of them contained, which could be assigned a quite well-defined place within the established culture sequence in terms of their shape and of details of manufacture.

Even with the increasing dependence upon radioactive clocks, there are factors (such as the atmospheric concentration on ^{14}C) that are climatically dependent. The archaeological context has on occasion been of significance not only for providing materials of approximately appropriate antiquity (for instance of old wood for tree-ring work), but in a few cases for providing material that can be fairly well dated by historical means. For instance those dating techniques that depend upon variations in the direction and intensity of the Earth's magnetic field have generally needed first to establish what these chronological variations have been before using this information to date sites of uncertain chronology. The initial documentation of that chronological variation has usually relied upon the finding of suitable samples of baked clay in secure contexts which could be dated by traditional archaeological (i.e. historical) means.

It is worth recalling also that Willard Libby's first clear clue that his assumption of secular constancy for the atmospheric concentration of radiocarbon was questionable came from what was essentially an archaeological source: from Ancient Egypt. The first test of the radiocarbon dating method (Libby 1955) had come from a plot of the 'absolute specific radioactivity' of selected 'samples of known age' against their historical age in years. All of these samples were from archaeologically secure contexts in Egypt, dated by the historical chronology for Egypt, ultimately based upon the records of the Ancient Egyptians themselves of the successive reigns (and their duration) of the pharaohs. Later, a fuller plot of radiocarbon determinations for comparable material indicated that something was wrong (Libby 1963):

> This plot of the data suggests that the Egyptian historical dates beyond 4000 years ago may be somewhat too old, perhaps five centuries too old at 5000 years ago, with decrease to 0 at 4000 years ago. In this connection it is noteworthy that the earliest astronomical fix is at 4000 years ago, that all older dates have errors, and that these errors are more or less cumulative with time before 4000 years ago.

The observation was a sound one, but Libby's assumption that the Egyptology rather than the physics was at fault was not. As is known, subsequent research has in general vindicated the Ancient Egyptian historical dates, whereas the variation in the atmospheric concentration of radiocarbon makes necessary the calibration of radiocarbon dates. One of the great triumphs in the field of chronology of recent years has been the establishment of the bristlecone pine dendrochronology by the late Charles Wesley Ferguson, and the pioneering dating programme by Hans Suess, with the production of the first calibration curve, about which we have heard in the course of this conference. It is particularly gratifying that the Irish bog oak dendrochronological sequence and the accompanying radiocarbon dating programme in the Belfast laboratory (see M. G. L. Baillie, this Symposium), has served to corroborate and confirm this work, so that for the Holocene period radiocarbon dating, appropriately calibrated, is the principal dating technique available to the archaeologist.

The impact of this technique in archaeology has of course been immense. The most striking consequence is undoubtedly the emergence of a 'world archaeology', where the culture sequence of each continent can be dated without any need for a pre-existing historical chronology. In a sense the prehistory of Australia, for instance, continued to the eighteenth century A.D., but the arrival of the first humans there, at least 40 000 years ago, and subsequent activity there can be approached with the same confidence as in Europe or Western Asia.

A further major consequence of the 'second radiocarbon revolution', if we may so term the impact of the calibration of the radiocarbon timescale, has been the revision of many of our ideas about cultural relations between Europe and the Near East. To choose one example, where our French colleagues have been particularly active, the megalithic monuments of Western Europe were once thought to derive from Mediterranean ancestors in the earlier third millennium B.C., and thus themselves to begin around 2300 B.C. in such supposedly peripheral areas as Iberia, France, Britain and Denmark. Yet we now know that the earliest of these structures were built in Brittany, where the radiocarbon laboratory at Gif-sur-Yvette has yielded uncalibrated radiocarbon dates as early as 3800 B.C., leading to calibrated dates of as early as 4600 B.C. Monuments such as Barnenez or Ile Longue can now be recognized as predating the pyramids of Egypt by more than 1500 years. They are quite simply the earliest well-preserved built stone monuments that are known.

With the establishment of a globally valid dating system for the Holocene, which is of course essentially an achievement in the field of geophysics, albeit reliant also upon dendrochronology and using archaeological material in the earlier stages, it might be thought that the archaeologist would have no further contribution to make to questions of geochronology. But yet this is not quite true. For there is at least the hope that, for some time ranges, global marker events may one day be used to allow certain dates to be established within a single year.

The first such 'global event' to gain widespread attention is the paroxysmal eruption of the volcano of Santorini (Thera) in the Aegean Sea sometime around the sixteenth century B.C. The dating problem was first posed by the archaeologists. Spyridon Marinatos, the excavator of the splendid Bronze Age site of Akrotiri on Thera, put forward the suggestion that it was precisely the erupton on Santorini (which had destroyed and buried the settlement of Akrotiri) that was responsible for the destruction of the Minoan civilization of Crete in around 1450 B.C. (Marinatos 1939).

Subsequent work called this into doubt, because the pottery buried on Thera (in 'Late Minoan Ia' style) was typologically earlier than the pottery found in the destroyed palaces of Crete ('Late Minoan Ib' style). Radiocarbon dating was brought into play, but the dates showed a bewilderingly wide range, suggesting that local geochemical distorting influences were at work (Nelson *et al.* 1990).

The fascination of this dating question from the geophysical point of view, however, is that the fine tephra from the eruption may have had global climatic effects, as has been postulated both for ice-core sequences in Greenland (Hammer *et al.* 1987), and for the Californian bristlecone pine (La Marche & Hirschboeck 1984). These have been used to suggest the much earlier date of 1628 B.C. for the eruption of Thera, a proposal which finds some support also from the Irish tree-ring date (Baillie & Munro 1988). It would indeed be fascinating if these two dating methods, which follow solar-dependent annual counts, could be used to give a precise date, within a single calendar year, for this archaeologically well-dated event. It would then be applicable to other archaeological contexts, for instance in Rhodes, Melos and Crete (Renfrew 1978; Marketou 1990), where traces of Santorini tephra from the Minoan eruption have been identified. There is the prospect, therefore, of using these correlations between disciplines, to offer a chronological precision within a single year (instead of the precision restricted to several decades offered by direct radiocarbon determination), possibly applicable in several parts of the world.

Unfortunately, however, it is difficult to establish that it was precisely the Santorini eruption which was causing such severe effects in 1628 B.C. in both Greenland and California. Volcanos much nearer home may be suspected, and at present there seems no easy way of deciding. Yet this is a good example of the sort of chronological precision which may, in just a few favourable cases, be obtainable.

CLIMATE AND ARCHAEOLOGY

When we turn to the interrelations between the fields of climatology and archaeology, matters are very much less clear. Indeed if one were to make a criticism of the programme of the present conference, it would be that, in its concentration upon the geophysical and astronomical sciences, it has almost entirely omitted the essential links between global climate upon the one hand and human history on the other. This has been a geophysicists conference, into which the life sciences (with the fleeting exception of tree-ring studies) have not been allowed to intrude.

The contributions to archaeology that we have been discussing have been, almost without exception, to chronology. The broader consequences of solar variability have only been cursorily explored, notably in the very interesting paper by H. Faure & M. Lerous (this Symposium).

Yet it cannot be denied that the significance of climate for the course of human history has been enormous: the flora and fauna upon which human populations depend for their subsistence are governed by climatic factors. Yet these climatic factors, while no doubt ultimately dependent at least in part upon solar variability, are more effectively expressed in terms of other parameters: above all in rainfall. But the relevant measure may not be an aggregate annual one. Seasonal measures of rainfall are needed, and of direct solar radiation during the growing season, and of extent of winter frosts. Moreover, these measures are needed locally. There can be no doubt that a regional approach is needed, so that the local climatic history may be reconstructed, and then the local vegetational history inferred and confirmed.

For it is not only that 'an army marches on its stomach'. Cultures and civilizations perish or flourish largely upon the basis of their subsistence productivity. Whether we are discussing hunter-gather régimes, or farming economies, it is the effects of climate upon subsistence which interest us. All other climatic effects are secondary. So if we are interested in human history and prehistory – which is, I take it, the principal reason for inviting an archaeological participation in this meeting – it is imperative to give due attention to such fields as geomorphology and crop plant geography, and above all to entertain the contribution of precise regional studies. For what we, and earlier cultures, experience on the ground is not global climate but the relevant microclimate.

The casual reader of the programme for this conference might suppose that the great advances being made in the study of global climatic variability (which in reality means mainly global variability in net annual solar radiation) have yielded some advances in our understanding of the principal changes in Holocene human history. Nothing could be further from the truth. For although, mainly from the study of oceanic temperature variation in the Pleistocene (from deep-sea cores), we know much more about the background to the palaeolithic culture sequence, the much more subtle effects of climate in the Holocene are little understood.

The best example is perhaps offered by the phenomenon of system collapse, for which three controversial examples may be cited here: the decline of the Indus Valley civilization in what is now India and Pakistan in the early second millennium B.C.; the demise of the Mycenaean civilization of Greece around the twelfth century B.C., and the collapse of the Classic Maya civilization of Mesoamerica around A.D. 900. In each case archaeologists have offered a bewildering range of conflicting explanations: plague, invasion, earthquake, revolt and of course climatic reasons. What I would like to highlight here is that in every case famine due to drought, and famine due to flooding and increased rainfall have been simultaneously offered as explanations (see, for example, Rhys Carpenter (1966) for the Mycenaean civilization). Nothing could more effectively illustrate the primitive state of the present stage in climatic studies as applied to the early past that explanations based upon (*a*) major rainfall increase, (*b*) major rainfall decrease, and (*c*) no significant change in rainfall may still be offered for each of the places and times under discussion (see Adams 1973, p. 23).

To say this is not to belittle the contributions that have been made, especially in the field of vegetational studies, mainly by palynology but also using other vegetational or climatic indicators such as land molluscs. We now know a good deal about the environment, in terms of geomorphology, flora and fauna, of many ancient sites during their *floruit*. But this often tells us little about the dynamic aspects, about such issues as repeated crop failure over a number of successive years, which might have prompted radical social changes. Environmental archaeology has only rarely developed techniques that allow it to operate on a timescale of individual years. Yet it is on a timescale of months, years and at the very most decades, that climatic and subsistence changes are experienced.

The annual variations under discussion in this Symposium are thus potentially of the highest importance for archaeology. But, as I have tried to stress, they have to be understood in terms of their local climatic effects. Much more detailed modelling is needed to give us the links between fluctuation in solar radiation, seen as cause, and local subsistence effects (including climatically dependent crop failure) seen as effect.

This, perhaps essentially negative, point is the main one that I wish to stress in this Symposium. If global effects are to be applied to human affairs, they must be applied with some vision, some understanding, of the relevant scale of analysis. We have to begin to envisage crop production in a local region of just a few hundred square kilometres, on a year by year basis. We have to forecast (or retrodict) both long-term trends in production of the relevant plants (including plants used by grazing animals), and catastrophic crop failures, and above all repeated crop failures. Until the modelling techniques are available for these undertakings there is little that will allow us to link global solar variation with the archaeology on the ground.

<h3 style="text-align:center">THE CAUSES OF ENVIRONMENTAL CHANGE</h3>

By way of example, to illustrate the complexity of the task facing the archaeologist who seeks to interpret the cultural data utilizing evidence for climatic change, I should like to cite the important question of local environmental change on the north Mediterranean littoral during the Holocene period. The focus of interest is upon the period following the inception of farming, i.e. on the Neolithic, Bronze Age and Iron Age periods, to use a rather outmoded terminology. The archaeologist naturally seeks to reconstruct the environment of the communities under study at the time they were flourishing. But this task is greatly complicated by several factors, not least the changes in the landform caused by erosion and deposition, which have materially changed the agricultural potential of the catchment area for many archaeological sites.

One of the most sustained and detailed examinations of this issue has been carried out by Gilman & Thornes. They aptly remark, on one aspect of the issues facing them (Gilman & Thornes 1985, p. 48):

> Our view is that fluviatile responses to climatic change are still so poorly understood, even under present regimes, that attempts to assert past climates from localised sedimentary sequences are still too primitive and conflicting to provide a...secure base....

The converse is of course also true that a precise knowledge of past global climate would not at present allow any adequate evaluation of local geomorphological effects.

To offer a specific example, some years ago I was excavating a major prehistoric site, Phylakopi, on the north coast of the Cycladic island of Melos in the Aegean. The site flourished during the Bronze Age: in the late Bronze Age, from around 1600 B.C. it was fortified, with a central administrative building or mansion from which came evidence of literacy in the form of a fragment of a tablet in the Minoan Linear A script. There were fresco fragments also showing signs of influence of the Minoan civilization of Crete. From around 1400 B.C. the

[263]

resemblances were rather with the Mycenaean civilization of Mainland Greece, with a sanctuary and indications of cult practice comparable with other found on the Mainland. Around 1100 B.C. the site was abandoned, never to be reoccupied, like many in the islands. This was the time of the decline and end of the Mycenaean civilization.

Whether the decline and end of such sites had something to do with climatic change is an issue that has been much debated, but it was not one which we could hope to answer from this single site. However, we could hope to determine at what date the fine natural harbour to the south of the site became silted up. Earlier geomorphological studies in Greece of Holocene soil erosion and deposition have established two major phases, the more recent of them, termed the Younger Fill by Vita-Finzi (1969) within the Holocene. This major depositional phase has been studied by Bintliff in several areas. Vita-Finzi put forward evidence to suggest that the episode responsible for the Younger Fill deposits which he studied were to be dated to the late Roman and early mediaeval periods, from *ca.* 200 B.C. He initially suggested that deposition of the Younger Fill was synchronous throughout the Mediterranean, thus emphasizing a climatic cause.

Geomorphological work on Melos by Davidson & Tasker (1982) suggested other possibilities. In the first place a well was examined in what are now the field immediately to the south of the site of Phylakopi, where the harbour had been situated in Bronze Age times. The material in the section may be described as colluvium, formed by the deposition of material washed down from the surrounding hill slopes. The pottery from the lower levels indicated that the hillslope wash had been initiated after the inception of the late Bronze Age and before classical times.

Buried organic matter in a colluvial section in the upper Phylakopi valley yielded a radiocarbon date which after calibration could be set at 1050–1150 B.C. (Davidson & Tasker 1982, p. 88).

These observations suggest that the processes leading to the deposition of what might be termed the Younger Fill began in Melos very much earlier than Vita-Finzi had suggested, indeed before the abandonment of Phylakopi in *ca.* 1100 B.C. They raise the possibility that, in Melos at least, and perhaps in other areas, the process was an anthropogenic one, arising in the first instance through over exploitation and in particular probably over-grazing leading to a phase of increased erosion in the late Bronze Age. Whether this was a process which contributed to the strong economic and agricultural recession seen at the end of the Bronze Age cannot yet be established. For it could be argued that the converse was the case, and that the abandonment of many fields as a consequence of this recession, and the failure to maintain the terrace systems that may already have made many hill slopes cultivable, may have led to increased erosion.

Now the point of offering this modest investigation to your attention on this occasion is to illustrate how far we are from integrating into the discussion a clear and reliable view of the impact of climatic change upon human occupation of Melos or indeed the Mycenaean world in general. The pollen evidence for Greece (Turner 1978), which is at least as good as for other Mediterranean lands (although no suitable deposits have been found on Melos itself), does not give clear cut evidence of a climatic episode that could be associated with a widespread episode of erosion and deposition such as might account for the Younger Fill.

Was there such a climatic episode? Or is the Younger Fill in fact to be dated differently in different areas, the complex product of a relation between anthropogenic and climatic factors?

It is of course precisely here that the investigations under discussion at this meeting should come to our aid. We are discussing major and widespread geomorphological changes that might be synchronous and that might be climatically determined. But the intervening arguments are lacking that might take us from the emerging data for variations in solar radiation, experience on a global scale, and climatic fluctuations, mainly variations in rainfall, experienced in south Greece and the Aegean islands.

My suggestion is that comparable difficulties will be found for most of the major changes in human settlement pattern or in agricultural production that archaeologists have wished to investigate in different parts of the world. The archaeological problems are there, involving changes of major significance in human history. The global climatic data are evidently improving in quality in a striking way, as the papers at this meeting have impressively documented. But where are the intervening arguments, the reliable frameworks of inference, which will take us from known global variation in radiation or temperature to specific climatic effects in individual regions? Until these are developed and adduced we cannot be said to be undertaking interdisciplinary discussions, as we should be doing. We are merely speculating, as archaeologists and geophysicists have done for more than a century.

References

Adams, R. E. W. 1973 The collapse of Maya civilisation, a review of previous theories. In *The Classic Maya collapse* (ed. T. P. Culbert), pp. 21–34. Albuquerque: University of New Mexico Press.

Baillie, M. G. L. & Munro, M. A. R. 1988 Irish tree-rings, Santorini and volcanic dust veils. *Nature, Lond.* **307**, 344–346.

Carpenter, R. 1966 *Discontinuities in Greek civilisation.* Cambridge University Press.

Davidson, D. & Tasker, C. 1982 Geomorphological evolution during the late Holocene. In *An island polity, the archaeology of exploitation in Melos* (ed. C. Renfrew & M. Wagstaff). Cambridge University Press.

Gilman, A. & Thornes, J. B. 1985 *Land-use and prehistory in south-east Spain.* London: George Allen and Unwin.

Hammer, C. U., Clausen, H. B., Friedrich, W. L. & Tauber, H. 1987 The Minoan eruption of Santorini in Greece dated to 1645 BC? *Nature, Lond.* **328**, 517–519.

La Marche, V. C. & Hirschboeck, K. K. 1984 Frost in trees as records of major volcanic eruptions. *Nature, Lond.* **307**, 121–126.

Libby, W. F. 1955 *Radiocarbon dating.* Chicago University Press.

Libby, W. F. 1963 The accuracy of radiocarbon dates. *Science, Wash.* **140**, 278–280.

Marinatos, S. 1939 The volcanic destruction of Minoan Crete. *Antiquity* **13**, 425–439.

Marketou, T. 1990 Santorini tephra from Rhodes and Kos; some chronological remarks based upon the stratigraphic evidence.

Nelson, D. E., Vogel, J. S. & Southon, J. R. 1990 More radiocarbon data for the last occupation of Akrotiri.

Renfrew, A. C. 1978 Phylakopi and the Late Bronze I period in the Cyclades. In *Thera and the Aegean World* I (ed. C. Doumas), pp. 403–422. London: Thera and the Aegean World.

Turner, J. 1978 The vegetation of Greece during prehistoric times: the palynological evidence. In *Thera and the Aegean World* I (ed. C. Doumas), pp. 765–773. London: Thera and the Aegean World.

Vita-Finzi, C. 1969 *The Mediterranean valleys: geological changes in historical times.* Cambridge University Press.

Zeuner, F. E. 1946 *Dating the past.* London: Methuen.

Phil. Trans. R. Soc. Lond A, **330**, 665–670 (1990)

Printed in Great Britain

Human biogeography and climate change in Siberia and Arctic North America in the fourth and fifth millennia BP

By W. R. POWERS AND R. H. JORDAN

Department of Anthropology, University of Alaska Fairbanks, Fairbanks, Alaska 99775, *U.S.A.*

This paper explores the relation between the geographic shifts in prehistoric hunting populations and changes in climate between 4500 and 3000 before present (BP) within the polar regions from the Yenisei River in Siberia to Greenland. We have chosen this time period because major human geographic changes occurred over much of northeastern Asia and northern North America, and because these changes appear to be linked, at least in part, to a palaeoclimatic fluctuations. The cultures under consideration have been termed the Early and Middle Neolithic (Syalakh and Bel'kachi) in Siberia and the Arctic Small Tool Tradition (with such local variants as Denbigh, Independence I, Pre-Dorset, and Sarqaq) in North America. Despite these terminological differences, these groups shared such a close similarity in their technology and adaptive patterns that they must have once shared a direct historical relation.

INTRODUCTION

The assumption underlying this paper is that there are linkages between palaeoclimatic fluctuations and such changes in the natural environment as sea-ice conditions, snowfall régimes and terrestrial spring icing conditions that directly affected the distribution and abundance of critical mammalian resources. These, in turn, affected the distribution and abundance of ancient hunting groups especially in the most extreme northern latitudes. On one hand, the northernmost distribution of all mammals, including humans, should be very sensitive to climatic and environmental fluctuations. On the other, humans adapt primarily through learned, patterned, symbolically transmitted informational and behavioural systems that anthropologists term culture. As such, the ability of humans to adapt to changing natural conditions is much enhanced, and more rapid and flexible than other animal species. Interspecifically, humans are probably the least sensitive species to climatic and environmental fluctuations, particularly if there are exploitable resource alternatives. Intraspecifically, however, human populations inhabiting the northernmost margins of the globe should, and probably were, among the most susceptible to such changes.

PALAEOCLIMATIC CONDITIONS

Palaeoclimatic data are best derived from a number of North American sources. Oxygen isotope data from Camp Century and Dye 3 in Greenland and the Devon Ice Sheet in Canada all indicate a mid-Holocene warming that peaked about 5000 before present (BP)† (Dansgaard *et al.* 1969; Dansgaard *et al.* 1982; Paterson *et al.* 1977). Average temperature

---

† All dates in this paper are presented in calendric time to provide a consistent and comparable chronolological framework. Radiocarbon dates have been recalibrated according to Pearson *et al.* (1986).

conditions declined thereafter but must have been above present average conditions through the first half of the fifth millennium BP. For example archaeological sites from north Greenland, dating between 4000 and 4500 BP, contain driftwood and seal bones, both unavailable regionally under current conditions given permanently frozen seas. Contemporaneous sites from Devon Island, High Arctic Canada, have also produced faunal evidence for warmer conditions (McGhee 1979, p. 37).

Data from the fourth millennium BP, however, register a period of increasingly colder conditions. Evidence for these conditions is registered by the oxygen isotope ratios in the Greenland and Canadian ice cores, and a 300 km southerly shift of the forest–tundra border west of Hudson Bay (Sorenson *et al.* 1971; Sorenson & Knox 1973). They are further corroborated by tree-ring data from western North America (La Marche 1974), palynological data from the Labrador–Ungava Peninsula (Short 1978) and Keewatin (Nichols 1975 with references), and global expansions of glaciers after 3300 BP (Denton & Karlen 1973).

HUMAN BIOGEOGRAPHY

About 5500 BP the Middle Neolithic Bel'kachi culture developed from the Early Neolithic Syalakh culture in the Lena Basin (Mochanov 1969). It spread westward to the Taimyr Peninsula and eastward to the Bering Sea. In general, the Bel'kachi culture is very similar to the preceding Syalakh culture. Camps were situated in the areas where the maximum return in game resources could be obtained: near the confluences of smaller and larger streams and on the shores of lakes with rich fish and migratory waterfowl resources. People lived in round skin tents warmed by small interior hearths. Sites on higher terraces also have bark-lined storage pits in which the bones of moose and large fish have been recovered (Mochanov 1969).

The Bel'kachi culture is characterized by the following: small bifacial triangular points; bifacial diagonally retouched willow-leaf points; microblades and cores; chipped burins and spalls; flaked and polished adzes; small, bifacial chisels; unifacial and bifacial, single- and double-ended endscrapers; concave sidescrapers; flake knives and scrapers; multifaceted burin-drills; bifacial knives; microblade insets with one longitudinal worked edge; perforators or gravers; and net sinkers. Bone tools comprise polished tips, awls, needles, and slotted arrows. The pottery is cord marked (Mochanov 1967, pp. 167–172, 1969, pp. 235–247)†.

About 4500 BP an aceramic variant of Bel'kachi spread east across north Alaska and entered the Canadian High Arctic and Greenland, a completely uninhabited area (figure 1). The strong similarities in the lithic technology and typology of the tool kits are grounds for calling the entire cultural phenomenon, from the Taimyr to Greenland, the Arctic Small Tool Tradition. We are, in effect, subsuming the Siberian data into a long-recognized North American tradition. In addition, we see the North American Arctic Small Tool Tradition originating in northeast Siberia as a complex, which was derived from Bel'kachi (cf. McGhee 1976). The ultimate origins of this development lie even deeper in the preceding Neolithic (Syalakh) and pre-Neolithic (Sumnagin) cultures of the Siberian North.

The only minor differences among early assemblages are the presence of pottery, and the distinctive multifaceted burin-drills in Bel'kachi that have not been reported in North America. An emphasis on delicately chipped, bifacial insets is distinctly Alaskan, whereas Eastern Arctic

† Russian and American terminology differs considerably for some tools of identical form and probable function. We have chosen the American terminology in this paper.

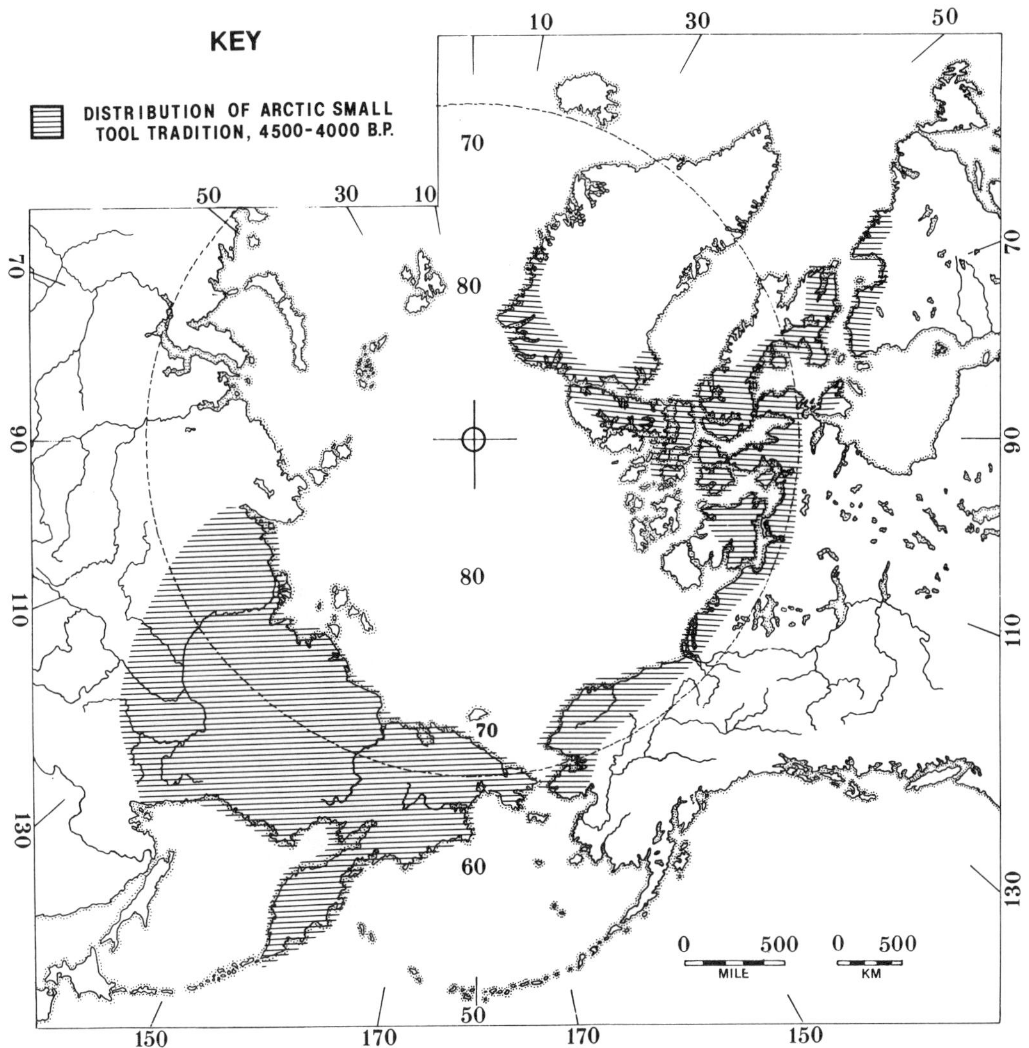

FIGURE 1. Distribution of Arctic Small Tool Tradition, 4500–4000 BP.

assemblages evidence a stronger emphasis on the stemming of bifaces. Yet the similarities far outweigh these differences. This has lead us to the conclusion that this tradition could not have evolved independently on both sides of the Bering Strait from such different antecedents as Syalakh and American Palaeoarctic.

In North America, the earliest manifestations are found in north Alaska and the High Arctic regions of Canada and Greenland. Soon after, they appear in such comparatively southern regions as Labrador and south Baffin (Anderson 1988; Cox 1978; Knuth 1967; Maxwell 1973, 1985; McGhee 1979). The nearly synchronous dates from the Eastern Arctic with those in Alaska and Siberia suggest an extremely rapid population spread. In the Eastern Arctic, where organic preservation is better, the data indicate that the subsistence economy was based on such terrestrial mammals as caribou and musk-oxen. Marine mammal hunting, particularly for ringed and bearded seals, also played an important role (Knuth 1967; McGhee 1979). Evidence for fishing has been recovered from early sites in Alaska and Greenland (Anderson 1988; Knuth 1967). Despite this broad subsistence base, sites generally consist of thin deposits and a variety of tent structures and occasional cache pits, indicating a seasonally nomadic existence.

The termination of the mid-Holocene warming had severe effects on the distribution of this tradition, particularly in northern Canada and Greenland where a southern extension of permanent pack ice must have occurred. Both these regions were abandoned during the fourth millennium BP (figure 2). Populations expanded south along the shores of East and West

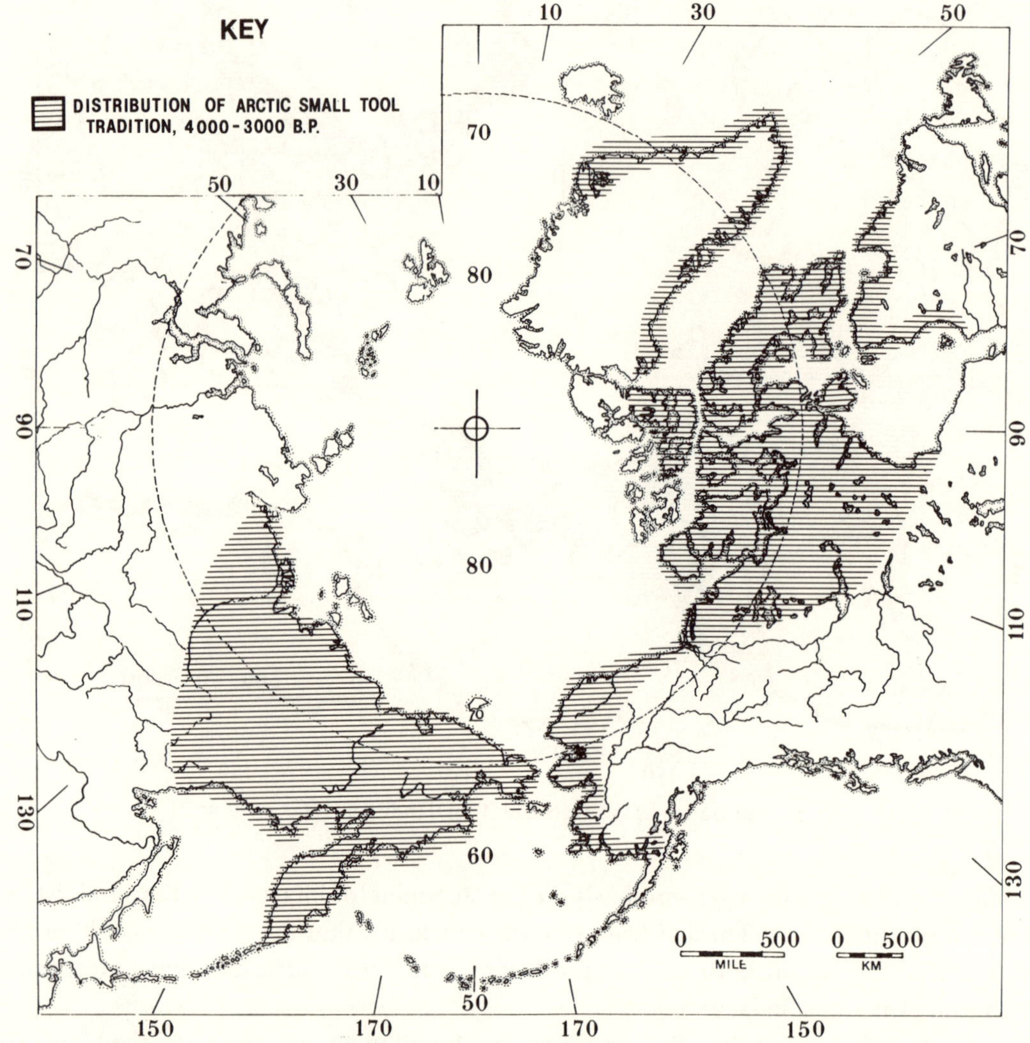

FIGURE 2. Distribution of Arctic Small Tool Tradition, 4000–3000 BP.

Greenland, along both shores of Hudson Bay and the broad expanses of the Barren Grounds as far south as Churchill, Manitoba and Lake Athabasca in northern Saskatchewan (Gordon 1975; Maxwell 1985; Moberg 1986; Nash 1969; Plumet 1976). Another southern geographic expansion as far south as the Alaska Peninsula apparently occurred during between 3500 and 3000 BP as well (cf. Dumond 1981; Giddings 1964).

Discussion

About 4500 BP the Arctic Small Tool Tradition spread into Alaska and the Eastern Arctic from Siberia. Although this was a period of generally warmer climatic conditions, this major population expansion cannot be correlated with changing climatic conditions. The archaeological data suggest that the most significant alterations may have been in the subsistence economy. The addition of seasonal marine mammal hunting in Arctic waters to an existing base of terrestrial hunting and fishing seems to have been central to this adaptive success. Geographic shifts in the fourth millennium BP, however, seem to be clearly linked to climatic cooling as the Canadian and Greenlandic High Arctic was abandoned and populations expanded their ranges south.

In addition to the reconstruction of human history, this paper suggests that northern archaeology has important applications to the historically oriented branches of the physical and biological sciences. For example, the changing geographic distributions of prehistoric hunting cultures can be used, albeit with caution, as proxy palaeoclimatic data if economic adaptations can be accurately evaluated. Moreover, well-preserved archaeological sites, commonly found throughout the north, are repositories of large quantities of environmental data that can provide new insights into changes in the Earth's climate and ecology.

We thank Ted Goebel for his assistance in drawing the figures. We also appreciate the travel support provided by Anne Shinkwin, Dean of the College of Liberal Arts, University of Alaska Fairbanks, which allowed us to attend the Royal Society meeting in London.

References

Anderson, D. D. 1988 Onion portage: the archaeology of a stratified site from the Kobuk River, Northwest Alaska. *Anthrop. Pap. Univ. Alaska* **22**, nos 1–2.

Cox, S. L. 1978 Palaeo-Eskimo occupations of the North Labrador Coast. *Arct. Anthrop.* **15**(2), 96–118.

Dansgaard, W., Clausen, H. B., Gundestrup, N., Hammer, C. U., Johnson, S. F., Kristinsdottir, P. M. & Reeh, N. 1982 A New Greenland deep ice core. *Science, Wash.* **218**, 1273–1277.

Dansgaard, W., Johnsen, S. J. & Møller, J. 1969 One thousand centuries of climatic record from camp century on the Greenland Ice Sheet. *Science, Wash.* **166**, 377–381.

Denton, G. H. & Karlen, W. 1973 Holocene climatic variations – their pattern and possible cause. *J. Quat. Res.* **3**(2), 155–205.

Dumond, D. E. 1981 Archaeology on the Alaska Peninsula: the Naknek Region, 1960–1975. *Univ. Oregon Anthrop. Pap. no. 21*. Eugene.

Giddings, J. L. 1964 *Archaeology of Cape Denbigh*. Providence, Rhode Island: Brown University Press.

Gordon, B. H. C. 1975 Of men and herds and Barrenland prehistory. *National Museum of Man, Mercury Ser. no. 28*. Ottawa: Archaeological Survey of Canada.

Knuth, E. 1967 Archaeology of the Musk-Ox Way. *École Practique des Hautes Études. Contributions du Centre d'Études Arctiques et Finno Scandinaves*, no. 5, Paris.

La Marche, V. Jr. 1974 Paleoclimatic inferences from long tree-ring records. *Science, Wash.* **183**, 1043–1048.

Maxwell, M. S. 1973 Archaeology of the Lake Harbour District, Baffin Island. *National Museum of Man, Mercury Ser. no. 6*. Ottawa: Archaeological Survey of Canada.

Maxwell, M. S. 1985 *Prehistory of the Eastern Arctic*. Orlando: Academic Press.

McGhee, R. 1976 Parsimony isn't everything: an alternative view of Eskaleutian linguistics and prehistory. *Can. archaeol. Ass. Bull.* **8**, 62–81.

McGhee, R. 1979 The paleoeskimo occupation at Port Refuge, High Arctic Canada. *National Museum of Man, Mercury Ser. no. 92*. Ottawa: Archaeological Survey of Canada.

Moberg, T. 1986 A contribution to paleoeskimo archaeology in Greenland. *Arct. Anthrop.* **23**, 19–56.

Mochanov, I. A. 1967 The Bel'kachi Neolithic Culture on the Aldan. *Soviet Archaeology* **1967**(4), 164–177. Moscow: Nauka. (In Russian.)

Mochanov, I. A. 1969 *The stratified Bel'kachi I site and the periodization of the Stone Age of Yakutia.* Moscow: Nauka. (In Russian.)

Nash, R. J. 1969 The Arctic Small Tool Tradition in Manitoba. *Occ. Pap. Dept. Anthropology, no. 2.* Winnipeg: University of Manitoba.

Nichols, H. 1975 Palynological and paleoclimatic study of the late quaternary displacement of the Boreal Forest-Tundra Ecotone in Keewatin and MacKenzie, N. W. T., Canada. *Occ. Pap. no. 15.* Boulder: Institute of Arctic and Alpine Research, University of Colorado.

Paterson, W. S. B., Koerner, R. M., Fisher, D., Johnsen, S. J., Clausen, H. B., Dansgaard, W., Bucher, P. & Oescher, H. 1977 An oxygen-isotope climatic record from the Devon Island Ice Cap, Arctic Canada. *Nature, Lond.* **206**, 508–511.

Pearson, G. W., Pilcher, J. R., Baillie, M. G. L., Corbett, D. M. & Qua, F. 1986 High-precision ^{14}C measurements of Irish oaks to show the natural ^{14}C variations from A.D. 1840–5210 B.C. *Radiocarbon* **28** (2B), 911–934.

Plumet, P. 1976 Archeologie du Nouveau-Quebec: habitats paleo-Esquimaux à Poste-de-la-Baleine. *Paleo-Quebec* no. 7. Laval: Le Centre D'Études Nordiques de L'université Laval, Quebec.

Short, S. K. 1978 Palynology: a Holocene environmental perspective for archaeology in Labrador-Ungava. *Act. Anthrop.* **15**(2), 9–35.

Sorenson, C. J. & Knox, J. C. 1973 Paleosols and paleoclimates related to late Holocene forest/tundra border migrations: MacKenzie and Keewatin, N. W. T., Canada. *Int. Conf. Prehistory and Paleoecology of Western North American Arctic and Subarctic,* pp. 187–203.

Sorenson, C. J., Knox, J. C., Larsen, J. A. & Bryson, R. A. 1971 Paleosols and the forest border in Keewatin, N. W. T. *Quat. Res.* **1**, 468–473.

Phil. Trans. R. Soc. Lond. A **330**, 671–677 (1990)

Printed in Great Britain

671

The bearing of phyto-archaeological evidence on discussions of climatic change over recent millennia

By J.-L. Vernet

*Laboratoire de Paléobotanique, Université des Sciences et Techniques du Languedoc,
Place Eugène Bataillon, 34060 Montpellier, France*

Phyto-archaeological data based on macro-remains studies, especially prehistoric charcoal, provide evidence concerning the changes in western Mediterranean vegetation during the last millennia. Comparisons are based on present vegetation levels as defined by Ozenda. From the last glacial period to the present time, differences between warm and cool vegetation were of about 8 °C in the south of France but less extreme in more southern regions. The late Pleistocene and early Holocene (Late Palaeolithic and part of Mesolithic) were a period of transition with pines and junipers. Then, the late Mesolithic and the early Neolithic are typically periods of good forestation. During the Neolithic period deciduous and holm oaks had a role of varying importance in all the present Mediterranean levels (thermo-meso- and supramediterranean). Man's influence on the vegetation became significant in the middle Neolithic (south of France) or earlier (south of Spain) and may be characterized by plants such as *Buxus sempervirens*, *Quercus ilex*, *Pinus halepensis* and heaths. The Chalcolithic, the Roman period and the Middle Ages are also periods during which Man's influence was important.

Introduction

Just like palynology, macro plant remains allow us to take an ecochronological approach to ancient Mediterranean floras and vegetations. This paper is especially concerned with prehistoric charcoal analysis. During the past 15 years this methodology has given many answers concerning Man's impact on vegetation. Prehistoric charcoal analysis is the study of wood collected by Man. Combustion generally facilitates the long-term conservation of wood. These processes take place in the Mediterranean area because woody vegetation components are abundant and grow either in forests or in various non-sylvatic formations. On the other hand, a large spectrum of taxa, up to 30 or more, is witnessed. This observation and others prove that wood was not selected by Man. Charcoal may be considered more as random samples than as selective collections. However, in for example fire places, there is a low number of taxa because of the short time they are used. Thus charcoal analysis may be considered as a complement of palynology by taking into account ecological and ethnological components of Man's environment. Charcoal analysis may be used to study the past 20 millennia including the development of modern man and the transition from hunting-gathering to agriculture. This time period concerns particularly the last glacial maximum and the post-glacial warming.

Climatic influences on vegetation

Charcoal analysis provides evidence that between 20000 and 11000 years before present (BP) the periglacial vegetation under the latitudes of southern France was very closely related to the present mountain vegetation (Vernet 1973, 1980*a*; Bazile-Robert 1981). The typical taxa were pines, *Pinus sylvestris*, birch, *Betula verrucosa*, and also *Hippophaë rhamnoides* and *Sambucus racemosa*. During periods characterized by a warmer temperature (dated from 20000, 16000, 13000 BP (Bazile *et al.* 1986)) thermophilous taxa expanded from their refuges. A particular refuge, on the northeastern Pyrenean slope during a cool period of the late glacial, the upper early Dryas dated from about 12000 years BP (Jalut *et al.* 1975) is especially emphasized. During the warm periods vegetation was quite similar to the present mesomediterranean or supramediterranean levels. Temperature variations would be estimated to be 6–8 °C (Vernet 1986*a*) according to the data (Sabatier & Van Campo 1984; CLIMAP 1976). Vegetation of the lower Mediterranean levels were replaced by forest-steppe or preforest communities mainly composed of the pines *Pinus sylvestris* (figure 1).

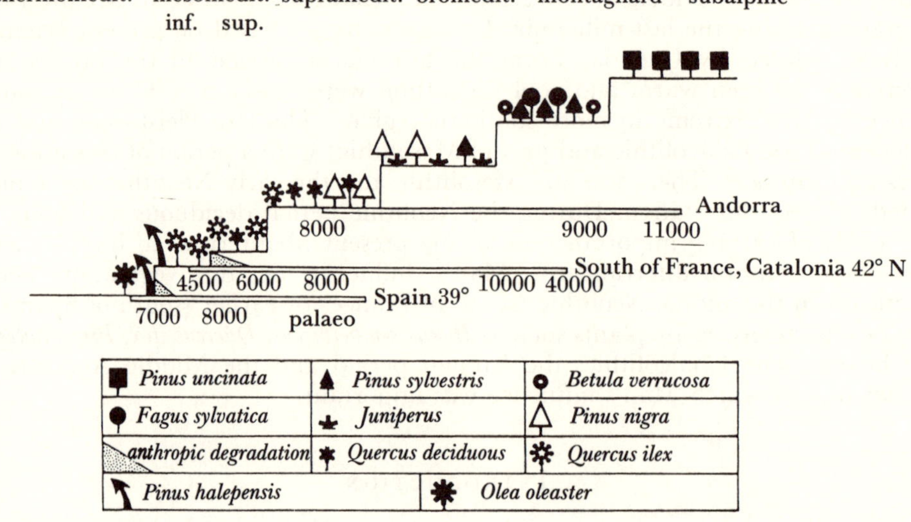

Pinus uncinata	Pinus sylvestris	Betula verrucosa
Fagus sylvatica	Juniperus	Pinus nigra
anthropic degradation	Quercus deciduous	Quercus ilex
Pinus halepensis	Olea oleaster	

FIGURE 1. Relations between climate and prehistoric vegetation based on present vegetations stages: the double line for Andorra, south of France and Spain, record the change of vegetation and climate from late Palaeolithic to post-Neolithic time (Vernet 1986*b*).

Another important climatic aspect may be deduced from birch ecology. The existence of this species (*Betula* cf. *verrucosa*) at lower levels during the last glacial up to the late glacial indicates moist summers. However, it is difficult to know if summer was the most rainy season. In this case, the climate cannot be assumed to have been Mediterranean. However, from 15000 to 10000 BP (late glacial) thermophilous periods were more frequent. This first transition period is correlated with a progresssive incoming of dry summers. A second transition phase took place between 10000 and 8000 BP with the occurrence of pines, mainly *Pinus sylvestris* and junipers, probably *Juniperus phoenicea*, *J. oxycedrus* and *J. communis*. This phase preceded the establishment of deciduous oak forest (late Mesolithic and early Neolithic) and may be related to oro- or supramediterranean vegetations levels.

This information raises some interesting points.

(i) The decrease and disappearance of birch during the late glacial from lower Mediterranean areas may be correlated with its extension in the regions under oceanic climate since 13000 BP (Van Campo 1984) as a result of the increase in temperature and moisture (change from a continental climate to an oceanic one in non-Mediterranean area). In other words, the late glacial may be regarded as a time of divergence between the Mediterranean and Atlantic bioclimatic history.

(ii) The oromediterranean *Juniperus* vegetation level has been recognized and exploited by man between 12000 and 8000 BP; the palynological optimum of this plant being generally dated from 13000 to 14000 BP in the south of France (Triat-Laval 1979; Jalut *et al.* 1982). Likewise, the beginning of deciduous oak vegetation (8000 BP) based on charcoal studies is correlated with the palynological dominance of this vegetation.

(iii) Finally, the transition from forest-steppes to deciduous oak forests (the early Holocene post-glacial 'climax') is the last major phytoclimatic division at the 42–44° N latitude. Later, the upper Holocene vegetations record only human influence.

HUMAN IMPACTS ON VEGETATION

Latitudes 43–44° N

Human action on the vegetation is recorded during middle Neolithic with the Chasséens (6000 BP, Vernet & Thiébault 1987). Vegetation opening is characterized by typical indicators such as *Buxus sempervirens* and an increase of *Quercus ilex*. Archaeological data from Dourgne in the Aude basin (Vernet 1980*a*; Guilaine *et al.* 1987) show us that hunting was losing importance in the prehistoric economy, and was being replaced by the breeding of sheep and goats. Later on, especially between 4500 BP and Roman times, breeding and human action on vegetation underwent different degrees of development as a result of Man's interference. Based on these observations we may consider that present 'garrigues' did not appear before the late Neolithic (*ca.* 4500 BP). During the Roman period and the Middle Ages an important change took place. During the first period olive trees and vineyards have taken a decisive role just as at present time, for instance in the Aude plains next to Narbonne, France. During the same period *Pinus halepensis* was increasing with the enlargement of deforested areas. Both beech and fir charcoal have been found in early Middle Ages excavations in alluvial plains. Accurate palynological evidence (Planchais 1982) has recorded beech disappearance after 1300 BP in the Languedoc plain of southern France. Therefore, riverside forests would have been cleared during the Middle Ages, between the tenth and the end of the twelfth centuries. This hypothesis is supported by the history of this period when new soils were constituted along the rivers (Beziers area). These land valorization was facilitated by the diffusion of the 'moulin à paissière' (a type of water-mill). This technique allows systematic irrigation especially in cultivated areas and natural meadows. We must also take into account the development in the making of tools used to cultivate the heavy soils, totally or partly unexploited until this moment.

In Languedoc, the 'silva' was definitely fixed in its major components between the tenth and thirteenth centuries. At this time, last glacial relicts disappeared from plains and a new soil organization was initiated (Durand & Vernet 1987).

Mediterranean area north boundary

The southern thermophilous components such as *Quercus ilex*, are less recorded or absent. At Choranche-Coufin an oak phase is recorded during Mesolithic and early Neolithic. During late Neolithic and posterior times the human activity is mainly revealed by the exploitation of ash. Charcoal ash maximum may be correlated with the occurrence of Bovidae bones (Thiébault

[275]

1983). Foliage exploitation for domestic animals known nowadays in this area had its origin in the Mesolithic. Another feature is the importance of *Taxus baccata* between 5000 and 2000 BP. This species has also been important between 6500 and 5000 BP, as shown by pollen diagrams (Clerc 1985). On the other hand, the data from the Sarrasins cave (Thiébault 1983) reveal the disappearance of *Taxus* and forest exploitation during about 1000 years. The exploitation would be realized in four stages:

(*a*) oak stage with *Fraxinus*, *Taxus*, *Abies*, *Acer* (late Neolithic);

(*b*) ash stage with the decrease of *Taxus*, the disappearance of fir, and the increasing of beech along with *Cytisus laburnum* (Chalcolithic, early Bronze Age);

(*c*) hazel stage with an increase of oak, and the disappearance of yew (middle and late Bronze Age);

(*d*) finally, a new oak stage with *Corylus* and *Fraxinus* is recorded (late Bronze and Iron Ages).

These charcoal stages would reflect a major exploitation from late Neolithic to early Bronze Age followed by a reforestation illustrated by hazel progression during middle and late Bronze Age until the forest's regeneration.

Thermomediteranean level

The northern and central regions of Catalonia in Spain are an area of transition between the charcoal model known for the south of France and more southern arid areas. In the Cova del Frare cave (Ros-Mora 1985; Ros-Mora & Vernet 1987) a post-glacial succession quite similar to the one for the Languedoc has been identified.

On this site a deciduous oak stage (early and late Neolithic) is followed by a *Quercus ilex* stage (Chalcolithic, Bronze Age). Pines (*Pinus halepensis*) were not recorded before early Bronze Age, whereas *Taxus baccata* has a relative importance during Neolithic. This may be related to the particular bioclimatic features of this Iberic Peninsula area.

The present thermomediterranean conditions of the region of Valencia show new sequences. In Cova de l'Or we have recorded an early Neolithic phase with *Quercus ilex* and scarce *Quercus faginea* (Vernet *et al.* 1983, 1987). These results are similar to the displacement observed in the last glacial. Climatic vegetation here is based on *Quercus ilex* formations and not on deciduous oaks, as in northern areas. Therefore, this scheme is more mesomediterranean than supramediterranean. Our conclusions are supported by pollen analysis from Padul at the foot of Sierra Nevada (Pons & Reille 1986). For these authors, the early Holocene climatic vegetation is a 'formation thermophile à *Quercus ilex* et *Pistacia*'. Moreover, wild olive (*Olea europaea* var. *sylvestris*) found in charcoal samples played an important role particularly from about 7500–8000 BP to the present time. In Padul, the beginning of the *Olea* continuous pollinic curve has been dated to 7840 BP (Pons & Reille 1986). Charcoal analysis in Cova de l'Or recorded an increase of olive and the begining of a *Pinus halepensis* curve at about 6800 BP. In Padul the increasing importance of *Pinus* pollens increasing may be related to this scheme. This history can be explained by the increasing human influence.

Pollen analysis and geomorphological studies in cave deposits agree with our results (Fumanal-Garcia & Dupré-Ollivier 1986) particularly with data upon vegetation's degradation. Our results also demonstrate the great importance of *Pinus halepensis* and *Olea europaea* var. *sylvestris* in the middle Neolithic. This evidence was corroborated by the Recambra charcoal diagram (Grau-Almero 1984). Here there is evidence of an early, middle and late Neolithic *Pinus halepensis* charcoal stage with *Olea europaea* var. *sylvestris*, scarce *Quercus faginea* and *Pinus nigra*. However, with the study from Recambra the post-Neolithic evolution is acknowledged. During the Chalcolithic and the Bronze Age the *Pinus halepensis* charcoal remains become more scarce and *Quercus faginea* and *Pinus nigra* disappear. These species were replaced by plants from *Rosmarino-Ericion* (for instance heaths as *Erica multiflora*, Leguminosae, *Anthyllis cytisoides*, etc.) and *Quercus coccifera*. This history, which culminates in the present vegetation, represents a regressive evolution in contrast to that of the northern boundary, where a cyclic evolution has been recorded.

In thermophilous area in the south of France, particularly in Provence, a progression of *Pinus halepensis* during the early Neolithic has been identified (Vernet 1971, 1980*b*). This evolution is similar to the thermomediterranean model, but may differ in the scarcity or absence of *Quercus ilex* and *Olea sylvestris*. In this area *Quercus pubescens* may be present and abundant both in coast regions and inland.

CONCLUSIONS

Charcoal analysis provides evidence of major climatic influences in the vegetation transformation from Palaeolithic to the early Neolithic (table 1). There was probably a decrease of glacial influences with latitude. From the Neolithic on, Man's influence dominates

TABLE 1. MAJOR EVENTS IN THE PREHISTORIC LATE GLACIAL AND POST-GLACIAL SOUTH OF FRANCE VEGETATION

(Note the climatic determinism before 8000 BP and the increasing of human influence during the past six millennia.)

year (BP)	Age			the first ruptures
2000 3000 4000	Iron Bronze Chalcolithic	}	4	growth of anthropization: regression of deciduous forests,
5000	Neolithic (final) (middle)	}	3	deciduous forests are cleared extension of green oak and box tree
				adaptations
6000 7000	(ancient) Mesolithic (final)	}	2	exploitation of deciduous forests no equilibrium rupture growth of deciduous forests
8000 9000	(middle)	}	1 b	Juniper phase
10000 11000 12000	Epipalaeolithic	}	1 a	Scots pine phase disappearance of birch in the Mediterranean region

Mediterranean vegetation changes. Middle Neolithic populations were the first responsible for vegetation degradation. Their influence was continued by the populations of the Chalcolithic, Bronze and Iron Ages, Roman and Middle Ages. Lastly, forest potentialities of the Mediterranean area are certainly important in mesomediterranean levels under subhumid climate and in the surrounding mountains. This does not apply to the thermomediterranean semi-arid and arid climates, where the degree of degradation is deeper.

REFERENCES

Bazile, F., Bazile-Robert, E., Debard, E. & Guillerault, P. 1986 Le Pléistocène terminal et l'Holocène en Languedoc rhôdanien; domaines continental, littoral et marin. *Revue Géologie dynam. Géographie phys.* **27**(2), 95–103.

Bazile-Robert, E. 1981 Flore et végétation des gorges du Gardon à la moyenne vallée de l'Hérault, de 40000 à 9500 BP d'après l'anthracoanalyse. *Paléobiologie continentale* **12**(1), 79–90.

Clerc, J. 1985 Première contribution à l'étude de la végétation tardiglaciaire et holocène du piémont dauphinois. *Documents Cartographie écologique* **28**, 65–83.

CLIMAP 1976 The surface of the ice-age earth. *Science, Wash.* **191**, 1131–1137.

Durand, A. & Vernet, J.-L. 1987 Anthracologie et paysages forestiers médiévaux, à propos de quatre sites languedociens. *Annls Midi* **99**, 397–405.

Fumanal-Garcia, M. P. & Dupré-Ollivier, M. 1986 Aportaciones de la sedimentologia y de la palinologia al conocimiento del paleoambiente valenciano durante el Holoceno. *Proc. symp. 'Quaternary climate in western Mediterranean', Madrid*, pp. 325–343.

Grau-Almero, E. 1984 *El hombre y la vegetacion del Neolitico a la edad del Bronce valenciano en la Safor, provincia de Valencia, segun el analisis antracologico de la Cova de la Recambra.* Tesis licenciatura, Facultad de Geografia e Historia, Valencia. (130 pages.)

Guilaine, J., Barbaza, M., Gasco, J., Geddès, D., Jalut, G., Vaquer, J. & Vernet, J.-L. 1987 L'abri du Roc de Dourgne, écologie des cultures du Mésolithique et du Néolithique ancien dans une vallée montagnarde des Pyrénées de l'Est. *Actes Colloque int. 'premières communautés paysannes en Méditerranée occidentals'.* CNRS, pp. 545–554.

Heinz, C. 1988 *Dynamique des végétations holocènes en Méditerranée nord occidentale d'après l'anthracoanalyse de sites préhistoriques: méthodologie et paléoécologie.* Thèse Université des Sciences et Techniques du Languedoc, Montpellier.

Jalut, G., Delibrias, G., Dagnac, J., Mardones, M. & Bouhours, M. 1982 A palaeoecological approach to the last 21000 years in the Pyrenees, the peat bog of Freychinède (alt. 1350 m, Ariège, south France). *Palaeogeogr. Palaeoclimat. Palaeoecol.* **40**, 321–359.

Jalut, G., Sacchi, D. & Vernet, J.-L. 1975 Mise en évidence d'un refuge tardiglaciaire à moyenne altitude sur le versant nord-oriental des Pyrénées. *C.r. hebd. Séanc. Acad. Sci., Paris* D **280**, 1781–1784.

Ozenda, P. 1975 Sur les étages de végétation dans les montagnes du bassin méditerranéen. *Documents cartographie écologique* **16**, 1–32.

Planchais, N. 1982 Palynologie lagunaire de l'étang de Mauguio, paléoenvironnement végétal et évolution anthropique. *Pollen Spores* **24**(1), 93–118.

Pons, A. & Reille, M. 1986 Nouvelles recherches pollenanalytiques à Padul (Granada), la fin du dernier glaciaire et l'Holocène. *Proc. symp. 'Quaternary climate in western Mediterranean', Madrid*, pp. 405–420.

Ros-Mora, M. T. 1985 *Contribucio antracoanalitica a l'estudi de l'entorn vegetal de l'home, del Paleolitic superior a l'edat del Ferro a Catalunya.* Tesi licenciatura, Univers. autonoma de Barcelona. (198 pages.)

Ros-Mora, M. T. 1987 L'environnement végétal de l'homme du Néolithique à l'âge du Bronze dans le nord-est de la Catalogne, analyse anthracologique de la Cova del Frare, St Llorenç del Munt (Matadepera, Barcelona). *Actes Colloque int. 'Premières communautés paysannes en Méditerranée occidentale'.* CNRS, pp. 125–129.

Sabatier, M. & Van Campo, M. 1984 L'analyse en composantes principales de variables instrumentales appliquée à l'estimation des Paléoclimats de la Grèce il y a 18000 ans. *Bull. Société botanique France* **131**, 85–96.

Thiébault, S. 1983 *L'homme et le milieu végétal à la fin du Tardiglaciaire et au Postglaciaire, analyse anthracologique de six gisements des préalpes sud-occidentales.* Thèse 3° cycle, Université Paris I. (215 pages.)

Triat-Laval, H. 1979 *Contribution pollenanalytique à l'histoire tardi et postglaciaire de la végétation de la basse vallée du Rhône.* Thèse Aix-Marseille III. (343 pages.)

Van Campo, M. 1984 Relations entre la végétation de l'Europe et les températures de surface océaniques après le dernier maximum glaciaire. *Pollen Spores* **26** (3–4), 497–518.

Vernet, J.-L. 1971 Analyse de charbons de bois des niveaux Boréal et Atlantique de l'abri de Châteauneuf les Martigues. *Bull. Museum Histoire naturelle Marseille* **31**, 97–103.

Vernet, J.-L. 1973 Etude sur l'histoire de la végétation du sud-est de la France au Quaternaire d'après les charbons de bois principalement. *Paléobiologie continentale* **4**(1), 1–90.

Vernet, J.-L. 1980*a* La végétation du bassin de l'Aude au Tardiglaciaire et au Postglaciaire d'après l'analyse anthracologique. *Rev. Palaeobotany Palynology* **30**, 33–55.

Vernet, J.-L. 1980*b* Premières données sur l'histoire de la végétation postglaciaire de la Provence centrale d'après l'analyse anthracologique. *C.r. hebd. Séanc. Acad. Sci., Paris* **291**, 853–855.

Vernet, J.-L. 1986*a* Changements de végétation, climats et action de l'homme au Quaternaire en Méditerranée occidentale. *Proc. Symp. 'Quaternary climate in western Mediterranean', Madrid*, pp. 535–547.

Vernet, J.-L. 1986*b* Ecologie préhistorique et étages de végétations inframontagnards entre les 45° et 39° parallèles en Méditerranée occidentale. *Société botanique de France, groupement scientifique ISARD, Colloque international de Botanique pyrénéenne*, pp. 81–90.

Vernet, J.-L., Badal-García, E. & Grau-Almero, E. 1983 La végétation néolithique du sud-est de l'Espagne (Valencia, Alicante) d'après l'analyse anthracologique. *C.r. Séanc. Acad. Sci., Paris* **296**, 669–672.

Vernet, J.-L. & Thiébault, S. 1987 An approach to northwestern Mediterranean recent prehistoric vegetation and ecologic implications. *J. Biogeogr.* **14**, 117–127.

Vernet, J.-L., Thiébault, S. & Heinz, C. 1987 Nouvelles données sur la végétation postglaciaire méditerranéenne d'après l'analyse anthracologique. *Actes Colloque 'Premières communautés paysannes en Méditerranée occidentale'*, CNRS, pp. 125–129.

Phil. Trans. R. Soc. Lond. A **330**, 679–681 (1990)
Printed in Great Britain

General discussion

S. K. RUNCORN, F.R.S. (*University of Newcastle upon Tyne, U.K.*). In examining whether there is evidence in the records of the Earth's climate for small variations with periods of about 200 years, the geophysicists suppose that the large fluctuation from year to year (of which we are very conscious) due to the instabilities of the atmosphere–ocean system may average out over long periods so as to reveal shorter-period changes in solar activity. There is an interesting analogy with the Earth's magnetic field, which varies considerably on a shorter timescale, yet over longer time spans the mean field is exactly that of a dipole aligned along the axis of rotation. Can we tell, from the archaeological record, whether climatic changes, resulting for instance in the migration of peoples, are short lived, or whether they might be produced by longer-term environmental changes?

J. D. EVANS, F.B.A. (*Institute of Archaeology, London, U.K.*). Archaeologists are not well placed for direct inferring of short-term fluctuations in climate or solar activity. They must depend upon archaeo-biological evidence and physical dating techniques (such as ^{14}C), most of which are not adequately sensitive to reveal 11-year periodicities, though they may sometimes pick up 200-year periodicities.

A. M. SNODGRASS, F.B.A. (*Museum of Classical Archaeology, University of Cambridge, U.K.*). At one extreme it could be held that the ending of the last glaciation is the latest climatic event for which it would be reasonable to expect to find broad reflections in the archaeological record. At the other extreme, however, one might start looking for archaeological effects that covered only three or four years, like the one towards the end of the Bronze Age that P. I. Kuniholm hinted (this Symposium). My own view is much closer to the former extreme than the latter; as an archaeologist I am very unhappy about linking up such short-term phenomena with major breaks in historical continuity. As Kuniholm said, it is not merely climatic change, but human perception of it, that has even the potential for affecting historical processes.

E. M. JOPE, F.B.A. (*The Queen's University, Belfast, U.K.*). While tree-rings (with their potential sensitivity to fluctuations in both climate and solar activity) are cognisant of single-year increments, there is often much uncertainty in relating the tree-ring sequence of the sample exactly to the archaeological or historical context or 'event'. The same difficulty applies to high-precision ^{14}C dating; however, in very favourable (but rather rare) circumstances it can occasionally be possible to date the building of a timber structure to less than $\pm 1\%$ (Jope 1986). But to date a structure is still not to relate to a climatic episode or solar activity.

Concerning Runcorn's point about relation of migration of peoples to climatic changes, I must stress that multiple factors are at work here: social, economic, epidemic, climatic, catastrophic, volcanic eruption (Pyle 1989). Such parameters (each in itself multivariate) might all be integrated under the concepts of chaos theory: a people approaching the threshold of instability in one parameter alone might be sufficient to disrupt life and lead to migration (a proposition that interested P. Foukal).

J. D. Evans, F.B.A. Many archaeologists have for long been ever more precisely quantifying their evidence. Much fruitful work has been done in recent years in the application of statistical techniques to interpretating archaeological data (e.g. in the study of wear on implements, or the changes down through sediment- or ice-cores), and some of this can have direct bearing on following short-term climatic or solar fluctuations.

A. M. Snodgrass, F.B.A. I strongly endorse Professor Evans's point about quantifying archaeological data. Archaeologists too have a large body of primary data that can be presented graphically in just the ways that scientific findings have been presented at this meeting; and some of us are increasingly presenting our data in such ways. This would then constitute a proper basis for correlating archaeological and scientific data, preferable to searching for one-off matches with unique climatic or historical events.

E. M. Jope, F.B.A. Much of the evidence nowadays produced from archaeologically motivated work is really archaeo-scientific data. It is the archaeological scientists' job to make the fullest possible use of such data, in harness with the more purely archaeological evidence, to intensify our understanding of the human past and the climatic and solar past. Some kinds of evidence are still not adequately used, such as soils, or sediment cores, and especially ice-cores; the latter can be very informative in relation to past human activity and climatic fluctuation, and they are particularly time sensitive (H. Oeschger & J. Beer, this Symposium), and note a recent example in Peru (Thompson *et al.* 1988): 'The ice cores from the Quelccaya ice cap contain a 1500-year record of the climatic variability that affected human activities in Southern Peru'.

G. Barker (*Leicester, U.K.*). It is difficult to fund archaeological work on the scale needed to yield meaningful archaeo-climatic data, especially in relation to Mediterranean lands.

G. Beran (*Wallingford, U.K.*). I'd like to draw attention to the potentially informative nature of hydrological studies, feeling (perhaps justifiably) that the topic had been underplayed at this Discussion Meeting.

Sir George Porter, P.R.S. What possibility is there for forecasting future climatic trends from archaeological or historical data? I ask this because there might be money in it!

E. M. Jope, F.B.A. It might be tried once, but archaeologists' futurology is not likely to be invoked for a second time.

This Panel, which is representative of archaeologists, warmly welcomed this opportunity to recognize Hans Suess's seminal influence on the subject; if he had not in earlier times stood doggedly by the actuality of his 'wiggles', we should not today be in the very favourable position over high-precision tree-ring calibration of the ^{14}C timescale, through more than 10000 years (Pearson & Stuiver 1986).

D. H. Tarling (*Geological Sciences, Polytechnic of the Southwest, Plymouth, U.K.*). It is interesting to note that, some 10 years or more ago, most explanations of the variation in production of ^{14}C in the ionosphere were attributed to variations in geomagnetic properties. For good reasons, particularly ^{10}Be, moderate and short-term solar fluctuations are now being considered as major factors affecting the short-term discrepancies between the dendrochronological and uncorrected radiocarbon ages, while it is still assumed that long-term (not less than 10^3 year) variations are associated with changes in the intensity of the main dipolar geomagnetic field. However, the truth probably lies somewhere between these extremes. The entire catalogue of ancient geomagnetic field intensity determinations, based on either archaeological or geological fired materials, is insecure (see, for example, Walton 1988) and, in any case, have a very confined distribution and hence relate more to local geomagnetic changes, rather than necessarily global changes. At the moment, the strongest evidence for long-term global geomagnetic field changes is that of the long-term fluctuations in the production of ^{14}C and other causes cannot be excluded. Conversely, it has recently been found that there are clear cyclicities in the modal rate of change in the direction of the geomagnetic field in at least northwestern Europe (U.K. and France) during the past 2000 years or so (Tarling 1988, 1989) based on Clark *et al.* (1988). The most prominent is a periodicity of 266 ± 27 years in the modal rate of solid angular change and this is also reflected in the rate of change relative to the axial geocentric dipole field direction. The relationship between rate of directional change and field intensity is not established and such results are still highly localized. None the less, archaeomagnetic studies do provide the only method of defining such geomagnetic variations during the past few thousand years and are required to establish the relationship to many interacting, little understood processes.

References

Clark, A. J., Tarling, D. H. & Noël, M. 1988 Developments in archaeomagnetic dating in Britain. *J. archaeological Sci.* **15**, 645–667.

Jope, E. M. 1986 Sample credentials necessary for meaningful high-precision ^{14}C dating. *Radiocarbon* **28**, 1060–1064.

Pearson, G. W. & Stuiver, M. 1986 High-precision calibration of the radiocarbon time-scale, 500–2500 B.C. *Radiocarbon* **28**, 839–862.

Pyle, D. M. 1989 Ice core acidity peaks, retarded growth and putative eruptions. *Archaeometry* **31**, 88–91.

Tarling, D. H. 1988 Secular variations in the geomagnetic field: the archaeomagnetic record. In *Secular solar and geomagnetic variations in the last 10000 years* (ed. F. R. Stephenson & A. W. Wolfendale), pp. 349–365. Dordrecht: Kluwer.

Tarling, D. H. 1989 Geomagnetic secular variation in Britain during the last 2000 years. In *Geomagnetism and palaeomagnetism* (ed. F. J. Lowes, D. W. Collinson, J. H. Parry, S. K. Runcorn, D. C. Tozer & A. Soward), pp. 55–62. Dordrecht: Kluwer.

Thompson, L. G., Davies, M. E., Thompson, E. M. & Liu, K.-B. 1988 Pre-Incan agricultural activity recorded in dust layers in two tropical ice-cores. *Nature, Lond.* **336**, 763–765.

Walton, D. 1988 The lack of reproducibility in experimentally determined intensities of the Earth's magnetic field. *Rev. Geophys.* **26**, 15–22.

Phil. Trans. R. Soc. Lond. A **330**, 683–684 (1990)
Printed in Great Britain

683

Closing remarks: astronomical

By J.-C. Pecker

Collège de France, Annexe, 3 Rue d'Ulm, 75231 Paris, Cedex 05, France

Over the past two days, we have covered many facets of the basic interactions between the solar activity and the Earth's climate. As an astronomer, I should perhaps first comment on the fact that solar activity is not the only astronomical or astrophysical phenomenon to influence physical conditions in the biosphere. Over a very long timescale of thousands of millions of years the evolution of the Sun from a pre-main-sequence star to a star of G type has not only fundamentally controlled the physical and chemical processes in the formation of the planets but has controlled their surface physical characteristics. Over timescales an order of magnitude less, the location of the Solar System in the Galaxy may have influenced life on Earth. For example it has been noted that when the Sun crossed the spiral arms of the Galaxy and their dense dust clouds, some catastrophies might have resulted; the disappearance of the dinosaurs could be accounted for by such phenomena, as was once suggested by Sir William McCrea, F.R.S.; but nearby supernovae, grazing comets, and on large meteorites might very well have played a decisive role in the evolution of species and of our Earth. On a smaller timescale, a million years, the variation in solar energy falling on the Earth, due to secular changes in the terrestrial orbit parameters (Milankovitch–Berger theories), would have caused climatic changes and have been shown to account for the successive ice ages of the Quaternary. While bearing this in mind the role of solar activity on the timescale of recent millennia, but also on shorter timescales, is of obvious importance to society and, as we have seen in this meeting, is only now being properly investigated.

The Sun is a complex generator of radiation, particles and magnetic fields sent far into space. Even if the total radiation emitted is constant, the amount of radiation received by the Earth changes with time: this change is complex and involved with a redistribution of energy emitted at different solar latitudes than to a real change in solar luminosity. We should not forget that our location, almost in the solar equatorial plane, favours the radiation coming from the solar low latitudes; the polar solar phenomena are strongly affected by foreshortening. These changes in the Earth's illumination may be function of the wavelength, and have various effects in different layers of the Earth's atmosphere. The Sun also emits particles of all energies. Some of them find their way through the magnetopause, giving rise to auroras, to magnetic storms, to ionospheric disturbances and the like. The possible climatological effects are yet obscure: the more energetic particles affect more directly the low terrestrial latitudes; they have a broader entrance door! Is therefore the Wolf number, used as a unique parameter for solar activity, suitable for all climatological studies? Clearly, spots of high latitude and of low latitude give rise to different particle paths; 'young' spots and 'old' spots are not surrounded by the same type of activity, of flares or prominences: why should they enter in a single activity index?

The Earth also is a very complex machine. Paths of particles in the geomagnetic field, propagation of phenomena and transfer of energy from ionosphere to troposphere, all of this involves difficult physics. Climate in polar zones is influenced by solar activity probably more than in equatorial zones. Averaging phenomena over the Earth's surface is not a satisfactory method. However, couplings exist and one cannot reduce the response of the Earth's atmosphere to the solar input to simple physics.

Actually, neither the understanding of solar activity, nor the physical description of the terrestrial response, are very far advanced. We are far from a completely causal detailed description of the interaction, even though the existence of this interaction seems fundamental to our discussion. This thought makes me quite cautious about the use (perhaps the abuse) of power-spectrum analysis of the time series. There is no doubt that the undecennal cycle of the Sun exists; and there is no doubt that the Sun rotates in about 27 days (though complicated by differential rotation). Finding in terrestrial phenomena the 11-year or the 27-day signature gives us some hint about the reality of the solar influence. But it should be considered as not much more than a hint! One has afterwards to look for one-to-one relations, say, between some type of flares and some type of meteorological change. And then one has to understand the physical mechanisms at work. And only then shall we be able to derive sound conclusions of a climatological nature.

The climate depends upon the solar activity. Vice versa, the climate is a good proxy, or should be a good one, to determine the past variation of solar activity. In this meeting, I must say that I have been a little disappointed about developments in this direction, possibly because of lack of understanding of all the physical mechanisms involved. In particular, the doubt about the Precambrian records (are they a signature of the solar cycle, as I would hope them to be, or of the tidal effects of the Moon?) is to an astronomer most frustrating, even though I do not doubt the importance of studies of varves *per se*!

Is the Earth, or are the other planets, such as Jupiter, influencing, within the Solar System, our star the Sun? This matter has hardly been touched upon. However, this field could be looked at: small, even very small tides, due to Venus and Jupiter, may well trigger the emergence, at the solar surface, of magnetic flux tubes if they are in a situation of near equilibrium.

A first conclusion, on the astronomical side, is at this stage obvious: to understand the solar terrestrial relations better, one should pursue regularly, unceasingly, the so-called 'routine' observations, from ground-based stations and from space, of solar phenomena of all timescales, including 'long' cycles (Gleissberg's cycle, or cycles covering several centuries, suggested by Link for example). One should also include in the solar programmes, studies, 'out of the ecliptic', of the polar regions of the Sun of which at present our knowledge is scanty because of our location with respect to the Sun.

A second, and more general conclusion, is clear. The sensitivity of human behaviour and of conditions of human life to even a small change in climatic conditions, is very large. The Sahel drought, and the Bangladesh floods are dramatic illustrations of this sensitivity. Therefore a good knowledge of solar–terrestrial physics is most relevant for mankind, a political need, as our understanding of the phenomena is necessary to predict them; and their prediction is necessary to help the populations to face climatic catastrophies. May this last conclusion be heard by decision-makers!

Phil. Trans. R. Soc. Lond. A **330**, 685–687 (1990)

Printed in Great Britain

685

Closing remarks: geophysical

By S. K. Runcorn, F.R.S.

Department of Physics, The University, Newcastle upon Tyne NE1 7RU, U.K.; University of Alaska Fairbanks, Fairbanks, Alaska 99775, U.S.A.; Imperial College of Science, Technology and Medicine, London SW7 2BZ, U.K.

The search for a connection between solar variability and the Earth's climatic variations has given rise to much controversy: given the great instabilities of the atmosphere and the relatively short span of accurate observations this was inevitable. Meteorologists, very conscious of the complexity of modelling the atmosphere, even assuming a constant energy input, have often been very critical of claims to have detected the solar cycle, and even hostile to the search for one. Now, as so often is the case in science, important evidence has come from an entirely different field; ^{14}C dating. One can therefore have sympathy with the distinguished meteorologist who, when informed of the 200-year period in the Sun inferred from this work and its possible effect on the climate, said 'Dear Professor Suess, you must understand that the Sun has nothing to do with the climate.' I trust that it is not an apocryphal story!

It is almost an invariable rule in geophysics that a phenomenon has more than one cause. The spectrum of ^{14}C variations shows lines of about 200 years and 2000 years. The ^{10}Be data appears to show the 200-year period and because both isotopes are produced by similar nuclear physics processes in the high atmosphere but find their way to the ground by different physico-chemical processes, we seem on firm ground in attributing the 200-year period to modulation of the cosmic-ray flux by magnetic field changes around the Sun with this period. The apparent absence of a 2000-year line in the ^{10}Be spectrum suggests that it is possible that in the C^{14} inventory a fundamental role is played by the oceans, which have such overturn times and store much of the CO_2.

In these discussions it has always been tacitly – although reasonably – assumed that the sunspot periods are accompanied by periodic variations in the solar bolumetric luminosity. This assumption has now been proved by the satellite measurements of the solar constant over recent decades.

Attempts to find in meteorological data evidence for the solar cycles has naturally focused hitherto on the 11-year and 22-year periods and has attracted, as we have pointed out, much criticism because the variations, which evidently result from instabilities of the atmosphere, are so great. But an analogy from another geophysical field may be useful. The geomagnetic field as observed over historical times presents a complex picture, many harmonics are required to describe it and its secular variation. This reflects the magnetohydrodynamics, a turbulence, of the Earth's core, but when averaged over thousands of years, the field is a dipole aligned along the axis of rotation and its moment depends on the power driving the dynamo process. Therefore, it is not unreasonable to expect that suitably averaged and appropriately selected meteorological data in certain areas of the globe will reveal periodic or quasi-periodic variations in the solar energy input to the atmosphere. The evidence seems now at last becoming more convincing. The analysis of time series goes back to the work of Udny Yule

[287]

(1927), the newer methods being elaborations of his autocorrelation one, but it is now possible to test the statistical significance of the periods found, an important theme in this Symposium.

The historical records of sunspots and the discovery of minima of solar activity are in agreement with the ^{14}C data in showing the existence of an approximate 200-year cycle. These minima are the Maunder (1645–1715 A.D.), the Spörer (1420–1530 A.D.), the Wolf (1280–1340 A.D.), the Oort (1010–50 A.D.) and a possible later one, the Dalton (1790–1830 A.D.). The correlation of these minima with cold winters in western Europe (Lamb 1982) aroused much interest and we now know that for the 11-year cycle at least the solar constant is greatest at the sunspot maximum, not immediately obvious perhaps as the dark sunspot areas are cooler. The quiet Sun is associated with a higher ^{14}C generation rate in the high atmosphere as fewer cosmic rays are deflected in the surrounding plasma of the Sun. The maxima of the Suess wiggles do correlate with the 'Little Ice Age' associated with the Maunder Minimum, i.e. the correlation has the correct sign.

The modulation of the sunspot cycle over 200 years might be expected to have a larger effect on the climate than the 11-year period, so it is reasonable to ask whether archaeological evidence can contribute to this study. Obviously, migrations of early peoples could be caused by slow climatic changes. The archaeologist ideally requires from the geophysicist annual records of mean temperatures and rainfall in the areas which he studies. It is possible that ^{18}O studies of tree rings may yet yield information about temperature. Leona Libby (1983) worked on that and published a record of temperature over the past few millennia for Europe, the last 200 years agreeing well with mean European temperatures based on the exhaustive compilations of recorded temperatures since Fahrenheit's invention of the thermometer. The mechanism by which the ^{18}O ratio in trees depends on temperature is obscure unlike that in the palaeo-temperature study of fossils and microfossils pioneered by H. C. Urey and C. Emiliani. A careful new approach is described in this Symposium by S. Epstein and R. V. Krishnamurthy. The results referred to above on European trees over recent millenia have been disputed and should be repeated. But the geophysicist would be much stimulated if the archeologist were able to point to certain migrations that seem best explained by climatic changes. We can look forward in this study of variations of climate and the Sun to as fruitful an interaction between archaeologists and geophysicists, as that during the development of ^{14}C dating. I well remember W. F. Libby coming to Newcastle and saying that in the previous few days he had been convinced by the Keeper of Egyptology, I. E. S. Edwards, in the British Museum that his ^{14}C dates were wrong by about 1000 years and he wished to know whether there was evidence that the geomagnetic field intensity had changed over these times sufficiently to alter the cosmic-ray flux into the Earth. This led to the need for an independent calibration of the ^{14}C dating method, which was achieved by Professor Hans Suess using tree rings and which has now led to the new evidence of solar variability, which has been the origin of this meeting.

The greenhouse effect points to the need for a better understanding of the atmosphere and of climatic change. By studying the effect of the changes of solar output with periods of 11 and 200 years on the atmosphere from the historic and archaeological records, we may hope to come to a better understanding of the Earth's atmosphere and how it responds to changing energy input. In geophysics we have to wait for Nature to do the experiment. But in assessing the influence of the greenhouse effect from observations over the past 150 years, the possibility that some of the change may be due to the 200-year cycle in the Sun must be considered. We showed in the Introductory remarks that the global increase in temperature over the second half of the last century and the first half of this – as strikingly illustrated by the frontispiece – correlates with the Sun's activity, as determined by geomagnetic disturbances. This implies that at least some part of the observed increase in temperature is due to increased solar energy output.

REFERENCES

Lamb, H. H. 1982 *Climate history and the modern world*. London: Methuen.
Libby, L. M. 1983 *Palaeoclimate, tree thermometers, commodities and people*. Austin, Texas: University of Texas Press. (142 pages.)
Yule, G. U. 1927 On a method of investigating periodicities in disturbed series, with special reference to Wolfer's sunspot numbers. *Phil. Trans. R. Soc. Lond.* A **226**, 267–298.

INSTRUCTIONS TO AUTHORS

(Philosophical Transactions series A: publication after July 1990)

1 General

Philosophical Transactions series A is published monthly. Separate issues contain original papers, 'Theme' articles, and the reports of Royal Society Discussion Meetings. The format of the journal is B5 (247 mm × 174 mm), single column.

2 Submission

Papers may be submitted (i) direct to the Editorial Office, The Royal Society, 6 Carlton House Terrace, London SW1Y 5AG, (ii) to the Editor or a member of the Editorial Board, or (iii) via a Fellow or Foreign Member of the Society. Three copies of the typescript (and of any figures, together with original drawings and prints) are required. The extra copies of any photographs should be prints rather than photocopies.

When sending their papers authors may, if they wish, suggest the names of referees, but such suggestions will not necessarily be adopted.

3 Copy

Papers should be clearly typewritten, **with double spacing throughout**, on one side of the paper only, with a margin of at least 3 cm all round; the sheets should be serially numbered and **securely clipped together**. Typescripts must be carefully corrected by authors before being sent in. Spelling should conform to the preferred spelling of the *Shorter Oxford English Dictionary*. Footnotes should be avoided.

Authors considering submitting papers on floppy disc should contact the Editorial Office before starting to write, for latest information on compatibility.

4 Title, Abstract

The title, which should be concise, should be typed on a separate covering sheet which should also bear the names of the authors and that of the laboratory or other place where the work has been done. Where the title is long, a short title suitable for page headings should also be indicated. Each paper must be accompanied by an abstract, which should not exceed 5 % of the length of the paper, and should give a precise and informative indication of its content.

5 Sectional headings

Papers should be divided into sections, described by short headings. Sections should be numbered and, when necessary, reference should be made to them in the text by use of the section sign (§) with the number, e.g. 'see §4'. Subsections should be lettered (*a*), etc., and sub-subsections numbered (i), etc. Papers that will exceed about 20 printed pages should include a list of contents.

6   Units and symbols

As far as possible the recommendations contained in *Quantities, units, and symbols* (1975, The Royal Society, £1.50) should be followed; in particular the International System of Units (SI) should be used whenever it is practicable to do so.

The use of symbols at the start of a sentence is deprecated and should be avoided. Special care is necessary in differentiating between handwritten symbols of comparable shape, e.g. $V\,v\,v$, $w\,W$, $s\,S$, $p\,\rho\,P$, $T\,\tau$. Marginal indications and differential underlinings should be used where necessary, the normal conventions being followed where applicable, e.g. $\sim\!\!\sim$ to signify bold characters. This is required for the printer's information even if the typescript has been prepared on a printer that implements such typographical distinctions.

Organic chemical formulae should be labelled by means of (unbracketed) bold arabic numbers.

7   Illustrations

Duplicate figures (e.g. Xerox or photographic copies) should be supplied with each copy. The author's name should be written on the back of all illustrations, and the number of the figure should also be shown there. Figures should be numbered in one sequence throughout the paper.

The position of each illustration should be clearly marked in the typescript thus:

Figure 2 near here

Line drawings

Long descriptions should appear not on the figures themselves but, much more conveniently for the printer, in the legends. Any labelling that is necessary for the understanding of a figure, e.g. the differentiation of curves, should be indicated lightly in pencil on the original drawings and exactly the same labelling should be inserted carefully in ink on the duplicate copies.

Where a graph is the subject of the illustration the description of the coordinates should be given on the duplicate copies.

All lettering of words should be in lower case except for proper names, where a capital should be used. Lettering for symbols should strictly follow the case and fount of type called for in the text. The printer's artist will insert these on the originals in a standard style of lettering and to a size to suit the reduction that will be made before printing. If an author is able to call on the services of an experienced lettering artist it is often preferable for heavily labelled figures, e.g. maps, to be completely lettered before submission. Adequate consultation between authors or their draughtsmen and the Editorial Office (telephone 01-839 5561, extension 229) will help to ensure satisfactory results. A leaflet on the preparation of illustrations for publication is available from the Editorial Office.

Legends

These should be typed with double spacing on a separate sheet at the end of the paper and should state concisely the points that the author wishes the reader to notice.

Figure legends should follow the style of presentation of information given below.

Figure 1. The course of oxidation of 2-methylpentane at 2.0 MPa and 800 K. (*a*) Non-sampling run: curve 1, pressure; curve 2, light transmission at 265 nm. Point A is the end of compression, B is the cool-flame reaction and C is the hot ignition. (*b*) Sampling run: curve 1, pressure; curve 2, light transmission at 265 nm.

Photographs

When it is essential to include photographs in a paper they should be carefully chosen to make the most efficient use of the space required. The area covered by the photographs should be restricted to the subject in question, or to a *minimum* representative area in photomicrographs, etc. This enables the photograph to be reproduced at the largest possible scale. The text area available in *Philosophical Transactions* series A is 212 mm × 135 mm.

Authors should supply unmounted glossy prints marked on the back with the authors' names and the number of the figure, and with top and bottom indicated. When lettering has to be inserted a rough set should be provided with the lettering clearly marked.

8   Tables

Tables, however small, should be numbered in arabic numerals and referred to in the text by their numbers (e.g. 'see table 3'), because it may not be possible to print a table in its immediate context.

The position of each table should indicated as in the following example:

Table 3 near here

Table headings should be brief, and will be printed in capitals and small capitals. Column headings should be in lower-case lettering except for the capital initial letters of proper names. The units of measurement and any numerical factors should be placed unambiguously at the head of the column, e.g. F/MHz, $10^{28}\,\sigma/\mathrm{m}^3$ or $q/(\mathrm{kJ\ mol^{-1}})$.

9   References

References to the literature cited must be given in double-spaced typing, in alphabetical order at the end of the paper. They should be arranged as follows.

1. Name(s) with initials of the author(s).

2. Year of publication of the **paper or book**.

3. The title of the paper.

4. Title of the periodical, abbreviated according to the principles of the *World list of scientific periodicals* (4th edn 1963–5), underlined to indicate italics. A booklet entitled *Short titles of commonly cited scientific journals* is available from the Royal Society at £2.00, including postage. When the correct abbreviation for a title cannot be deduced it should be given in full.

5. Volume number underlined thus 24, preceded where applicable by the series number in parentheses.

6. First and last page numbers of the paper.

7. When the title of a book is cited the place of publication, the name of the publisher, and the number of the edition should be given. A page, section or chapter number is nearly always necessary.

The reference to a paper will then be printed as in the following examples:

Hill, A. B. 1953 The mechanics of active muscle. *Proc. R. Soc. Lond.* B **141**, 104–117.
Taylor, G. I. 1930 Recent work on the flow of compressible fluids. *J. Lond. math. Soc.* **5**, 224–240.

and to an article in a multi-author work or to a book:

Penrose, R. 1979 Singularities and time-asymmetry. In *General relativity: an Einstein centenary survey* (ed. S. W. Hawking & W. Israel), pp. 581–638. Cambridge University Press.
Marchbanks, R. M. 1975 Biochemistry of cholinergic neurons. In *Handbook of psychopharmocology* (ed. L. L. Iversen, S. D. Iversen & S. H. Snyder), vol. 3 (*Biochemistry of biogenic amines*), pp. 247–326. New York and London: Plenum Press.

References in the text are made by giving the author's name and date of publication, e.g. (Brown 1965). Such reference is usually placed in parentheses unless the name of the author is part of the sentence, in which case the year only is required in parentheses. Where two or more papers published in any one year by the same author are cited, each paper should be distinguished by a small letter, *a*, *b*, etc., placed after the date, e.g. (Brown 1965*a*). Where there are more than two authors to a paper it should be cited thus: (Brown *et al.* 1978) unless there are good reasons for including all the authors, up to five, at the first mention. All the authors should, however, be included in the list of References. References to books should be to the latest editions.

References by serial number (e.g. A. N. Other (8)) are not permitted.

10 Proofs

Great care is necessary in checking proofs to ensure that all misprints are detected. Authors should note that systematic emendations may have been made to their typescript in accordance with the normal style of the Society's journals. If any changes are necessary to proofs every effort should be made by substituting matter of similar length to avoid extensive rearrangement. Authors are warned that they are liable for the cost of excessive alterations to their proofs.